E-COMMERCE, E-BUSINESS AND E-SERVICE

Communications in Information Engineering

ISSN 2333-2115

Volume 1

PROCEEDINGS OF THE 2014 INTERNATIONAL CONFERENCE ON E-COMMERCE, E-BUSINESS AND E-SERVICE (EEE 2014), HONG KONG, 1–2 MAY 2014

E-Commerce, E-Business and E-Service

Editor

Garry Lee
Information Engineering Research Institute, Bellflower, CA, USA

CRC Press
Taylor & Francis Group
Boca Raton London New York Leiden

CRC Press is an imprint of the
Taylor & Francis Group, an **informa** business
A BALKEMA BOOK

Published by:
CRC Press/Balkema
Schipholweg 107C, 2316 XC Leiden, The Netherlands

CRC Press/Balkema is an imprint of the Taylor & Francis Group, an informa business

Visit the Taylor & Francis Web site at
http://www.taylorandfrancis.com

and the CRC Press Web site at
http://www.crcpress.com

Typeset by V Publishing Solutions Pvt Ltd., Chennai, India

ISBN: 978-1-138-02646-9 (Hbk)
ISBN: 978-1-315-76222-7 (eBook PDF)

E-Commerce, E-Business and E-Service – Lee (Ed)
© 2014 Taylor & Francis Group, London, ISBN 978-1-138-02646-9

Table of contents

E-Commerce, E-Business and E-Service – Lee (Ed)
© 2014 Taylor & Francis Group, London, ISBN 978-1-138-02646-9

Preface

2014 International Conference on E-Commerce, E-Business and E-Service (EEE 2014) was held on May 1–2, 2014, Hong Kong.

Electronic commerce is a type of industry where the buying and selling of products or services is conducted over electronic systems such as the Internet and other computer networks. Electronic commerce draws on technologies such as mobile commerce, electronic funds transfer, supply chain management, Internet marketing, online transaction processing, electronic data interchange (EDI), inventory management systems, and automated data collection systems. Modern electronic commerce typically uses the World Wide Web at least at one point in the transaction's life-cycle, although it may encompass a wider range of technologies such as e-mail, mobile devices, social media, and telephones as well. Electronic commerce is generally considered to be the sales aspect of e-business. It also consists of the exchange of data to facilitate the financing and payment aspects of business transactions. This is an effective and efficient way of communicating within an organization and one of the most effective and useful ways of conducting business.

EEE 2014 focused on four topics: (1) E-Commerce, E-Business and E-Service; (2) Internet Computing and Information Services; (3) Computer science and Technology and (4) ICT for Business and Management. Some papers were highly evaluated by the scientific committees and the reviewers. For example, "E-business Models for Financial Services and Internet Banks" by M. Sahut and "Entrepreneurial Learning in a Secure e-Commerce Course through Creating Competitive Real-World Sites" by Ryan Garlick were highly evaluated. We expect that the conference and its publications will be a trigger for further related research and technology improvements in this importance subject.

EEE 2014 was both stimulating and informative with an interesting array of keynote and invited speakers from all over the world. Delegates had a wide range of sessions to choose from. The program consisted of invited sessions, technical workshops and discussions with eminent speakers covering a wide range of topics. This rich program provided all attendees with the opportunity to meet and interact with one another.

Finally, our sincere thanks to the support from the technology program committee for designing the conference web page and also spending countless days in preparing the final conference program in time for printing. Sincere thanks to our organization committee for their several months' hard work in sorting out manuscripts from various authors. Also thanks to other staffs assisting in the various stage of the editorial work, especially the international committees of our conference.

We hope that EEE 2014 was successful and enjoyable to all participants. We look forward to seeing all of you next year at EEE 2015.

E-Commerce, E-Business and E-Service – Lee (Ed)
© 2014 Taylor & Francis Group, London, ISBN 978-1-138-02646-9

Organizing committee

KEYNOTE SPEAKER

Gerald Schaefer, *Loughborough University, UK*

GENERAL CHAIRS

Dehuai Yang, *Huazhong University of Science and Technology, China*
Minli Dai, *Suzhou University, China*

PUBLICATION CHAIRS

Garry Lee, *Information Engineering Research Institute, USA*
Mark Zhou, *Hong Kong Education Society, Hong Kong*

IERI INTERNATIONAL COMMITTEE

P. A. Pereira Correia, *University of Vic, Barcelona, Spain*
JM. Sahut, *IPAG Business School, Paris, France*
D. Liginlal, *Carnegie Mellon University, Doha, Qatar*
Ryan Garlick, *University of North Texas, Denton, TX, USA*
H.C. Chen, *National Chin-Yi University of Technology, Taichung, Taiwan*
Wei-Siang Chiew, *Multimedia University, Cyberjaya, Malaysia*
M. Hoshi & V. Klyuev, *University of Aizu, Japan*
R. Contreras-Espinosa, *University of Vic, Barcelona, Spain*
R. Meeds, *Qatar University, Doha, Qatar*
I. García Medina, *Glasgow Caledonian University, Glasgow, UK*
Samini Subramaniam, *Multimedia University, Cyberjaya, Malaysia*
Z. González, *University of Vic, Barcelona, Spain*
Ming-Tsung Tsai, *Southern Taiwan University, Taiwan*

Section 1: E-Commerce, E-Business and E-Service

E-Commerce, E-Business and E-Service – Lee (Ed)
© 2014 Taylor & Francis Group, London, ISBN 978-1-138-02646-9

The study of administrative division reform in China on the basis of the E-government

Weidong Shi
Key Laboratory of Tourism and Resources Environment in Universities of Shandong, Taishan University, Taian, China

ABSTRACT: Based on electronic government connotation and denotation, and on the basis of in-depth analysis, from the analysis of the trend of the global government reform, electronic government is the motivation and means of reform, to explore the era of electronic government system reform, the necessity of China's administrative division level. E-government is put forward under the perspective of China's administrative reform strategy: reduce the administrative level, enlarge the amplitude of the management under the overall direction, return to the constitution "provincial, county (city), towns" three levels of the administrative management, to cancel a prefecture level, the prefecture implementation of provincial management system of county directly.

Keywords: E-government; Administrative division reform; China

1 THE CONNOTATION AND CHARACTERISTICS OF E-GOVERNMENT

E-government, as a term can refer to the use of information technology, especially Internet technology, with more convenience, customer—oriented, cost—benefit such as different and better ways to provide public services [1]. Public administration academia believes that E-government is an application of information technology which has prompted the Government to manage more democratic, open, service tools or means, similar to the electronic management category. Often related to the information age "mode of organizational change significantly related to the current management form in the field of public administration organization pattern consistent[2]. However, as the information technology and equipment to popularize and apply, in the field of government management and service provide a deepening dependence on information technology and equipment, the understanding of the connotation of the theory and practice of E-government continues to expand and deepen, the understanding also become more comprehensive and objective, showing the theoretical definition of focus has shifted from previous focuses on technology application gradually to focus more on how to use technology to make the government management and Public governance revolution.[3] E-government has gradually become the government management mode of the information age, the essence of which is the introduction of E-government that is the centralized management of the industrialization model, layered structure, the physical operation of big government in the economy, via the Internet into a new type of management system, to adapt to the virtual, global, based on the knowledge of digital economy, but also to adapt to the fundamental shift of social operation. The implementation of the E-government affairs is essentially a profound revolution of government management mode which is the alternative paradigm of traditional government management mode.

E-government has the following marked characteristics (see table 1). Firstly, offer round-the-clock service. E-government is a cross-platform combined with a virtual government, through the network to build up a cross-regional, across time and space remote administration, which can provide 24-hour online service for the society and the public to.

Secondly, the governance of the government is to achieve the standardization administration business. In decision-making, through the network investigation, live, consulting, public application, such as ask implementation strategy in the people, improve the level of decision-making; In performing link, establish application system power, achieve standardization of procedures, reduce the influence of artificial factors, and to improve the quality and efficiency of work; In the results section, by email-sending, text messaging, online inquiry, the measures of the public, to realize the execution result immediate notice.

Table 1. The comparison of traditional government and E-government.

	Traditional government	E-government
Spatial characteristics	Regional	Super-regional
Management hierarchy	Decentralized management	Centralized management
Management mode	The government entity management	System program management
Structure of the organization	Vertical stratification structure	Flat structure of radiation

Third, the integration of public service. Users only need a website to solve all the problems by themselves.

Finally, the government's flat organization structure tendency. As the government's growing business information technology application system, gradually formed the across different departments, business office, across application system of network coordination office system, the government management services changes from scattered, fragmented pattern to the centralized integration, one-stop mode, creating a seamless integration of the government.

2 THE ERA OF E-GOVERNMENT, THE NECESSITY OF ADMINISTRATIVE DIVISION HIERARCHY SYSTEM REFORM IN CHINA

2.1 *The global government reform trend*

Starting in the 1980s, the western developed countries such as Britain, the United States have experienced a called "reinventing government movement" of the government management reform, its purpose and mission are to explore to adapt to the information society and future government management mode to meet the need of economic globalization. Local decentralization reforms become a worldwide trend. The reason for this is that: after all, the international competition of economic globalization is more and more fierce, need to arouse the enthusiasm of the governments at all levels to improve their own competitiveness, to local government decentralization is conducive to mobilize the enthusiasm; Too much at the same time management level with foreign simplicity and effectiveness of government difficulties, affect the exchange with the world.

2.2 *E-government provides motivation and means for China's administrative system reform*

Directly in the internal management of government information technology penetration, including documents, file, data and other kinds of management information collection, production, delivery and so on all electronically, with the advantages of synchronic and sharing. It not only saves administrative costs, improves the administrative efficiency, and directly to the government management system and management methods, program brings revolutionary change.

With the development of world new technology revolution and the improvement of the economic foundation in China, with the computer as the core of the modern office equipment widely used in the administrative management, government information network system is increasingly perfect, office automation and modernization level increased significantly, which in time and space for the administrative performance improvement provides a strong material conditions and technical support.

The government set up E-government network, Internet, reforming the administrative process, is to the government for the information-based tide, to deal with in the aspects of conception, function, system challenge and an important step. The governments to actively respond to the citizen's demand on the Internet, and take corresponding measures, fairly and efficiently achieve its needs and interests. Government administrative processes of transformation through the E-government affairs system, can guarantee the standardization of the administrative behavior, overcome the randomness, and reduce the management and decision-making errors as much as possible. Government also can communicate with public information on the Internet; listen to the opinions and feelings, to better serve the public. Relying on the network the government administrative process reform also perform their functions to build a modern government, implementing management and service platform for the behavior. At the same time, it also gathers the government function transformation, system innovation, mechanism innovation and management methods of a variety of value. Through electronic networks, the government administrative management and the operation of the public service principle, implementation mechanism and the specific means are significantly different from the traditional government operations. Faced with the challenge of the new environment, the government administrative operation mechanism must appear new changes, changes in technology inevitably reflected in the management of the scope and level, to the old and the reality does not adapt the administrative system reform is imperative.

Anyhow, in the operation of the E-government affairs, to people's administrative culture leads to a remodeled drive, to achieve administrative culture by ethical type to the legal theory of type transformation, by controlling to participated in the transformation, by type led to a service-oriented transformation, from plain to the conversion of the apparatus of transformation, from peers to autonomy, conversion from closing to opening, from loose to the transformation of the communist party.

3 THE E-GOVERNMENT AFFAIRS UNDER THE PERSPECTIVE OF CHINESE ADMINISTRATIVE REFORM STRATEGY

China's administrative division system is a special part of China's administrative system, and it is a kind of government management on the basis of the regional special institutional arrangement, its structure by certain social economic conditions and specific government institutions and constraints. As super structural administrative division system, after a leap in productivity change is bound to make the corresponding change. The level of the administrative divisions and amplitude is the basic framework of administrative division system, is the foundation of the organization system of local government. Scientific and reasonable to determine the administrative divisions of the margin of management level and management is the core of the strategic exploration of administrative area.

Since the 1980s, no matter in developed countries or the emerging industrialized countries, or developing countries, they all have raised a hot wave of administrative system reform. Now countries around the world generally adopt the system of administrative divisions of the triangle, more is given priority to with four government three layer division, only a few countries for more than 3. Such as, Germany is a federal, state, county, township; France has a "central, prefecture, provincial, town"; Italy is also "the central, prefecture, and provincial, city (town) level management, etc." At present, most provinces and autonomous regions in China is a "central provinces (autonomous regions), city, county, township" five management system, Hainan, such as Beijing and Shanghai municipality directly under the central government are basically is a "central provinces (municipalities directly under the central government), county (district), rural" 4 class management system, normal operation and management efficiency is very high.

Known from the analysis of the above, the era of E-government to reduce the administrative level, and to expand the management scope is the overall direction of Chinese administrative management system reform at present stage.

China's current five government, level 4 administrative division system, management level, the administrative power concentrated on the top, order upload downward, information flow level is more, add to the bureaucracy, to a certain extent, contributed to the body harmony is swollen, unfavorable to establish flexible and efficient administrative system. Should return to the constitution "provincial, county (city), the villages and towns" level 3 of the administrative management level, cancellation area, city, county to the cities and counties in partition the system of direct management counties (province), is the best choice of compression on the macro level.

For zoning at the provincial level, more popular is "province", the opinion of many scholars and officials is redrawn provinces and regions, will be divided into about 50 provinces in China. The division of province in China, began in the Yuan dynasty, Ming formation, finalize the design in the late Qing dynasty, during the period of the republic of China, the provinces have already formed their respective characteristics of the local customs and culture context. Increase of around 20 provincial administrative division units, under the existing national cadre personnel management system is counterproductive to increase financial burden; At the same time, from thousands of years of Chinese history, from the administrative districts around the world can always see clearly, senior administrative division adjustment is mainly based on the country's political stability, it is the embodiment of the nation's highest interests. Throughout most of the Chinese and foreign history, in the political golden period, there is almost no "big" of provincial visible level of administrative zones for the importance of national stability and unity. On the other hand, with the development of traffic, communication, especially the advent of the era of E-government, province is fully capable to manage the existing number of county.

Prefecture administrative level is the most problematic. Prefecture administrative zones between China 's provincial and county level city, area, etc., some cities under the jurisdiction of only 1–3 counties. As table 2 shows, if the prefecture level is canceled, its management made county for 89.1, but if lost municipal district, the management range is about 70. Provincial part, it seems, no middle level or more, but if you consider the development of modern science and technology, transportation, communication technology, and management technology revolution and the transformation of government functions and the amplitude can be perfectly acceptable. In addition, provincial and county is between "government vs." relationship, rather than the villages and towns government directly to people, more over the establishment

Table 2. China's administrative division hierarchy range in 2012.

	Country	Provinces	Prefecture level	Counties	towns
Total number	–	34	333	2,852	40,446
Secondary administrative area averages	34	10.4*	8.6	14.2	–

Note: According to the law of the People's Republic of China on administrative division Jane copies of 2013, China map publishing house, in July 2013, the related data calculation; Take quotient 32 provincial administrative region, in the calculation of the two special administrative region is not taken into account.

of the socialist market economy development in China, is China's future more play to the government, enterprises and non-profit organizations (NGOs) the power of the third domain development gradually, some of the government functions are gradually transfer, the general trend is the general function of government, is more of a macro guidance function. "Made county", therefore, is the trend of The Times, have made it clear that the central government's recent planning in advance, should now with the E-government to promote accelerated.

In terms of county level administrative unit, it is the most stable level administrative units in China; many counties have a history of more than one thousand years. From the protection of Chinese cultural and historical style and features and basic stability point of view, county-level administrative region pattern is not easy to break. Chinese people formed in the long history of social, economic and cultural "channel" and "network" is more a county, the county is one of China's prefecture identities and sense of belonging of unit directly.

Village and township level administrative division is the biggest change, removal and polymerization is the most frequent primary division unit, change is the dominant factor of economic development. In recent years, with the deepening of the reform, too small scale of villages and towns, from the perspective of economy and the government streamlining, across different levels of the villages and towns removal, removal rate are positively correlated, and economic development, namely the more the more developed areas should develop aggressively.

REFERENCES

[1] Holmes D. E-Gov: E-Business Strategies for Government. *London: Nicholas Brealey Publishing,* 2001:2.

[2] Bellamy C. & Taylor J A. Governing In The Information Age. *Buckingham: Open University Press,* 1998:37.

[3] West D m. Digital Government: Technology and Public Sector Performance. *New Jersey: Princeton University Press,* 2005:13.

E-Commerce, E-Business and E-Service – Lee (Ed)
© 2014 Taylor & Francis Group, London, ISBN 978-1-138-02646-9

The role of Facebook as an interactive communication device for business

I. García Medina
Glasgow Caledonian University, Glasgow, UK

P.A. Pereira Correia, Z. González & R. Contreras-Espinosa
University of Vic, Barcelona, Spain

ABSTRACT: In the last century, the advances in technology have brought enormous changes for consumers, markets and marketing. From the early years of this century, information technologies have been invading the conventional market, evolving into what it is today. One of the factors that enable this technology wave is social media. As the possibilities of expression are increasing in social media, consumers also increase their ability to influence others through their own experiences. Some cases, particularly Facebook, have seen a spectacular growth. There are few studies on the phenomenon, so it is intended, in this study, through empirical observation and theoretical analysis to analyze the reactions of individuals in social media (particularly in Facebook) and its confluence with the organizations.

1 THE CONSOLIDATION OF SOCIAL NETWORKS ON A DAILY BASIS IS INCREASINGLY INTENSE. ACCORDING TO HAMEL (2007), "THE ADVANTAGE OF COMPETITIVENESS COMES FROM A TECHNOLOGICAL INTRODUCTION AND THE CONSTANT ADAPTATION ALONG TIME TO THE TECHNOLOGICAL EVOLUTION INSERTED IN THE STRATEGY OF THE ORGANIZATION." HAMEL (2007)

These new technologies allow the connection and interaction between individuals and are basically composed of three fundamental forces: computers and mobile phones, the low cost network connection and open source (free software) Rosales (2010).

Social media have a low cost level and are bias free Rosales (2010); therefore, they are the future of marketing. The connections between friends through social networks like Facebook also help companies to know the market better.

The value of something is not what it costs, but what we give up to get it. Many companies still have not integrated new technologies into their work processes, even when they assume that they will depend on it; they prefer to continue operating as usual since old ways still work for them. The opportunity cost that companies are paying for this attitude will be realized when competitors start taking their clients or when they recognize they are unable to win a proposal because they are not in a closer position to other candidates Tasner (2010).

The intention of the study is to understand the Facebook contribution as a marketing communication device for business and find out how participants interpret their experiences and the way they structure the social world in which they live, in order to build knowledge, understanding behavior and human experience.

2 THEORETICAL FRAMEWORK

Internet is now an indispensable communication tool for companies covering a wide range of functions, as a way of advertising, to educate or to communicate, as a way of interact, to construct and to support relations or simply as a way of entertainment, García (2011) and Peri (2009).

Nevertheless, many organizations simply have created electronic versions of printed materials. In many cases, the organizations refuse to use the whole potential of the collaborative tools (blogs or forums of discussion), with the fear of losing the control of the published information about the organization and the direct relationship with the public Peri (2009) and Jenkins (2011).

In marketing, the frequent communication plays an important role, but the technology and other factors have altered the way in which the consumers check the information and where they choose to check it. The rapid diffusion of Internet of broad band and the social media have led the professionals of marketing reflecting about the challenges that they need to consider and the changes occurred in the traditional practices (Kotler et al. 2009).

The web offers a simple way of spreading the ideas instantaneously to potentially millions of people, understanding and telling histories related to their problems Scott (2010). The social networks bring again to the world the art of the storytellers, with contents such as movies, advertising, commercial information, or presentation of the company. A good story is funny, sensational and emotional, has full of sense, is didactic, is not an intruder, since it works for invitation and favors the connection, participation and interactivity of his addressees Godin (2009) and Núñez (2009).

In this circle of social media are the social networks characterized by their social structure composed by people or organizations connected for one or more types of relations, which share values and common aims. They are inside websites such as Facebook, Twitter, MySpace, Tuenti (Spanish community) or LinkedIn, where the people connect with their friends (acquaintances offline and online or just online). In terms of organizational marketing, it represents an opportunity because the companies interact with the clients, through profiles, groups and pages (public profiles of organizations) Li & Bernoff (2011) & (Argal, et al. 2008).

The social networks are characterized principally by the opening and tolerance between the participants due to the fact that the relationships are horizontal and marked by expectations, confidence and loyalty. In spite of the fact that the major networks are orientated towards the relations (MySpace, Facebook, Twitter), they also can have a professional or political slope (LinkedIn, Xing). With an orientation for the relations, Facebook is an emergent space of convergence of the preexisting mass media, collaborative heterogeneous cultures and participating audiences, convergence that approaches the interrelationship between the audiences in the attempt of finding in the social network the characteristics of the new media cultures, (Adaime, et al. 2010).

Some authors, specifically Zarella (2010), emphasize from between the social media the Facebook Page, as a platform of spreading products and services or just a platform useful for promotion, principally for new products and to keep the reactions concerning the brand.

Customers use social media to analyze products, services and brands independently of the presence of the company. With a reduced cost, the companies must promote their participation in the social networks, which can generate a great financial impact in their business Zarella (2010).

3 METHODOLOGY

This research required a major theoretical work of conceptual understanding of the elements that comprise it, their historical evolution and their practical application, and a qualitative analysis about the behavior, reactions and attitudes of individuals towards organizations in the social network.

In this sense, the theoretical analysis in the field of social networks supported the propositions discussed in the empirical analysis, which is based on a random sample of individuals and representative companies.

The literature review allowed contrasting, updating, and complementing relevant data. Collecting information on the object of study portrayed in the theoretical framework gave way to the analysis of the different methodologies and to the discovery of relevant data that helps us understand the actual status of the research on our object of study.

Thus, the methodology follows a qualitative interpretative and exploratory study of Facebook case, which identifies the topic that the study wants to explore guided by the theoretical lines, a kind of primary theory which provides a framework for their investigation.

The modes of analysis of gathering, analyzing and interpreting qualitative data followed the propositions derive from theoretical framework analysis as a grounded theory which is a research method that seeks to develop theory that is grounded in data systematically gathered and analyzed. According to Martin and Turner (1986), grounded theory is "an inductive, theory discovery methodology that allows the researcher to develop a theoretical account of the general features of a topic while simultaneously grounding the account in empirical observations or data." The major difference between grounded theory and other methods is its specific approach to theory development—grounded theory suggests that there should be a continuous interplay between data collection and analysis.

The data process collection used different empirical materials to support the observation and interviewing methodology thorough telephone, Skype, electronic devices (table and computers) and face-to-face approaches following in some cases a structured interview and in other cases a semi-structured interview supported both by an interview guide with closed and open-ended questions.

The analyzed universe consists of a randomly unrepresentative group of consumers living in Portugal, particularly in the cities of Funchal (Madeira) and Lisbon, and by company representatives established in Funchal (Madeira), in order to understand if the theoretical arguments are also verified in these regions regardless of their particular characteristics, especially the geographical and demographic.

The choice of these cities for the empirical analysis is due to the fact that Lisbon is a very important city in Portugal, since it is the capital, and Funchal is the capital of an island territory marked by its geographic separation from European territories, where social networks are especially important to overcome the physical distance from their customers.

Embracing the exploratory purpose of the study, the selected sample is illustrative and not representative. It is a non-random sample according to the snowball sampling model, where the interviewee was asked to suggest another person for the study. The first person is chosen by the researcher following the criteria that may be more useful or representative for the study, which follows the model purposive (judgmental) sampling Babbie (2007).

The sample aggregates a small group composed of 15 Portuguese companies divided into four groups according to the number of employees the company has, and a larger group of 150 individuals. The period of observation was between October 2011 and January 2012.

The sample of companies was from the Madeira hotel sector due to the international characteristics of their customers.

The 4 company groups were divided according to the criteria of the recommendations of the Commission of the European Union May 6, 2003, based on the number of employees. In this sense, we have a group of micro companies (up to 10 employees), another of small business (10 to 50 employees), another of medium-sized enterprises (between 50 and 250 employees) and the last group of large companies (more 250 employees).

The contents of the interviews were analyzed following systematic procedures and an objective description of the data in the form of words, citations and transcripts of their responses, allowing a clearer understanding of the subject in study.

4 RESULTS

Proposition 1: The social network, as a means of communication is a social transformation of the users that integrates multiple communication tools rather than a simple technological change.

The majority of individual users agree that somehow other communication tools can be framed within the platform. The users give importance to the communication network as a multifunctional tool due to the concentration and consequent usability of the different communication tools an easy and fast platform to interact with friends and companies which implies a new behavior attitude.

Although the proposition has been validated by individual users, it wasn't totally confirmed by the company representatives, as only a minority sees Facebook as a tool that integrates other tools.

Thus, we can say that social networks are a social transformation of the users rather than a technological change.

Based on these theoretic findings we look to verify empirically, that the main reason for participation in social networking is social communication, but there are other interests related with the companies, their products and services. These business-related interests are increasingly important.

Proposition 2: The Individuals who participate in the social networks mainly for social reasons also accept communication with companies.

For the Facebook users group, the social relations are the main reason for using it, but they also give importance to the information about products/services.

Most of the participants are "fans" of a company, particularly the younger users, to follow the companies, their products and services, to stay updated, obtain support and read comments or replacing other communication tools, it further points out with "likes" about the published content and have a predilection for frequent communication.

So the emergence of social networks and the concentration of various communication tools into a single platform have captivated individual users and have revolutionized the way people communicate and share information, but have not been revealed as a purely technological change but, above all, as a cultural change.

Proposition 3: Should companies consider social media as a key element of their marketing strategy and not just as a channel of their traditional communications strategies and their participation on the social networks has changed the way they manage the communication with their audience.

The participants companies' representatives agreed with theoretical finds. Additionally sharing information interactively with clients could hardly be done through traditional channels. This bidirectional communication management is a key success factor of this type of communication strategies. Before the advent of social networks, organizational communication was traditionally characterized as unidirectional, the company transmitted the message and the audience was limited to hear it without having the option of responding, allowing the organization to control the message.

The bi-directionality between the company and fans is also a way to segment marketing activities and communication, an information source about needs and desires.

The evidences continue to show us that companies will pass through the biggest change of its short history. The models of network collaboration and innovation can provide managers with new possibilities to arouse the human potential, whose success lies in challenging the knowledge/experience that companies have accumulated over the years.

In recent years the marketing activities have been more extensive, going beyond the sales financial transaction. On nowadays the basic objective of trade relations goes far the simple buy/sale transaction, comprise also the establishment of stable and lasting relationships with the audience, which are beneficial for both parties and lead to an increase on trust and consequently the willingness to help each other, reflecting directly on the reduction of cost and time to complete the transactions.

So, currently organizational survival increasingly depends on socialization, sharing interests and activities with the audience and linked to the incorporation of digital technologies in their activities, especially those related to social networks. Technology emerges as a support for the satisfaction of social connection, a transformation of the communication between people and companies and making it much more dynamic and transparent.

Proposition 4: Within the social networks, Facebook is dominant.
The majority of the companies' representatives and individuals highlight three reasons to why Facebook is dominant:

1. Free communication media between people separated geographically.
2. Its capacity for mobilization and amplification of individual power.
3. The interaction and trust building between consumer and company.

Proposition 5: The adoption of Facebook technology is not sufficient for the integration in the online social movement.
The analysis of the interviews shows that the mere adoption of Facebook technology is not enough and that leads to a number of operational changes that go beyond the technological use and more closer to management procedures and users behavior which comprises a strategic approach to monitoring the consumers behavior, their needs and desires and that characterizes the company's behavior and interaction with his "fans" in the social network.

5 CONCLUSIONS AND IMPLICATION OF FINDINGS

Organizations should consider the way they communicate with their target audience and consider social networks and mobile technologies as a new way of expanding the business, adapting to this new consumer not contemplated by the traditional marketing and communication media.

The study presents a qualitative analysis of the behavior, the reactions and the attitudes of individuals towards organizations, with the aim of understanding which are the social factors that contribute to sustainable competitive advantage to organizations and support strategies and future actions.

The social aspects are a part of the experience in the Facebook community and also of the shopping experience. So it is important to monitor these behaviors in Facebook or other networks to perceive users of social networks, and consequently define marketing and communications actions to transform fans into customers.

Relevant factors come associated with tacit knowledge of the organization, particularly those related with learning and social interaction of the organization and organization knowledge about virtual communities. To a higher coexistence of these factors, the more difficult the replication is, and the higher the strengthening of the hypothesis of sustainability of competitive advantage.

Thus, it is important to note that monitoring and learning through experience in the network as part of culture and organizational routines are specific factors that may become a competitive advantage by being relevant factors associated with tacit knowledge of the organization.

For all this, we consider relevant and recommend further research on this phenomenon of vital importance to understand the new communication patterns between consumers and businesses, which will facilitate the economic development of the business and lead us to discover a new marketing paradigm.

In addition, it is recommended an extensive analysis involving research over time, so that the impact of social networks on the competitive position of the company can be investigated. Therefore, we propose a long-term research that allows, on one hand, to analyze the evolution of the various factors related to the social networks along the life of the organization, and on the other hand, to determine the effectiveness of communication/advertising, the integration in the marketing department, the network use efficiency, and results evaluations, but above all, the employees' behavioral changes which should lead to a more transparent and collaborative operational model.

Future research should focus on deepening knowledge on how to perform in social networks and their strategic potential for companies.

REFERENCES

Adaime, I. 2010. *El proyecto facebook y la posuniversidad. Sistemas operativos sociales y entornos abiertos de aprendizaje*. Editorial Ariel.

Argal, M., et al. 2008. *Web 2.0 y empresa. manual de aplicación en entornos corporativos*. Madrid: Asociación de Empresas de Internet.

Babbie, E. 2007. *The practice of social research*. USA: Thomson Wadsworth.

García, I. 2011. Marketing digital multimedia: nuevos formatos y tendencias. *Revista Géminis* 1: 37–45. Brasil.

Godin, S. 2007. *Permission marketing: turning strangers into friends and friends into customers*. Pocket Books, UK: Simon & Schuster UK Ltd.

Godin, S. 2009. *All marketers are liars: The power of telling authentic stories in a low-trust world*. USA: Penguin Group.

Godin, S. 2010. *Linchpin: Are you indispensable? How to drive your career and create a remarkable future*. UK: Piatkus Books.

Hamel, G. 2007. *The future of management*, Harvard Business School Press.

Jenkins, H. 2008. *Convergence culture: Where old and new media collide*. New York: New York University Press.

Kotler, P., Kartajaya, H., & Setiawan, I. 2011. *Marketing 3.0*. Madrid: LID Editorial Empresarial.

Li, C., & Bernoff, J. 2011. *Groundswell: Winning in a world transformed by social technologies*. Harvard Business School Press.

Martin, P.Y. & Turner. B.A. 1986. Grounded Theory and Organizational Research. *The Journal of Applied Behavioral Science*. 22: 141–157.

Núñez López, A. 2009. *Será mejor que lo cuentes*. España: Empresa Activa.

Peri, P. 2009. *Branding corporativo*, Colección de Libros de la Empresa, Santiago, Chile.

Rosales, P. 2010. *Estrategia Digital. Cómo usar las nuevas tecnologías mejor que la competencia*. Barcelona: Ediciones Deusto.

Scott, D. 2010. *The new rules of marketing and PR*, John Wiley & Sons, Inc.

Scott, D. 2011. *The new rules of marketing & PR: How to use social media, online video, mobile applications, blogs, news releases, and viral marketing to reach buyers & PR: how to use social media, blogs*. New Jersey: John Wiley & Sons; 3rd Edition.

Tasner, M. 2010. *Marketing in the Moment. The Practical Guide to Using Web 3.0 Marketing to Reach Your Costumer First*. New Jersey: Pearson Education, Inc.

Zarella, D. 2010. *The social media marketing book*, Canada: O'Reilly Media Inc.

E-Commerce, E-Business and E-Service – Lee (Ed)
© 2014 Taylor & Francis Group, London, ISBN 978-1-138-02646-9

E-Business models for financial services and internet banks

J.M. Sahut
IPAG Business School, Paris, France

ABSTRACT: Internet is not simply one more distribution channel among the multi-channel strategies used by the financial industry; it is fostering new "E-Business Models" such as Internet-primary banks. However, in spite of its strong development potential, this type of bank has often achieved a weak breakthrough onto this market and shows modest financial results. The goal of this article is to study the "E-Business Model" of Internet-primary banks and to determine if it can perform better than the "Business Model" of a traditional bank.

1 INTRODUCTION

Internet has given a new dimension to the convergence and the deconstruction of the value chain in the financial sector. Convergence has taken place on three levels:

- convergence of offers; by widening their product range, banks and insurance companies have entered into direct competition,
- convergence of the sub-sectors of the financial industry; banking, insurance and asset management activities increasingly overlap,
- financial institutions and non-finance actors have become more closely linked.

Previously, deconstruction came mainly from the offer side, i.e. from the emergence of new entrants (e.g. consumer credit). With Internet, it comes from the demand side: customers can choose the best supplier depending on their preferences (e.g. real estate loans, online brokering). With the appearance of banks which mainly sell their services by Internet (Internet-primary banks), the major competitive advantage of traditional banks—a network of local branches—has been diminished for certain types of customer. These customers have been attracted by the prospect of accessing their accounts and carrying out bank transactions 24 hours a day, seven days a week without having to go anywhere, and sometimes with a better quality of service than was offered by a bank branch. Moreover, the vast majority of Internet-primary banks charge lower account administration fees than those charged by traditional banks. This has often been used as an argument to attract customers in a nearly saturated market.

The goal of this article is to study the "Electronic Business Model" of Internet-primary banks and to determine if it can outperform the "Business Model" of a traditional bank. After having defined the Business Model and E-Business Model (E-BM) concepts, we will analyze the E-BM of online banks as an economic model through the study of its revenue sources, costs incurred, and how it creates value for customers. Then, we will question its strategic development prospects. Lastly, we look at Internet's impact on performance in the case of both traditional institutions as well as Internet-primary banks.

2 BUSINESS MODELS IN THE FINANCIAL SECTOR

2.1 *Business models and electronic business models*

Many papers have already been written which attempt to clarify this concept due to its dynamic dimension (George & Bock, 2011). We will look at how this concept has been applied to electronic services: the "Electronic Business Model" (E-BM).

One of the first workable definitions of this concept was provided by Timmers (1998). He defines a BM as being:

- the architecture for product, services and information flows including a description of the model's various business actors and their roles;
- a description of the potential benefits for each business actor involved;
- a description of the sources of revenue.

Other researchers have since expanded this idea. Linder & Cantrell (2000) state: "It's a rich, tacit understanding about how all the pieces work together to make money". These authors confirm this vision of BMs by the fact that 62% of the company directors they interviewed found it difficult to describe their BM over and above its success. While for Loilier & Tellier (2001), a BM can be likened to a firm's value creation method.

In fact, defining what a BM is can be a difficult exercise because this concept is associated with dynamic dimensions such as value creation, competitiveness and organizational change. Porter (2001) in particular described this concept as "fuzzy", "superficial", and "theoretically difficult to grasp". Magretta (2002) specifies that the usual error with respect to BMs is to regard them as being a strategy: "Business modeling is the managerial equivalent of scientific method—you start with a hypothesis, which you can test in action and revise when necessary". For this author, the BM describes in system form, how the firm manufactures and sells a service or a product. Equally, for Afuah & Tucci (2003), the BM is a basket of activities which allows a business to earn money in a sustainable way. Along the same lines, Demil et al. (2004) approach it as an intermediate concept used to make strategies operational. They define the BM as all the choices which a firm makes in order to generate revenue. Whatever definition you choose, it is important to dissociate the concept of BM from that of strategy and to regard it as a dynamic concept which is constantly reconsidered according to market conditions, technology, regulations, inter-company relations, etc.

The development of ICTs, and specifically the Internet, has generated new activities resulting either from technological innovation (for example Internet portals), from the destruction of value chains (e.g. loan comparison sites), or from new channels (e.g. online brokers). These activities then brought about the appearance of new BMs or redefined those which already existed (Applegate, 2001). For example, Internet strengthens the Research and Development function (a support service according to Porter's value chain concept) by helping in the collective design of products between sites and participants in the value system, by listing the concepts accessible to all the branches of the business and by giving access in real-time to all the sales and services databases. Also, Internet makes it possible to reduce order transmission times by automating both customer and supplier contacts and enables a truly integrated management system to be put in place. The impact of such practices on work efficiency creates an additional added value for the firm (Porter, 2001).

The term "Electronic Business Model" (E-BM) has since appeared to qualify the BM of these new activities. As the concept is derived from that of BMs, it is also difficult to define and can be understood in different ways (Casadesus-Masanell & Ricart, 2010). This is why many authors have tried to define it using a heuristic approach, in other words starting from their observations of the market. The result has been a wide diversity of studies which cover both the number of E-BMs identified and their characteristics: Timmers (1998) counted eleven different types, Loilier & Tellier (2001) counted five and Rappa (2004) nine.

Even if many similarities are identified, convergence between these typologies started to take place in 2001 with the analysis of value creation. Indeed, Timmers (1998) has a vision which focuses on the internal dynamics of E-BMs and on their interaction with the environment. As for Mahadevan (2000), he prefers a "macro" vision in which E-BMs depend on the types of relations which exist between actors from the same market. Applegate's article (2001) marks a transition by putting forward a very precise classification of E-BMs concerning value creation for the E-BM (sources of differentiation, revenue and costs incurred), but the value created for customers is not a central issue. It wasn't until Novak & Hoffman's article (2001) that the different dimensions of E-BMs were brought closer together. They present "Customer Model Integration" in which the definition of an E-BM is jointly linked to both value models for customers and revenue models.

In particular, Novak and Hoffman identify twelve value models. However, their revenue or value models created for customers cannot be regarded as exhaustive for two primary reasons:

- Given the difficulty of defining only one typology of E-BM, it is possible to present as many models as there are combinations of sources of revenue; the appearance of a new E-BM means the existence of new original combination of sources of revenue, even the creation of a new element for generating revenue;

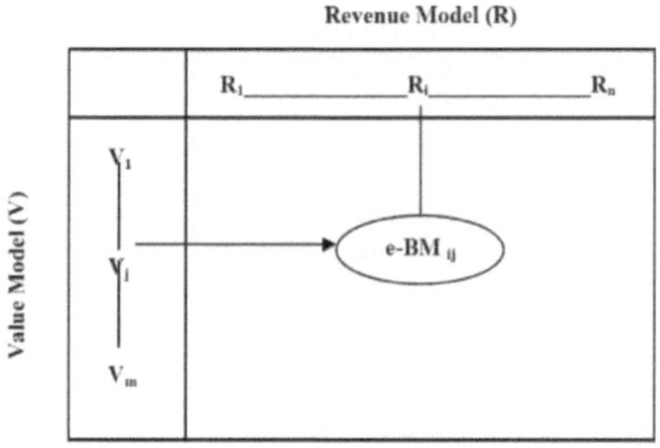

Figure 1. Values, revenues and customer model integration.
Source: Adapted from Novak and Hoffman (2001).

- It would seem that revenue sources other than those presented by Novak and Hoffman can exist, for example the sale of customer data (e-mail, addresses, purchasing behavior…), a purchase made in a store following the consultation of the product on Internet, etc.

Moreover, the type of activity strongly influences the structure of an industry (Casadesus-Masanell & Ricart, 2010). As a result E-BMs are very different from one sector to another. Hereafter, we will look at the financial sector and at the different E-BMs which characterize it, and more specifically, at Internet banks.

2.2 *Typology of E-Business Models for financial services*

Due to their apparent contradictions, it is difficult to come to any firm conclusions on the positioning of Internet-primary Banks compared to more general E-BMs. On the other hand, if one focuses on the specific sector of financial services, a greater convergence of E-BMs can be clarified by studying the work of Horsti et al. (2005), Zott & Amit (2010), and George & Bock (2011).

Between these different E-BMs for financial services, there exist several forms of exchange, partnership and strategic alliances. In particular, portals offer virtual banking agencies, aggregators and financial actors in general a space to advertise and be referenced. Moreover, certain portals develop co-branding strategies (coproduction) by coming together with certain producers to produce or sell joint services. Thus while calling into question the comparative advantages of traditional actors, the financial service sector on Internet has become very competitive and obliges actors to innovate by creating alliances, partnerships or mergers making it possible to build up new comparative advantages, to catch up with competitors, or to break into new markets. The movement which recently saw a consolidation of online brokers in Europe illustrates this phenomenon. Between 2002 and 2006, 22 brokers were taken over or merged in France (Les Echos, n 19783, 30/10/2006).

Table 1. Typology of E-BMs for financial services.

Type	Characteristics	Revenue created	Value model
Vertical portal Financial portal [MUY 01] Portals [SAH 01]	They mainly offer general or specialized information services in several fields. Some have created "personal finance spaces", like Yahoo with "Yahoo Finance", which proposes, in addition to informational services, transaction services (credit card, account aggregation, advice about stock market investments, etc).	Advertising, Affiliation, Commission, Sale of products and services.	Brokerage; Content; Search; Communication; Community; Affiliate, Transaction.
Aggregator [MUY 01] Aggregators – Brokers – Quoters [SAH 01]	These are sites whose role is to act as online intermediary between different actors. "Quoters", contrary to brokers, do not carry out the transaction, they are infomediaries which bring business to virtual agencies, or compare offers for consumers.	Brokers: Sale of products and services. Quoters: Commission, Advertising, Affiliation.	Brokers: Content, Transaction. Quoters: Content, Brokerage.
Speciality manufacturer [MUY 01] Suppliers [SAH 01]	These are producers of financial services (like Visa, equity funds, traditional banks, etc) which distribute them through their own network, or external networks (resale or co-branding).	Commission, Sale of products and services.	Outsourcing, Transaction.
Company web sites [MUY 01] Virtual agencies (banks, insurance, or broking on line) [SAH 01]	These are online banking, investment or insurance services. The sites with the best performance offer, in addition to advanced information and transaction services, customer relations management services (tools to help in decision-making, online advice, development of personalized products, etc). The main difference with brokers/aggregators is that they are not satisfied to aggregate the existing offer. They mainly sell products of their brand.	Product sales and services, Commission, Advertising.	Outsourcing, Transaction, Content, Brokerage, Community.

Moreover, it is difficult to understand the financial equilibrium of E-BMs because not only have they brought about new commercial concepts but, above all they have provided a new logic for value creation. In the start-up phase, companies in the E-Business field generate large negative free cash flows for a certain length of time, before turning round and progressing exponentially. This type of cycle which is more marked than for traditional industries can be explained in particular through the theory of the network economy (Shapiro and Varian, 2003). In fact, the main specificities of E-BMs which develop in an economy based on networks show that:

- profitability is only achieved by developing the use of a good or product (one speaks about "an experience good") because the consumers are not motivated to buy or carry out transactions on line when they have doubts about the quality of the products or the operation's level of security. Once the "reputation" of the E-BM has been made, this type of psychological barrier decreases;
- profitability is determined by "the attention" paid to the E-BM. In fact, Internet provides an important mass of information with rapid, permanent and inexpensive access. Many BMs therefore try to benefit from this by setting up an informational website. Thus, gradually, competition in the physical markets has moved onto Internet. However, there are so many websites for companies selling the same products that saturation has been reached: far from increasing their visibility, these companies have provoked an "attention shortage". It is therefore imperative for an online bank wishing to "grab the attention" of customers from other banks already offering Internet Banking services, to propose superior value to that offered by its competitors. This requires big investment (in the development of services and the acquisition of customers) and explains the high level of negative free cash flows suffered by E-BMs at the beginning of their life cycle;
- profitability depends on the technological infrastructure and the capacity to give added value to the offer on Internet. In fact, contrary to a BM, E-BMs are characterized by the high costs of technological infrastructure (computers, software, data-base servers, computer maintenance, network equipment, etc.). This technological infrastructure can be the basis for the acquisition of a competitive advantage as soon as it improves capacity for storing, researching, sorting, filtering and sending out information (Applegate, 2001). It increases, at the same time, the value of the information itself (and the service provided to the customers in general). But the high fixed cost that technology represents cannot be amortized until a critical number of customers has been reached. However, such an acquisition takes a long time to achieve.

Lastly, we can notice that the profits expected by Internet Banking services are more indirect, in other words they come from efficiency to the detriment of productivity in value. Even before the appearance of Internet banks, Rowe (1994) had noticed that the IT expenditure of French banks was increasing faster than productivity. He saw this phenomenon as confirmation of "the Solow paradox".

Should we therefore call into question Internet's productivity and its contribution as a channel of distribution? Will the traditional banks continue to perform the same with or without Internet? What are an online bank's real prospects for profitability?

3 E-BUSINESS MODEL, INTERNET BANKS AND PERFORMANCE

After having analyzed different E-BMs, their sources of revenue, their costs and their method of creating value, we study the profitability of Internet banking services.

The few studies which have looked into this aspect are divided into two categories: those interested in the introduction of Internet banking services on the profitability of traditional banks (known as click-and-mortar), those comparing the performance of Internet banks with other banks.

The first studies tried to show whether or not the introduction of Internet banking services in traditional banks in the USA increased their profitability (Egland et al, 1998; Carlson et al, 2000; Furst et al, 2002). But, these services represented too small a proportion of the activity to really influence the profitability of these banks. Similarly, Sullivan (2000) showed that the multichannel banks of the 10th Federal Reserve District do not appear more profitable on average when they have a transactional website. In Italy, Hasan et al. (2005) demonstrate a positive relationship, over the 1993–2001 period, between Internet adoption and the profitability of click-and-mortar banks.

More recently, Degado et al. (2007) studied the impact of the adoption of Internet Banking services on the performance of 72 commercial banks in Spain over the period 1994–2002. They conclude that the effects of this adoption take time to appear and result in a fall of overhead expenses. One need 1.5 year to notice a significant increase in ROA (Return On Assets) and three years for ROE (Return On Equity). In this context, Internet is used more as a complementary channel than as a substitute to the physical branches.

These contrasting conclusions are debatable because they are based on average results obtained at the beginning of the development of these services. Thus they depended more on the customers' adoption of Internet than on the real contribution of Internet Banking services on the overall profitability of the bank. In fact, these studies have mainly highlighted the problems of measuring the profitability of these services.

The second wave of studies tried to free themselves from these limits by defining a broader measurement of the performance of Internet banks compared to that of the other banks. We are going to concern ourselves mainly with this second type of study. But beforehand, we will look at performance measurement in the banking sector.

3.1 *How to measure profitability?*

Demirguc-Kunt & Huizinga (2000) highlight that *the approach to profitability in banking and finance is characterized by its complexity and its multiform aspect.* They explain their analysis through five main points:

- The merging of "raw material", "money deposited", "final product" and "money loaned": the fungibility of money makes it more complicated to calculate profitability due to the difficulty of dissociating resources from their uses;
- The impossibility of establishing provisions of profitability in the short term because of the existence of several uncertainties which are part of bank-customer relations (loan prepayment, litigation, change of address, etc.);
- The difficulty in establishing profitability per product because traditional banking is based on linked product sales which have high indirect costs;
- The strong regulation (or commoditization) of some products: an innovation in the banking sector cannot be patented and can easily be copied;
- The strong constraint of rigidity of costs in banking which are mainly "overhead costs" and indirect. Defining the profitability of a product, a customer segment or a center of responsibility (branch, area, etc.) is a complicated task and depends on method of indirect cost allocation.

For all these reasons, banks performances are assessed ex-post starting from general accounting indicators such as the level of deposits, the losses on loans, the ROA (Return On Assets) or the ROE (Return On Equity). But, in the case of a comparison of the performance of newly created Internet banks with that of traditional banks, these indicators (ROA and ROE) are not very relevant because the net income can be negative (because the activity is starting up), and they do not take into account other elements such as "possible market power" (Shepherd et al., 1999).

Among the other methods used, the most famous is the profit efficiency model (DeYoung, 2005). This method allows us to identify between the technology-based scale effects and the technology-based experience effects, showing whether the profitability gaps evaporate as Internet banks grow larger, gain experience and capture economies of scale.

For Cyree et al. (2008): "Profit efficiency indicates how well management produces outputs for a given input mix along with other market characteristics and is measured as the distance from the best-practice frontier". Moreover, this methodology captures "technology-based experience" and "technology-based scale" effects.

3.2 *Do Internet Banks give a superior performance?*

On the basis of the idea that "experience" could be a determinant of cost reduction and production efficiency, DeYoung (2001) presented a first comparison of the ROA between newly chartered banks (newly created traditional banks) and Internet banks (Internet-primary banks) between 1997 and 1999. He notes that the Internet-primary banks show significantly lower profits than those achieved by the newly chartered banks because of difficulties generating deposit accounts and higher non-interest expenses. The gap is very wide during the first two years but is reduced quickly thanks to technology experience effects. The growth rate of the Internet-primary banks declines to meet that of the newly chartered banks. The banks then progress at the same rate except for the deposit-to-asset ratio. The maturity effects are similar for the two types of bank. Lastly, Internet-primary banks, just like the newly chartered banks, can only reach the same profitability (ROA) as traditional banks at the end of approximately 10 years of activity (DeYoung, 2001).

In a second study, DeYoung (2005) confirms these results by using the profit efficiency model. He shows that the Internet-primary bank startups tended to underperform the branch bank startups over the period 1997–2001 in the USA. This seems to call into question the viability of the Internet bank E-BM. One can conclude from this that the success of an Internet bank is only possible if it reaches a sufficient level of economy of scale and has efficient management practices, particularly for cost management.

The more recent results from Cyree et al. (2008), which study the performance of Internet-primary banks and newly chartered traditional banks from 1996 to 2003, provide more details of these performance gaps with certain conflicting conclusions to the studies of DeYoung (2001, 2005).

Their univariate analysis shows that Internet-primary banks have lower ROA, ROE, loan losses, and net-interest margin, compared to newly chartered traditional banks. But, they indicate that the Internet-primary banks are more profit efficient than the newly chartered traditional banks. In fact, several elements can justify the performance gap of Internet banks with traditional banks especially when starting up (EFMA, 2013):

- Incompressible structural costs: These are expenses inherent to all banking activities and are mainly composed of high fixed charges and costs of IT development. In the case of Axa Banque, these IT costs account for 30% of total operating costs,
- A high turnover of advisers and the difficulty of arranging schedules for hot line,
- A very high cost of customer acquisition: For Axa Banque, this cost must not exceed 300 euros per customer recruited to be profitable. This stumbling block can only be overcome with a diversification of customer recruitment methods, loss leaders, and precise customer targeting.

On the other hand, other factors are favorable to the development of online services including:

- The specific characteristics of customers who use on line banks: 37% are on average expert users in Europe. This clientele is autonomous in decision making, they are mainly men, managers, with a high income who subscribe regularly to financial products on line;
- Internauts are more profitable than non-internaut customers. In the case of Axa Banque, their internaut customers bring in 15% more revenue than other customers as soon as they join and achieve the target revenue after 18 months instead of 30 months on average;
- The potential of productivity for online banks is higher than that of the traditional bank: in the case of Axa Banque, the productivity by employee is 405 customers (which is a higher performance than the average French bank) with a productivity potential which can reach double that of a traditional bank;
- The structure of the net banking income from an online bank is very different from that of a traditional bank: for example, Axa Banque draws 70% of its net banking income from payment methods. The remaining 30% comes from the banking offer (credit, savings and stock exchange transactions) which have strong development potential;
- The backing of a big group means that certain costs can be minimized by benefiting from a phenomenon of material synergy (compensation operations, debit and credit cards, etc.) and intangible synergy (experience, notoriety, etc.). For example, the Midland Bank provided First Direct with a back office of 7 000 ATMs in Great Britain which was the basis of its success. The risk of the cannibalization of channels, competition between the online bank and the mother company (traditional bank) is low because the banks attract a different clientele.

In fact, these studies and research show that the success of an online bank is mainly conditioned by:

- Building a strong competitive advantage which lies in access to customers and a good understanding of their behavior and tastes. Many banks believed that the competitive advantage depended on the technology used; however, technological advances are very quickly copied. This advantage makes it possible to control the cost of acquiring a new clientele. The support of an institution delivering financial services (insurance, supermarkets, banks, etc.), the creation of an attractive loss leader (savings accounts with a high interest rate like ING Direct, etc.) or an innovating service (for example: online brokers at the end of the 90s, etc.) are the main means of building up a competitive advantage;
- The difficulty of maintaining continuous investments without making profits forces certain online financial operators to merge with each other. This was specially the case of certain online brokers who, when they started dealing on the stock exchange, were able to collect large funds. But, as their period of investing at a loss lengthened, they had difficulties accessing other investments. The only way out for those in most difficulty was to merge. One finds several examples in the case of online banks such as the absorption of ZeBank by Egg (which gave it access to the French market), the merger of First-e (Ireland) and Uno-e (Spain), etc;
- A strong brand which inspires trust from potential customers and which also helps reduce the cost of acquiring new customers.

These conditions highlight the difficulty of creating an online bank for actors from outside the banking sector.

4 CONCLUSION

In conclusion, we have shown that the dynamic of the financial services sector on Internet is intensifying and that a certain structuring of the sector can be observed. Indeed, the actors are looking for new "revenues-created values" combinations following the example of online brokers, such as Schwab, who have become information providers for the portals, and are trying to reach a critical size via mergers and strategic alliances.

However, the shape of an E-BM specific to online banks hasn't clearly emerged. Despite the strong development potential of online banks, their survival as a specific E-BM can be questioned because of their weak breakthrough onto the international and national markets and their modest financial results. The most successful actors have concentrated on market niches, like ING Direct in Europe.

One can question the capacity of online banks, as E-BMs, to remain banks which are virtual, independent and generalist. In this case, it is extremely probable that they will have to develop their own physical sales network (like Schwab) or enter into partnerships with traditional establishments. Otherwise, they are likely to disappear or be absorbed by traditional banks and become gradually integrated into their multi-channel strategy.

REFERENCES

Afuah, A., & Tucci, C. 2003. Internet Business Models and Strategies. New York: McGraw-Hill.

Applegate, L.M. 2000. Emerging e-Business Models: Lessons from the Field. Boston: Harvard Business School Press.

Berger, A.N., Demsetz, R.S., & Strahan, P.E. 1999. The consolidation of the financial services industry: Causes, consequences, and implications for the future. Journal of Banking & Finance 23(2–4): 135–194.

Casadesus-Masanell, R., & Ricart, J.E. 2010. From Strategy to Business models and onto Tactics. Long Range Planning, 43(2–3): 195–215.

Carlson, J., Furst, K., Lang, W., & Nolle, D. 2000. Internet Banking: Markets Developments and Regulatory Issues. Office of the Comptroller of the Currency, Economic and Policy Analysis. Working Papers 2000-9.

Cyree, K.B., Delcoure, N., & Dickens, R. 2008. An examination of the performance and prospects for the future of internet-primary banks. Journal of Economics and Finance. June.

Delgado, J., Hernando, I., & Nieto M.J. 2007. Do European Primarily Internet Banks Show Scale and Experience Efficiencies? European Financial Management 13 (4): 643–671.

Demil, B., Lecoq, X., & Warnier, V. 2004. Le business model: l'oublié de la stratégie? 13th AIMS Conference, Normandie. 2–4 June.

Demirguc-Kunt, A., & Huizinga, H. 2000. Financial structure and bank profitability. World Bank, Policy Research Working Paper Series 2430.

DeYoung, R. 2001. The financial progress of pure-play internet banks. Bank for International Settlements, Monetary and Economic Department. BIS Papers n 7: 80–86.

DeYoung, R. 2005. The Performance of Internet-Based Business Models: Evidence from the Banking Industry. Journal of Business 78 (3): 893–947.

Essayan, M., Rutstein, C., & Wetenhall, P. 2002. Activate and Integrate: Optimizing the Value of Online Banking. Boston Consulting Group.

EFMA (2013), Banking in a digital world: Entering the next level. EFMA Studies. October

Egland, KL, Robertson, D., Furst, K., Nolle D.E., & Robertson, D. 1998. Banking over the Internet. Office of the Comptroller of the Currency. Currency Quarterly 17: 25–30.

Furst, K, Lang W.W., & Nolle D.E. 2002. Internet banking. Journal of Financial Services Research 22: 95–117.

George, G., & Bock, A.J. 2011. The Business model in Practice and its Implications for Entrepreneurship Research. Entrepreneurship Theory and Practice, 35(1): 83–111.

Hasan, I., Zazzara, C., & Ciciretti, R. 2005. Internet, Innovation and Performance of Banks: Italian Experience. Unpublished manuscript.

Hensman, M., Van den Bosch, F.A., & Volberda H. 2001. Clicks vs. Bricks in the Emerging Online Financial Services Industry. Long Range Planning Journal 34: 33–235.

Hernando, I., & Nieto, M.J. 2007. Is the Internet delivery channel changing banks' performance? The case of Spanish banks. Journal of Banking & Finance 31(4): 1083–1099.

Horsti, A., Tuunainen, V.K., & Tolonen, J. 2005. Evaluation of Electronic Business Model Success: Survey among Leading Finnish Companies. Proceedings of the 38th Annual Hawaii International Conference on System Sciences, Volume 7.

Linder, J., & Cantrell, S. 2000. Changing Business Models: Surveying the Landscape, Accenture Institute for Strategic Change.

Loilier, T., & Tellier, A. 2001. Nouvelle Economie. Net organisations. Paris: EMS Eds.

Magretta, J. 2002. Why Business Models Matter? Harvard Business Review, May: 90–91.

Mahadevan, B. 2000. Business Models for Internet-based e-Commerce: An anatomy. California Management Review, 42(4): 55–69.

Novak, T.P., & Hoffman, D.L. 2001. Profitability on the Web: Business Models and Revenue Streams. eLab Position Paper. Owen Graduate School of Management. Vanderbilt University. January: 9–18.

Porter, M.E. 2001. Strategy and the Internet. Harvard Business Review, June.

Rappa, M. 2004. The utility business models and the future of computing services. IBM System journal, 43(1):32–42.

Rowe, F. 1994. Des Banques et des Réseaux: Productivité et Avantages Concurrentiels. Economica: 246–247.

Sahut, J.M. 2001. Vers une révolution du secteur bancaire? La Revue du Financier. n 131: 34–38.

Shapiro, C., & Varian, H.R. 2003. Information rule. Ethics and Information technology. 5(1): 51–73.

Sullivan, R.J. 2000. How has the adoption of internet banking affected performance and risk in banks? Federal Reserve Bank of Kansas City. Financial Perspectives. December: 1–16.

Timmers, P. 1998. Business models for electronic markets. Electronic Markets 8(2): 2–8.

Zott, C., & Amit, R. 2010. Business model Design: An Activity System Perspective. Long Range Planning, 43(2–3): 216–226.

E-Commerce, E-Business and E-Service – Lee (Ed)
© 2014 Taylor & Francis Group, London, ISBN 978-1-138-02646-9

Localization for a high context culture: An exploratory study of cultural markers and metaphors in Arabic E-Commerce websites

D. Liginlal & M. Rushdi
Carnegie Mellon University, Doha, Qatar

R. Meeds & R. Ahmad
Qatar University, Doha, Qatar

ABSTRACT: Researchers claim that arabs have an affinity for high context but low content. This paper aims to examine how well arabic E-Commerce websites have adapted their site designs and content strategy to conform to the needs of a high context culture. First, based on a set of cultural dimensions commonly considered important for localization of webpages, we analyzed a purposive sample of 39 arabic E-Commerce websites. Given that visual communication is an important facet of a high context culture, we then conducted a qualitative study of nonlinguistic metaphors gathered from 530 arabic E-Commerce websites across 22 arab countries. Finally, we sampled a set of linguistic metaphors used in arabic language websites to examine how attuned they are to a high context culture. The results of these studies, which examine three facets of a high context culture, provide initial insights for web designers, arabic language translators, and managers of global companies who aim to expand their businesses into this fast developing region.

1 INTRODUCTION

1.1 *A cultural journey through commercial Arabia*

The very mention of the Arab world and its rich cultural history brings to one's mind visions of Arab traders on camels, checkered headdress, veiled women, incense, magic carpets, calligraphic art, dhows, and couscous. Given that the Arab world spans 22 countries across a large geographic region, misconceptions about the culture are prevalent. Consider Turkey and Iran, which are part of the Middle East. Iran's official language is Persian and Turkey's official language is Turkish. Indeed, the Arabic language is the defining element of the "Arab world." Nevertheless, there are cultural and historical differences between different Arab countries. For example, the "Abaya" (ladies' wear) and "Thawb" (men's dress) are traditional dress only in some Arab countries. Similarly, the Qatari tea, Karak, would be good in advertising for drinks on a website in Qatar, but Moroccan tea would be the corresponding alternative in Morocco. Statues in Islamic culture are not preferred, but pictures of nature are. Colors are also culture specific. For example, on a happy occasion black would not be used. All these aspects of culture influence art, and in turn, the design of a website, particularly in the use of images, metaphors, and even colors. Islam has influenced Arabic calligraphy, and abstract artistic expressions are widely seen in Islamic art and architecture.

1.2 *Culturally rich history, but web design still lags behind*

It is natural to assume that the long and rich tradition of Islamic calligraphy has strongly influenced Web design in the Arab world. However, many studies have shown evidence to the contrary. Khashman and Large (2010) used a widely adopted cultural dimension model attributed to Hofstede (1991) to examine the design characteristics of Web interfaces from Arab countries. Their results from content analysis of a sample of 15 government Web portals suggest that Hofstede's model of culture does not seem reflected in the Arabic sites studied. Another study (Marcus & Hamoodi, 2009) of university websites from three different Arab countries concluded that the sites only partially exhibited characteristics of Hofstede's cultural model. Furthermore, studies on Arabic website design have mainly focused either on the differences that exist between Arab and Western countries (Al-Badi & Mayhew 2004, Al-Badi 2009) or on domains other than E-Commerce (Khashman & Large 2009, Marcus & Hamoodi 2009). Very little research has focused attention on cultural issues related to the E-Commerce websites of Arab countries (Liginlal & Pasi, 2012). The available ones are all based on Hofstede's (1991) model of culture. We aim to fill this gap in the literature.

1.3 *Implications to E-Commerce in the Arab world*

With the radical growth of E-Commerce in recent years, many companies have plunged into the global market without adequately understanding the cultural implications of their actions or without the right tools and models to do a cultural analysis. As the non-English-speaking population on the Web increases at a rapid rate, companies that respond positively to the need for multilingual sites will thrive. The localization of a webpage considers not just the need to bridge this language gap, but also the use of several cultural markers such as graphics, navigational elements, layout, color, and metaphors. Faiola and Matei (2005), in a study that explored issues related to Web designers' cultural cognitive styles and their impact on users' responses, also concluded that users perform information-seeking tasks faster when using Web content created by designers from their own cultures.

Therefore, this paper focuses on the interplay between Web design, culture, and language. We integrate three facets of Hall's (1976) model of high-low context cultures. First we report on the results of an exploratory study of bilingual Arab E-Commerce websites to understand how well these sites reflect the dominant characteristics of a high context culture. Then, we extend our study to 530 randomly selected Arabic language websites to specifically explore the use of visual communication elements. Finally, we selected a purposive sample of 50 websites encompassing multiple business categories and multiple Arab countries and studied the linguistic metaphors used in the home pages of these sites. We conclude with a discussion of the implications for Web designers and language translators and the expected contributions to promoting E-Commerce in the region.

2 LOCALIZATION IN ARAB E-COMMERCE WEBSITES

2.1 *Investigating localization using Hall's model*

Culture is considered an important determinant of website usability. Website localization is the process of modifying an existing website to make it accessible, usable, and culturally suitable to a target audience (Payne, 2007). The majority of studies on cultural adaption of websites have relied on Hofstede's (1991) five cultural dimensions of power distance, individualism/collectivism, uncertainty avoidance, masculinity/feminity, and long/short term orientation. Riding and Rayner (1998) have defined cognitive style as "an individual's preferred and habitual approach to organizing and representing information." Nisbett and Norenzayan (2002) have suggested that cultural variations in cognitive processes are tied to cultural differences in basic assumptions about the nature of the world; these basic assumptions are specifically related to differences in methods of reasoning, for instance the holistic vs. analytical reasoning differences between East Asians and Westerners.

Hall (1976) considered Arab countries as high context cultures. As Hall (2000) mentions, it is important to look at meaning and context together with the content. Further, greater confidence is placed in the nonverbal aspects of communication than in its verbal aspects (Wurtz, 2005), particularly signs, symbols, and metaphors. For our first study, we adopted the following guidelines based on the results of Wurtz (2005).

(i) Visual communication on Web sites is expressed in general through layout and the use of images, photographs, and animation to provide context. Also integration of multimedia effects, animation, and interactivity will essentially provide a sense of human presence. (ii) High context cultures are believed to be collectivist in nature, which implies they value group welfare over the goals of individuals. The imagery chosen in websites for high context audiences should reflect characteristics such as family values, with consequent expectation that pictures depicting groups of individuals and widgets facilitating group interaction would be prevalent on such websites. (iii) Expectations of inequality and power differences are also attributed to high context cultures (Marcus & Gould, 2000). (iv) Diversity in layout and color schemes is expected for high context cultures, given their parallel thinking patterns. Further, emphasis on the aesthetic experience is expected in websites directed at high context cultures.

2.2 *A localization study of Arabic E-Commerce websites*

Country-specific websites were chosen using Google search with specific keywords such as online shopping, fashion, clothing, travel, and bookstores. The final sample for the study was selected based on two criteria: first, each site was bilingual, and second, E-Commerce transactions were facilitated by the sites. Only the home page of each website was chosen for further inspection.

The analysis of the resulting 39 website designs focused on their functional, structural, visual, and graphical elements such as navigational schemes, menus, images, presence of search engines, site maps, color, gender in images, use of animation, etc. The term "cultural markers," as defined by Barber & Badre (1998), apply to these elements when certain attributes are most prevalent and preferred within a particular cultural group. The description and frequency of occurrence of these were determined by two judges. Before the analysis, these judges were trained on unsampled websites to identify any ambiguity in the analysis process. The results of classifying the cultural markers were compared, and all cases of disagreement were resolved through discussion with a third researcher.

2.3 *Results of the localization study*

It is believed that in high context cultures, the use of animation, especially in connection with human images, is more prominent than in low context cultures (Wurtz, 2005). The first interesting observation of our exploratory study is that a majority of the sites studied use both real and animated human images. Moreover, most of the websites used explanatory images rather than just aesthetically pleasing images of the products.

The results of a chi-square test, with a null hypothesis that 50% of the websites will have the cultural marker present, are summarized in Table 1. Thirty-two websites used detailed navigational cues typical of users expecting a high level of support instructions and detailed navigational cues at every step of their navigation of the website. We could not identify many websites with structured data presentation methods such as tabular columns or messages corresponding to authoritative figures that would indicate a strong hierarchy in information presentation. Company hierarchy information was present in a clear majority (79.5%) of the websites ($\chi^2 = 13.6$, df = 1, $p < 0.001$). Similarly, links to local sites were also noted in statistically half the sites (50.0%, $\chi^2 = 0.0$, df = 1, $p > 0.05$). One airline site provided a link to its government website, clearly depicting an affinity for high context. The majority of sites had privacy statements (79.5%, $\chi^2 = 13.6$, df = 1, $p < 0.001$), which are also considered to indicate a hierarchy. Also supporting this is the use of featured images depicting the products in use by individuals, which typically denotes a high context culture. From the above results, the power distance dimension of Hofstede's model (i.e., the extent to which less powerful members expect and accept unequal power distribution within a culture) (Nouf & Andre, 2010) was partially confirmed in our study.

Table 1. Results of chi-square tests for localization.

Variable	% Present	Chi square
Company hierarchy info	79.5	13.56***
Simple menus	92.3	27.92***
Easy access to info	66.7	4.33*
Live chat	15.4	18.69***
Newsletters	41.0	1.26 (ns)
Links to local sites	50.0	0.00 (ns)
Simple colors	74.4	9.26**
Use of images	94.9	31.41***
Privacy statement	79.5	13.56***
Clear gender roles	83.3	13.33**
Female images	41.0	1.26 (ns)
Guided navigation	79.5	13.56***

* = $p < 0.05$; ** = $p < 0.01$; *** = $p < 0.001$; all df = 1.

Twenty-eight websites used monotonous colors, and a high text-to-image ratio was not observed in most sites. Images were self-explanatory in relation to the products on display, and descriptions were of low information density. The website of a high context culture is characterized by the presence of many sidebars and menus that open onto new windows for each new page (Wurtz, 2005). In our study, 32 websites were characterized by linear navigation, and the menu levels of all the websites had a depth ranging from one to three. Most websites did not open a new window for every new page.

Arab websites are expected to use more images of men and animation (Nouf & Andre, 2010) than of women. Clear gender roles were apparent on most of the sites (83.3%, $\chi^2 = 13.3$, df = 1, $p < 0.001$), but statistically half the sites contained female-themed images (41.0%, $\chi^2 = 1.3$, df = 1, $p > 0.05$). Linear navigation of the pages also denotes a masculinity aspect in the website design (Reinecke et al., 2010). This seems to be prevalent in the sites, because they support linear navigation and the use of explanatory images of products. The website design in most cases can be described as "simple" and unsophisticated.

The simple menus measure contained little variance. Simple menus were found in 92.3% of the sites ($\chi^2 = 27.9$, df = 1, $p < 0.001$). Information was deemed easy to access in two-thirds of the sites (66.7%, $\chi^2 = 4.3$, df = 1, $p < 0.05$). Most of the sites used simple color schemes (74.4%, $\chi^2 = 9.3$, df = 1, $p < 0.01$). Overall, the Arabic websites satisfied most of the presence test for cultural markers conforming to a high context culture.

3 VISUAL AND LINGUISTIC METAPHORS IN ARAB E-COMMERCE WEBSITES

3.1 *Why investigate metaphors?*

The prominence given to symbolism in high context cultures implies that visual communication on websites is emphasized through layout and the use of images, photographs, and animation (Wurtz, 2005). Images particularly carry many subtle cultural messages besides conveying rich information about the product or company. The symbolism of icons and the signs they portray differ by culture. An image of a house to represent the home page, a letterbox indicating mail, and specific culturally sensitive images in logos are all examples that highlight the need for caution. Choosing a wrong color for a company logo or as the background of a banner message may turn away potential customers.

The word metaphor is often associated with a style of writing, used as a literary figure of speech for a descriptive effect. Metaphors are also based on embodied human experiences. Many studies have shown that the use of metaphors is pervasive in all domains of our everyday life. People use them to grab attention, win arguments, structure abstract concepts, and sell their products. In the globalized market of the 21st century, which is characterized by information overload, the use of metaphors can make one product stand out from the barrage of information about others that are competing for customers' attention. The power of metaphors lies in the visual images and meaningful associations they invoke in customers' minds. Metaphors present in their visual form on websites also help bridge the gap between virtual and real life, thus enhancing human experience.

Visual metaphors are widely used in E-Commerce to present messages in a succinctly and grab customer's attention. Like linguistic metaphors, the understanding and effectiveness of visual metaphors depends on the social and cultural factors of e-customers. With the emergence of multilingual websites that are marked by the use of linguistic and visual metaphors, it is crucial to understand how visual metaphors are "translated." Scholars argue that visual metaphors may not be universal (De Mooji, 2010). Although numerous studies show that linguistic metaphors, to be comprehensible and effective, need to be translated into the language of the customer, there seem to be no studies that investigate the use of visual metaphors, especially their use in E-Commerce.

Metaphors "play powerfully yet silently in the unconscious minds of consumers" (Zaltman & Zaltman, 2008). Over the past few decades, the importance of communicating meaning through visual resources has increased. Consumers on the Web have shorter attention spans. The use of metaphors helps speed up the process of scanning a web page. An example is our association of an exclamation mark with error. Metaphors help users understand, remember, and enjoy the entities and relationships of computer-based communication systems (Aykin, 2004).

People differ in cultural background, values, experiences, beliefs, and religious preferences. For a metaphor to be efficient, it must assimilate these elements. Therefore, translation has to go beyond just the written word. During the process of assembling a multilingual website, translating languages is just as important as localizing metaphors—that is, modifying them to accommodate a user's cultural background. Failure in localizing visual metaphors may lead to "cases of intercultural miscommunication" (Aykin, 2004).

3.2 *Results of our study of visual communication elements in E-Commerce websites*

We examined our earlier sample of websites for the presence of nonlinguistic metaphors and other user interface elements that corresponded closely to "Arab culture." We noted a predominant use of nonlinguistic metaphors (we counted up to 63 different metaphorical elements) that are in line with the Western genre. Of 39 websites, only four used the color red, which is seen in Arabic culture as symbolizing danger or evil, out of context (Barber & Badre, 1998). The color green, which is considered holy in the region, was used prominently by only one website; all other websites predominantly used the colors white and blue. Six websites reflected a strong Islamic orientation, and none of the sites included religious words such as "Inshaa Allah," and only three websites exhibited calligraphy, which is considered an essential aspect of Arabic culture, on their home pages. Ten of the 39 websites had images that might be considered unholy in Islam.

To extend our study, we manually inspected the home pages of 530 E-Commerce websites from 22 Arab countries encompassing multiple domains, such as tourism, fashion, and retail. The countries include Egypt, Jordan, Qatar, Bahrain, and Saudi Arabia. These websites are really designed for consumer interaction in the Arab region and not the world at large. Our results demonstrate that little attention was given to redesigning the visual elements in these websites to accommodate the culture of the Arabic world. Of these 530 websites, only 40% had an Arabic version. Of those that had an Arabic version, none had localized their iconic metaphors to embrace Arabic culture. Rather, these websites had Western metaphors. Western cultural values embrace both the European and American traditions, but their efficiency is questioned in the eastern part of the world. Despite common beliefs, graphics are not universal. Although the widespread use of the same road signs across countries has given many designers the notion that graphics speak a universal language, "it is a misconception to think that visuals are universally understood across cultures" (De Mooij, 2010).

Figure 1 helps summarize the status of visual metaphors in multilingual websites in English and Arabic. Arabs represent a distinct linguistic,

Figure 1. A sampling of visual metaphors from three E-Commerce domains.

cultural, and religious group, but no effort seems to have been made to adapt visual metaphors to accommodate them. Although we plan to conduct a study in which we will design visual metaphors that are more culturally congruent and test their effectiveness, we can conclude based on this analysis that the lack of adaptation of visual metaphors opens up an opportunity to design and test the effectiveness that their "translation" might bring.

3.3 *Why investigate the role of language?*

In a high context but low content culture, the subtle use of language plays a major role. The use of proverbs, well-known sayings, and metaphors is popular in almost all Arab countries. Most E-Commerce websites in Arab countries are first created in English. Translation of these sites into Arabic is not straightforward. It is necessary to consider whether there exist direct translations of all phrases and metaphors. The sensitivity of some messages must also be considered. For example, "every man for himself" might be culturally sensitive in a high context culture that is correlated with a collectivist dimension, but local alternatives to such phrases and to other content such as content designed for humor often need to be found. The different dialects of the Arabic language pose other unique problems. This makes it necessary to ensure that Modern Standard Arabic is used in these websites.

Almost all Arab countries have invested heavily in Internet infrastructure to deliver various services electronically to citizens, businesses, and other governments. Language plays a prominent role in the adoption and acceptance of these services. It is of particular significance for the services consumed by citizens, because they benefit the most from digital government and a great number of citizens in the region prefer the use of the Arabic language for consumption of government services. Language stimulates a positive feeling toward the use of these services in addition to enhancing perceived usefulness, perceived social presence, and trust. Therefore, examining the role of language richness is of significance to e-government. The appropriate language style for the target audience, the selection of content to carry over, ensuring that images with text are also converted, and adapting technical language without changing the meaning are all challenging tasks.

3.4 *Results of our investigation of linguistic elements in Arab E-Commerce websites*

The linguistic aspect is a natural and significant extension of the previous results, given the strong relationship between language, culture, and identity. Our study determined that most sites' bilingual versions are mirror images of each other, and no specific attention seems to have been given to cultural sensitivity. For example, a few websites depicted both Arabic and English product versions on the same page with no description of the product being sold. One particular instance of an English version of a website being dynamically translated into Arabic was also observed. Our study determined that the use of Modern Standard Arabic is prevalent in these websites. Although all of the links pointing to the Arabic language page did take us to an Arabic version, the following issues were noted. In most cases, the Arabic versions were only similar looking pages with Arabic subtitles, but all the merchandise names and descriptions were still in English. The one Kuwaiti website that had everything in Arabic had grammatical mistakes and did not use any figurative or metaphorical language. One fashion website only had headings such as "dresses" and "shoes," with pictures and no textual description. The English pages of the websites, in general, seemed to be more informative than the Arabic ones.

Some illustrative examples of Arabic language metaphors we extracted are shown in Table 2. A preliminary sociolinguistic analysis by an expert concludes that these linguistic metaphors reflect no significant characteristics native to Arab culture. All metaphors that we found are those that can be understood universally.

In conclusion, our exploratory study has revealed that the majority of sites that offer an Arabic version are literal translations that fail to capture the real meaning or richness of the expression and detail of the Arabic language. The Arabic language pages are only premature versions of the English language pages. They are not really

Table 2. Examples of Arabic language metaphors.

Examples of linguistic metaphors in Arabic	*Meaning in English*
بيت أحلامك	The house of your dreams
للمصرف كلاعب دولي	The bank as a global player
عجلة الإنتاج والصناعة	Wheel of production and industry
سير بورصة	Stock movement
مشاعرك بحرية	Nautical feeling
المدينة الوردية	Pink City
رؤيتنا هي "الجمال هو حجر الزاوية في الثقة بالنفس"	Our vision is "Beauty is the cornerstone of self-confidence"
الاتجاه الذي تسير فيه هذه الشركة	The direction of this company
شعر كثيف ولامع؟	Thick and shiny hair
استضافة بياناتك	Hosting your data
برحلة موسيقية	In a musical journey
ثورة المعلومات	The information revolution

authentic pages that reflect either Arab culture or the pristine form of the Arabic language. Our pilot study did not investigate e-government sites; however, the proposed research aims to include a wider spectrum of domains.

4 CONCLUSIONS AND FUTURE RESEARCH

"Think globally, act locally" is considered the key to winning consumer trust in E-Commerce. However, there is a dearth of studies aimed at understanding the extent to which E-Commerce websites are localized for Arabic language and culture. Many researchers have argued that culture affects all facets of our lives and that websites are no exception to the influence of culture. Researchers also claim that Arabs have an affinity for high context but low content.

Our exploratory study of the prominent E-Commerce websites of businesses based in Arab countries has revealed that a majority of these bilingual sites, i.e., those that offer an Arabic version, are designed for a high context culture. The study also determined that the Arabic pages of these sites are literal translations that neither capture the real meaning nor the richness of expression and detail of the Arabic language. This is significant considering that the volume of E-Commerce transactions in this region is increasing rapidly. In Qatar, the build-up to the 2022 FIFA World Cup, along with a wave of investment in IT products and services, makes it the fastest growing market in the region. One important outcome will be the recognition of Qatar as a regional leader in localized digital content. To achieve this outcome, the national plan considers as significant drivers the increased creation and production of Arabic digital content, increased content sourcing from the region, and the digitization of cultural artifacts. Further, the plan emphasizes the need to address an unmet demand for more Arabic digital content along with increased preference for browsing in Arabic. The results of our research shed light on both linguistic and nonlinguistic factors that make Arabic E-Commerce websites more effective.

The results of the exploratory study reported in this paper may be considered preliminary. The authors are actively involved in a major funded research project that investigates the use of linguistic metaphors in Arabic language E-Commerce websites. This larger study has been undertaken with a larger sample size and the support of a number of sociolinguistic and E-Commerce experts. The study aims to build an Arabic language corpus and establish content patterns for E-Commerce websites. The project aims to create and test a model that connects linguistic richness with linguistic perception, social presence, and trust. Also, the project proposes to develop and disseminate a set of best practices for Arabic E-Commerce websites.

ACKNOWLEDGMENT

The authors thank the Qatar National Research Fund for supporting this study through a National Priorities Research Project.

REFERENCES

Al-Badi, A. & Mayhew P. 2004. Designers Perspective of Website Usability: The Cultural Dimension. *IADIS International Conference* WWW/Internet, Madrid, Spain.

Al-Badi, A. 2009. Designing Usable Localised Websites: Differences between Western and Arab Cultures. *IBIMA Business Review*, Vol. 2, No. 2, pp. 9–19.

Aykin, N. 2004. *Usability and internationalization of information technology*. Mahwah, NJ: Lawrence Erlbaum.

Barber, W. & Badre, A., 1998. Culturability: The Merging of Culture and Usability. Paper presented at *4th Conference on Human Factors & the Web*, Baskin Ridge NJ, USA.

De Mooij, M. 2010. The Hofstede model. *International Journal of Advertising*, 29(1), 85–110.

Faiola, A. & Matei, S.A. 2005. Cultural cognitive style and web design: Beyond a behavioral inquiry into computer-mediated communication. *Journal of Computer-Mediated Communication*, 11(1). http://jcmc.indiana.edu/vol11/issue1/faiola.html.

Hall, E.T. 1976. *Beyond culture*. New York, Doubleday. Hall, E.T. 2000. Context and meaning. In L.A. Samovar & R.E. Porter (Eds.), *Intercultural communication: A reader*, 9th ed. (pp. 34–43). Belmont, CA: Wadsworth Publishing Co.

Hofstede, G. 1991. *Cultures and Organizations*: Software of the Mind, McGraw-Hill, London.

Khashman, N. & Large, A. 2010. Cultural design analysis of Arabic websites. *Human Factors in Computing Systems*—CHI conference, Atlanta, USA.

Khashman, N. & Large, A. 2009. Investigating the design of Arabic web interfaces using Hofstede's cultural dimensions: A case study of government web portals. *Proceedings of CAIS*, pp. 1–5.

Liginlal, D., & Pasi, S.J. 2012. Cultural adaptation and Arab e-commerce: An exploratory study. *IADIS International Conference, eSociety*, Berlin.

Marcus, A. & Gould, E.W., 2000. Cultural dimensions and global Web user-interface design: What? So what? Now what?. *In Proceedings of the 6th Conference on Human Factors and the Web*, Austin, Texas, USA.

Marcus, A. & Hamoodi, S. 2009. The Impact of Culture on the Design of Arabic Websites. In Proceedings of the 3rd International Conference on Internationalization, Design and Global Development: Held as Part of HCI International, pp. 386–394.

Nisbett, R.E. & Norenzayan, A. 2002. Culture and cognition. In H. Pashler & D.L. Medin (Eds.), Stevens' *Handbook of Experimental Psychology: Vol. 2: Cognition* (3rd ed., pp. 561–597). New York: John Wiley & Sons.

Nouf, K. & Andre, L. 2010. Investigating the design of Arabic web interfaces using Hofstede's cultural dimensions: A case study of government web portals. *Proceedings of CAIS*, 1–5.

Payne, N. 2007. *Culture and Website Localization*. TranslationDirectory.com.

Reinecke, K., Schenkel, S. & Bernstein, A. 2010. Modeling a user's culture. in *The handbook of research in culturally aware information technology: Perspective and models*, IGI Global press.

Riding, R. & Rayner, S.G. 1998. *Cognitive styles and learning strategies: Understanding style differences in learning and behavior*. London: D. Fulton Publishers.

Würtz, E. 2005. Intercultural Communication on Web sites: A Cross Cultural Analysis of Web sites from High Context Cultures and Low Context Cultures. *Journal of Computer Mediated Communication*, 11(1), 274–299.

Zaltman, G., & Zaltman, L.H. 2008. *Marketing metaphoria: What deep metaphors reveal about the minds of consumers*. Harvard Business Press.

E-Commerce, E-Business and E-Service – Lee (Ed)

Entrepreneurial learning in a secure E-Commerce course through creating competitive real-world sites

Ryan Garlick
University of North Texas, Denton, TX, USA

ABSTRACT: In Spring 2013, the Secure E-Commerce course curriculum was revised to have both undergraduate and graduate teams creating functional real-world sites while trying to compromise the security of the other team during and after the development process. Outcomes from the course included a now functional E-Commerce site currently generating revenue for the ACM chapter at the University of North Texas. This paper presents the results of the process of active, entrepreneurial learning in E-Commerce from a pedagogical, motivational, and functional standpoint, with notes toward future course design and implementation.

1 INTRODUCTION

As a now well established subject area in computer science and business curricula, E-Commerce education has evolved rapidly. Active, experiential learning using industry tools has taken an important role [Ngai, 2005], as well as 'service' or 'community' based education with students working in conjunction with established businesses [Tabor, 2005]. For a thorough history of the development of E-Commerce courses, see [Abrahams & Singh, 2010]. We present a novel approach that guides students in two competing teams through the entire entrepreneurial approach to creating a secure online business [Garlick, 2013].

1.1 *Course details*

CSCE 4560/5560 Secure Electronic Commerce is a blended undergraduate/graduate course at the University of North Texas. In the Spring of 2013, 21 Undergraduate students and 13 graduate students enrolled. The class was divided into graduate and undergraduate teams, who then chose group membership within their team as detailed in Section 2.2. The goal for the course was to move beyond the typical case studies or 'service learning' of providing services to businesses in the community and create E-Commerce sites from the ground up. This necessitates the selection of product, platform and infrastructure, security provisions, business and graphic assets and the essential entrepreneurship quality of learning how to implement solutions to problems as necessary. In this way, the course was designed to give students the necessary skills to implement a secure E-Commerce site with exposure to marketing, entrepreneurship, advertising, business, fulfillment, software development, and payment gateways—with an over-arching focus on security.

1.2 *Security*

Students were encouraged to actively probe the defenses of the opposing team's site during the development process, with a few caveats: First, if a security breach was possible, nothing destructive should be done until the final presentation, and then only as a demonstration that could be reversed. Secondly, although other possibilities were discussed, only application-layer attacks were permitted. Exploiting weaknesses in the shopping cart software and database were promoted over DDOS or other network level attacks. Cross-site scripting, SQL injection, and several other attack vectors were presented. Lastly, social engineering was encouraged to promote an awareness of creating a secure development environment and team.

2 CLASS FORMAT

2.1 *Entrepreneurial learning*

Course content was presented in the order it would be needed for site development via real-world examples from live sites the instructor is involved with through outside consulting. Examples of the administrative side of shopping carts, phpMyAdmin, Google Analytics and Adwords, and Amazon feeds were shown as demonstration. Purchasing domains and SSL certificates was also covered. The instructor purchased hosting services and security certificates as needed since there were restrictions on using school equipment, but expenses were minimal.

2.2 *Student groups*

Students from each team (undergraduate and graduate) chose to join one of the following groups:

Project Management—a single student from each team, an effective project manager was critical to the process (as discussed in Section 3.4). He or she was held responsible for group progress, organizing meetings, and implementing project-tracking and collaboration software.

Security—the largest group in each team, security team group members were responsible for ensuring the security of their site throughout the development process as well as probing the other team's site for security holes.

Payment—the payment group was responsible for incorporating the payment gateway in the site. This team was also responsible for researching and implementing Bitcoin and Paypal as payment methods.

Database/Backup—responsibilities for this group included setting up and securing the site's database and scheduling regular backups. This team was also tasked with restoring the site in the event of a security breach.

Business—The business group was tasked with search engine optimization, advertising, email marketing campaigns, Google Analytics, and setting up an Adwords account for potential pay-per-click advertising.

Graphics—This group created all site graphics, product images, logos and templates.

Cart Software—The shopping cart group loaded products into the cart, implemented custom attributes, managed the shopping cart software, and customized code necessary for the implementation of each site.

Overlapping group requests at the start of the semester necessitated a few students being placed in their second choice group, but the members of each group quickly took to their roles, shared contact information and established independent meeting times. Workdays were also provided during class times for inter-group coordination.

2.3 *Sites chosen*

Students spent the initial class sessions researching E-Commerce platforms and selecting a site to create. The undergraduates chose to sell t-shirts and other merchandise for the ACM chapter in the department. Their selected platform was Drupal Commerce.

The graduate team developed a site for selling drone photography services to augment an autonomous UAV research project in the Electrical Engineering department. Their chosen platform was WooCommerce for WordPress.

3 RESULTS

3.1 *Evaluation methods of student work*

There were no exams or homework assignments in the class, and performance was judged based on milestone presentations at mid-semester and during the traditional final exam period. These were oral presentations and students were told to focus on presenting their contribution to the project and to strive for completion over adding additional features. Students were made aware they would be evaluated on measureable results contributed to the site, and not just research of potential solutions.

Each group within the team presented during the milestones, and grades were assigned based on individual performance. Feedback on individual student performance from the project manager was also significantly taken into account when assigning grades.

3.2 *Site status*

Overall course results were encouraging, particularly among the undergraduate team, which completed the site and received their first order before the class had concluded. The project manager for the undergraduate team was highly organized and efficient, leading to effective teamwork and progress.

The graduate team created a visually appealing site with significant front-end programming work to determine service availability based on user location. The site was not quite functional by the time the class completed, however.

Due primarily to the unfinished status of the graduate site, no students took over development and management once the course was complete. The undergraduate site was turned over to the ACM chapter where it continues to operate and earn revenue for the chapter.

3.3 *Security related outcomes*

As part of the final presentations from each team, security groups presented the results of attempting to compromise their cohort's site. Undergraduates were given a 2x grade modifier for successful attacks on the graduate site.

Although both teams used security tools to scan the other site for vulnerabilities and attempted to use known exploits against PHP and mySQL, no application exploits were successfully used against either team. This was attributed largely to the use of popular platforms with large community involvement in patching vulnerabilities. Both teams were security conscious throughout the development process, and the security groups were responsible for implementing defensive measures including monitoring for intrusions and unauthorized modifications.

The only successful attack was a social engineering exploit committed against the instructor during the first week of the project. The undergraduate team successfully convinced the instructor that a fake Gmail account belonged to the project manager of the graduate team, and initial login details from the hosting provider were sent there. The only factor mitigating the instructor's embarrassment was the unheeded advice by the graduate team to change their login credentials frequently. It was a learning experience for everyone.

3.4 *Course comments from students*

Students completed the standard course evaluation form at the conclusion of the semester. Reaction to the course format and content was largely positive. A few comments noted that the overall size of the undergraduate team led to groups too large for effective collaboration. In the future, limiting the size of teams to facilitate communication and workload will be addressed. Having about 2 people per group (security, cart, etc.) seemed to be the ideal size. A few students preferred the traditional model of an instructor choosing the platform and giving step-by-step guidelines, but most seemed to find their role and perform well on the team.

Anecdotally, the reaction from the students at the end of the semester was very positive, with two students indicating their desire to create E-Commerce sites for organizations they were associated with. The course material seemed effective at spurring the entrepreneurial spirit among some students. The existing sites also create an opportunity for students to create side jobs with managing and fulfilling orders.

Several students indicated that they appreciated the real-world nature of the class, and were more motivated to learn the material by actually having to put the skills into practice.

As mentioned previously, selection of an effective project manager was critical to the process. As with most group projects, team cohesiveness and delegation is essential. Having someone that can lead a team on both technical and personal details is critical.

3.5 *Pedagogical notes*

The unique nature of the course led to several initial observations. With no exams or homework, the course is essential plagiarism-proof. Students were encouraged to use existing modules for their platform of choice when available, but a significant amount of customization had to be done to each site. As each project was new, code, graphics and security had to be implemented from scratch.

The traditional approach of presenting a topic and subsequently testing students on the subject was turned on its head. Often during the creation of a site, an unanticipated problem would come up and the learning paradigm switched to identifying the problem, followed by researching and implementing a solution—a much more entrepreneurial approach.

3.6 *Conclusions and future implementations*

If the course is taught in the same format in the future, projects would be arranged prior to the class, possibly with community and business input to minimize setup time and expense, and provide direction for the design of the site, products, and fulfillment. Partnering with students from the Murphy Center for Entrepreneurship at UNT and co-listing a course to provide additional support for product creation, integration and fulfillment is also a possibility. Even with large teams, designing, implementing and securing the sites was more than enough work for the Computer Science students without worrying about designing t-shirts to sell online. Without sacrificing the entrepreneurship component, outsourcing some of the implementation of product fulfillment would be considered in the future.

REFERENCES

Abrahams, A.S. & Singh, T. 2010. An Active, Reflective Learning Cycle for E-Commerce Classes: Learning about E-commerce by Doing and Teaching. *Proceedings of ISECON*, Vol 27 (1308).

Garlick, R. SoMiC 2013. Conference Presentation, Security on the Move and in the Cloud. University of North Texas. www.cse.unt.edu/somic/.

Kaltenecker, N. & Hoerndlein, C. & Hess, T. 2013. The Drivers of Entrepreneurial Intentions-An Empirical Study among Information Systems and Computer Science Students.

Ngai, E.W.T. & Gunasekaran, A. & Harris, A.L. 2005. The Maturing of E-Commerce Education in Our Curricula. *Journal of Information Systems Education*, Vol 16(1).

Tabor, S.W. 2005. Achieving Significant Learning in E-Commerce Education Through Small Business Consulting Projects. *Journal of Information Systems Education*, Vol 16(1).

E-Commerce, E-Business and E-Service – Lee (Ed)
© 2014 Taylor & Francis Group, London, ISBN 978-1-138-02646-9

Estimation of figure of merit in ultra-wideband radio channel for location-based E-Service

Juan Li
College of Information Science and Engineering, Ocean University of China, Qingdao, China
College of Computer and Communication Engineering, China University of Petroleum (Huadong), Qingdao, China

Hao Zhang
College of Information Science and Engineering, Ocean University of China, Qingdao, China

Xuerong Cui & Chunlei Wu
College of Computer and Communication Engineering, China University of Petroleum (Huadong), Qingdao, China

ABSTRACT: With the development of information technology, there have been a growing number of location-based E-Services. UWB (Ultra-Wideband) technology is considered to be the most suitable for indoor wireless positioning and IEEE802.15.4a is the first radio ranging and positioning physical layer of IEEE standards. In order to let the sender know the quality of the ranging, FoM (Figure of Merit) is added in this protocol, but how to produce FoM is not given. On the basis of analyzing the statistical energy block characteristics of received signal, a method based-on the joint of Maximum Slope and Skewness was proposed to estimate the FoM in UWB radio channel. The simulation finds that this method can provide reference for accurate ranging and positioning, and can improve the ranging accuracy about 30%.

1 INTRODUCTION

With positioning techniques, operators can provide new E-Services such as asset tracking, emergency caller location, navigation or other location-based information services.

UWB (Ultra-Wideband) is a new wireless communications technology which has generated great interest due to the many potential applications such as precise location and high bandwidth data transition. The former application can be used in the location-based E-Services and IEEE802.15.4a is the first radio ranging and positioning physical layer of IEEE standards. UWB technology is ideal for precise position,[1] even in NLOS (Non-Line of Sight) or dense multipath fading environments which is because of the following reasons.

UWB can provide many advantages for precise location. Firstly, the frequency components of UWB signal may be able to penetrate obstacles and then provide LOS (Line-Of-Sight) location signals. Secondly, the pulses of UWB are very short, which makes sub-nanosecond to nanosecond high time resolution possible. Lastly, the wide signal band means a very low power spectral density, which can reduce interference to other radio frequency wireless communication systems.

Common positioning algorithms in location-based E-Services can be classified into range based positioning and non-range based positioning. For example, TOA (Time of Arrival)[2] and TDOA (Time Difference of Arrival)[3] are the kind of range based techniques, while RSS (Received Signal Strength) and AOA (Angle-of-Arrival)[4] are the kind of non-range based positioning. Range based positioning is very suitable to be used in UWB technology, because it can take advantage of the high time resolution with very short UWB pulses. Accurate TOA estimation is very important to any ranging methods, but always exists some challenging problems[5,6] due to the severe environments, e.g. inter-symbol interference, multi-path fading, thermal noise or reflection interference. The quality of received signal is different, and therefore, in IEEE802.15.4a, the physical layer notifies its application how good each range measurement is via a parameter called the FoM (Figure of Merit)[7] which indicates how accurate and reliable a range value is. However, how to produce the value of FoM is not proposed in IEEE 802.15.4a. In [8], the authors had proposed the concept of PSF (pattern stability factor) including FoM and frequency-domain correlation patterns, to characterize the stability of UWB. The concepts of time-domain correlation patterns, pulse fidelity, and frequency-domain pattern stability were also proposed.

By using this feedback, location receiver can dynamically adapt the ranging results to determine the location of RF-tag. For example, in the multiple reference base-stations, different weights will be set to different base-stations according to the received FoMs, and even some worst ranging results will be discarded, and then the greater positioning errors can be avoided because of the large ranging error.

In this paper, a method based on the parameters of Skewness and Maximum Slope is proposed to estimate FoM in UWB radio channel which can be used in low-complexity systems, thus improving system ranging positioning accuracy.

2 FOM IN IEEE802.15.4A

In IEEE802.15.4a, FoM also is called as confidence level, which is the probability that the detected arrival time of the leading edge of a signal will deviate from the true arrival time by at most the confidence interval. The FoM consists of Confidence Interval Scaling Factor (CISF), Confidence Interval (CI), Confidence Level (CL) and an extension bit, which is shown in Figure 1.

Effective CI, which consists of CI and CISF, is equal to the multiplication of CI and CISF. According to IEEE802.15.4a, because there are 4 CISFs (0.5 CI, 1 CI, 2 CI, 4 CI) and four CIs (100 ps, 300 ps, 1 ns, 3 ns), there are 4*4 = 16 effective CIs, that is, 0.05 ns, 0.15 ns, 0.5 ns, 1.5 ns, 0.1 ns, 0.3 ns, 1 ns, 3 ns, 0.2 ns, 0.6 ns, 2 ns, 6 ns, 0.4 ns, 1.2 ns, 4 ns and 12 ns. Corresponding to the 16 effective CI, the binary frames respectively are 00 00, 00 01, 00 10, 00 11; 01 00, 01 01, 01 10, 01 11; 10 00, 10 01, 10 10, 10 11; 11 00, 11 01, 11 10, 11 11, where the former 2 bits are CISF, and the later 2 bits are CI.

CL is the probability that the arrival time estimation of pulse rising edge less than the effective CI, and CL = 0 means no FoM. According to the format of IEEE802.15.4a, there are 8 CLs, that is, 0%, 20%, 55%, 75%, 85%, 92%, 97%, and 99%. The binary frame formats respectively are 000, 001, 010, 011, 100, 101, 110, and 111.

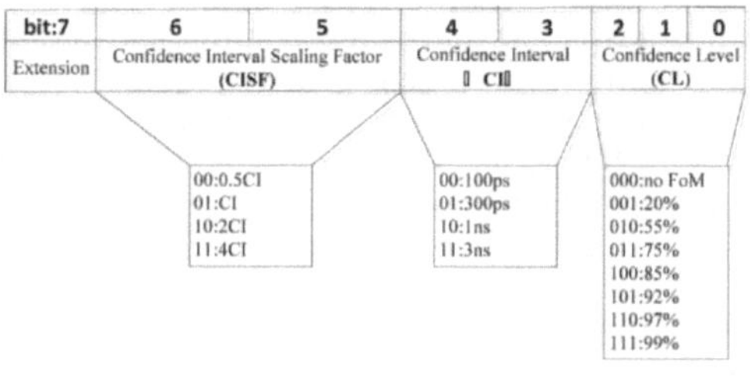

Figure 1. Format diagram of FoM.

Although, IEEE802.15.4a gives the format of FoM, how to calculate or produce the CISF, CI, and CL is not given.

3 CHARACTERISTICS OF UWB RECEIVED SIGNAL

The basic principle[6] of energy detection method is shown in Figure 2, that is, after the LNA (Low Noise Amplifier), the received UWB signal is squared, and then is input into an integrator whose integration period is T_b. Because of the inter-frame leakage due to multipath signals, the integration duration is set to $3T_f/2$, Thus the number of signal values for energy detection is $N_b = (3T_f)/(2T_b)$. The integrator output can then be expressed as

$$z[n] = \sum_{j=1}^{N_s} \int_{(j-1)T_f + (c_j + n - 1)T_b}^{(j-1)T_f + (c_j + n)T_b} r^2(t)dt, \qquad (1)$$

where $n = 1, 2, ..., N_b$ is the sample index with respect to the start of the integration period and N_s is the number of pulses in each symbol. Because N_s is set to 1, thus the integrator output is

$$z[n] = \int_{(c+n-1)T_b}^{(c+n)T_b} r^2(t)dt. \qquad (2)$$

If there is only noise, $z[n]$ is a distribution of centralized Chi-square, while if a signal is present, $z[n]$ is a distribution of non-centralized Chi-square. The variances and mean and values of noise and signal values are,

$$\mu_0 = F\sigma^2,\ \sigma_0^2 = 2F\sigma^4, \qquad (3)$$
$$\mu_e = F\sigma^2 + E_n,\ \sigma_e^2 = 2F\sigma^4 + 4\sigma^2 E_n, \qquad (4)$$

respectively, where E_n is the signal energy of the nth integration period, $F = 2BT_b + 1$ is the number of degrees of freedom and B is the signal bandwidth.

Statistical characteristics of Kurtosis, Skewness, Mean Square Error, and Maximum Slope for different Signal to Noise Ratios (SNRs), channel, and integration period have been investigated via simulation in IEEE802.15.4a channel.

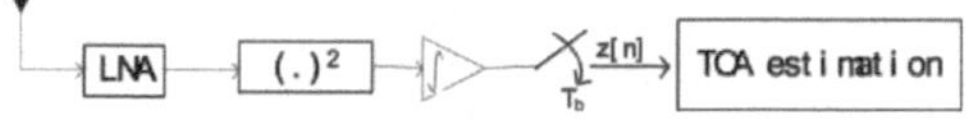

Figure 2. Diagram of the energy detection method.

3.1 *Kurtosis*

The Kurtosis is given by the second and fourth order moments and is calculated as,

$$K = \frac{1}{(N_b - 1)\delta^4}\sum_{i=1}^{N_b}(x_i - \bar{x})^4 - 3 \tag{5}$$

where σ is the standard deviation and $\bar{x}$ is the mean of energy blocks. The Kurtosis equal to 3, if the energy blocks is a standard normal distribution, and then, k is often defined as $K = k - 3$, that means excess Kurtosis, so that the Kurtosis of standard normal distribution has a value of zero. Positive Kurtosis means the "peaked" distribution of energy blocks, at the same time, negative Kurtosis means a "flat" distribution of received energy blocks. If there is only noise or the SNR is very low and the degrees of freedom of the Chi-square distribution are large enough, z[n] is a Gaussian distribution and Kurtosis is zero. On the other hand, K tends to increase, as the SNR increases.

3.2 *Skewness*

The Skewness can be expressed as

$$S = \frac{1}{(N_b - 1)\delta^3}\sum_{i=1}^{N_b}(x_i - \bar{x})^3 \tag{6}$$

where σ is the standard deviation of the received energy blocks, and $\bar{x}$ is the mean value. The Skewness is zero if energy blocks are normal distribution; actually any symmetric data will have a Skewness of zero. If the Skewness is negative value, the data will skewed left, while positive values the data will skewed right. Skewed left means that the left tail is longer than the right tail, while skewed right indicates the opposite. S will equal to zero nearly, if there is noise only, or very low SNRs, and F is large enough. Along with the SNR increases, S tends to increase.

3.3 *Standard deviation*

The Standard Deviation is widely used to measure the variability of values. It indicates that how much dispersion or "variation" there is from the expected or value mean. The Standard Deviation used in the article is given as (7).

$$D = \sqrt{\frac{\sum_{i=1}^{N_b}(x_i - \bar{x})^2}{N_b - 1}}. \tag{7}$$

3.4 *Maximum slope*

Skewness and Kurtosis cannot account for the propagation time of UWB signal, so the slope of the energy values is considered in the FoM estimate as an alternative measure. These energy blocks values are divided into $(N_b - M_b + 1)$ groups, with M_b values in each line group. The slope for each line group is calculated based-on a least squares line fitting. The Maximum Slope (M) is given by

$$M = \max_{1 \le N_b - M_b + 1} slope\{linefit\,(z[n], z[n+1], \ldots, z[n + M_b - 1])\} \tag{8}$$

For example, in Figure 3, there are eight energy values, which are divided in 5 groups with 5 corresponding slopes, and in each group there are $M_b = 4$ energy value to be fitted to calculate each slope, at last the maximum slope is selected to use to estimate FoM.

In order to examine the statistical characteristics of Kurtosis, Skewness, Standard Deviation, and Maximum Slope, the CM1 (residential LOS environment) and CM2 (residential NLOS environment) channel models of the IEEE802.15.4a standard were used. 1000 channel realizations were simulated and sampled at $F_c = 8$ GHz for each SNR value. The Gaussian pulse with second derivative is employed with $T_b = 4$ ns, $T_c = 1$ ns, $T_f = 200$ ns, and $N_s = 1$. Each simulation has a uniform TOA distribution within (0, T_f).

The four parameters were calculated, and the obtained statistical results were shown in Figures 4 and 5. These figures show that the characteristics of the four parameters with respect to the SNRs are similar for the two CM1 and CM2 channels. Furthermore, Figures 4 and 5 show that both Skewness and Kurtosis increase as the SNRs increase, but Skewness changes more rapidly. On the other

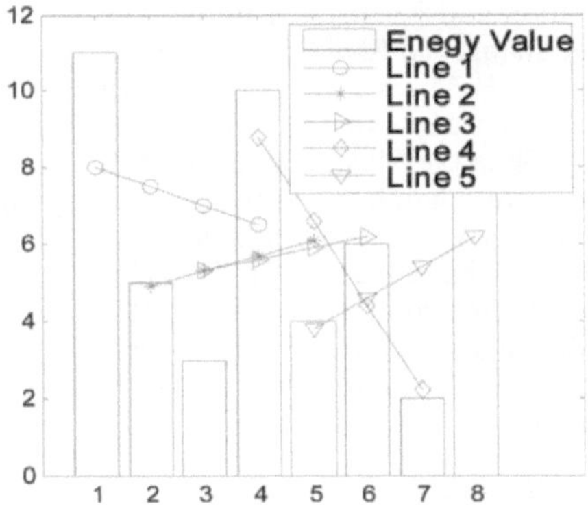

Figure 3. Diagram of the maximum slope of the received energy value.

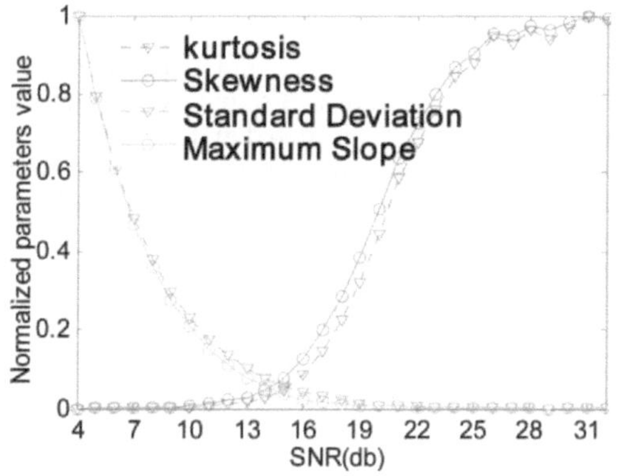

Figure 4. Four normalized statistical parameters change with SNRs in CM1.

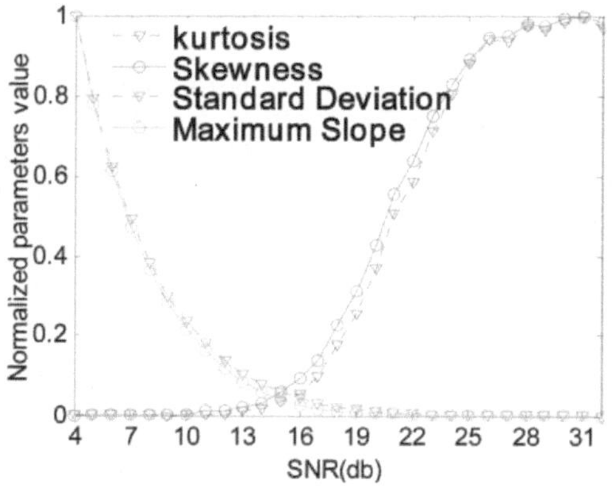

Figure 5. Four normalized statistical parameters change with SNRs in CM2.

hand, the Standard Deviation and Maximum Slope decrease as the SNR increases, while Maximum Slope changes more rapidly. Since Maximum Slope and Skewness change more rapidly than Standard Deviation and Kurtosis, therefor, they can reflect better changes in SNR. Thus, they are more suitable for FoM estimation. Furthermore, when the SNRs are less than 15 dB, Skewness changes slowly while the Maximum Slope changes rapidly. On the other hand, the Skewness changes rapidly but the Maximum Slope changes slowly when the SNRs are higher than 15 dB. Therefore, no single parameter is a good measure of SNRs change over a wide range of received energy values.

Thus, the following joint metric (K) based-on Maximum Slope (K) and Skewness (S) is proposed

$$J = S - K. \tag{9}$$

where M is the Maximum Slope and S is the Skewness of received energy values.

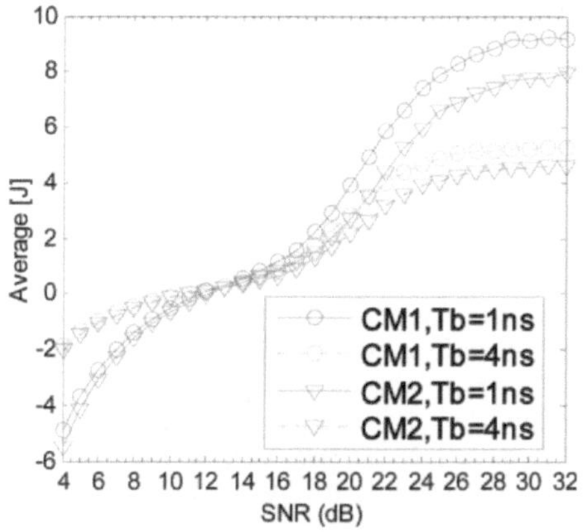

Figure 6. Relationship of the Joint parameter J and the SNR.

In order to verify that the proposed metric J is suitable for FoM estimation, that is to say, it is sensitive to both high and low SNRs, 1000 channel realizations were simulated for many SNR values in CM1 and CM2 IEEE802.15.4a channel. In the simulations, because the received signal is random, the J values are not unique for one SNR, but in order to get Figure 6, the average J value with respect to each SNR were calculated for each integration period and channel model. Because there were 29 SNR values simulated, there were 29 *J-SNR* pairs for each diffrent integration period and channel model. Figure 6 shows that for a large range of SNR values, J is a monotonic function and that J is more sensitive to the changes of SNR than any other single parameter. Thus, J can be used to describe the quality of communication channel, that is to say J is suitable for FoM simlation.

4 FOM ESTIMATES AND USED IN POSITIONING

The method of FoM estimation and used in positioning includes 5 steps, as shown in Figure 7.

5 RESULTS AND DISCUSSION

In this section, the simulation was examined in the CM1 and CM2 of IEEE 802.15.4a channels using TOA positioning method with 5 positioning base-stations. During positioning, the base-station with the worst FoM was discarded, and the other 4 base-stations were given 4 different weights according to their own FoMs.

Figures 8 and 9 show the location root mean square error of the two methods of FoM based

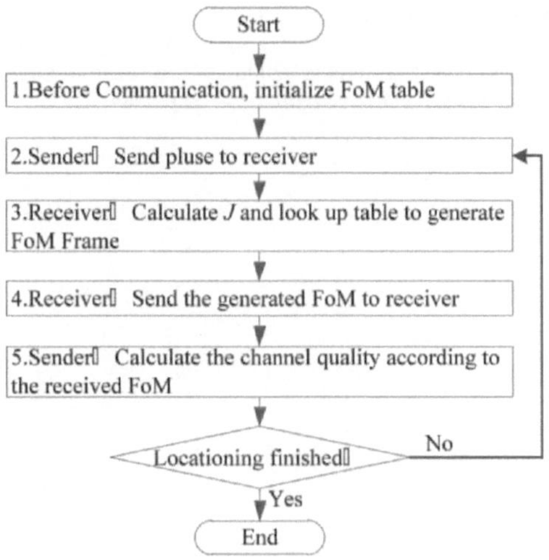

Figure 7. Flow chart for FoM estimation and applied in positioning.

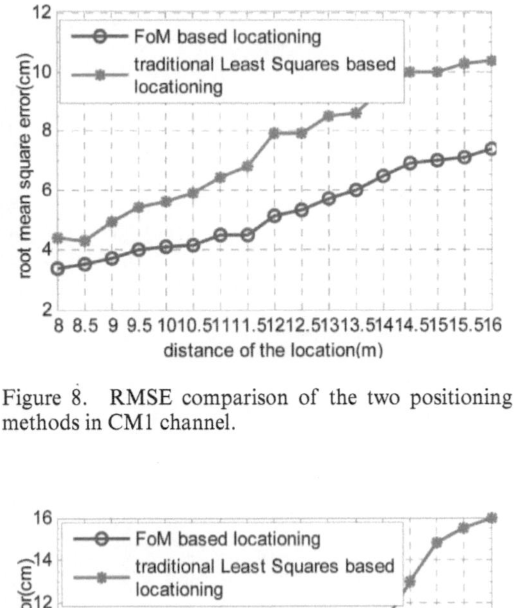

Figure 8. RMSE comparison of the two positioning methods in CM1 channel.

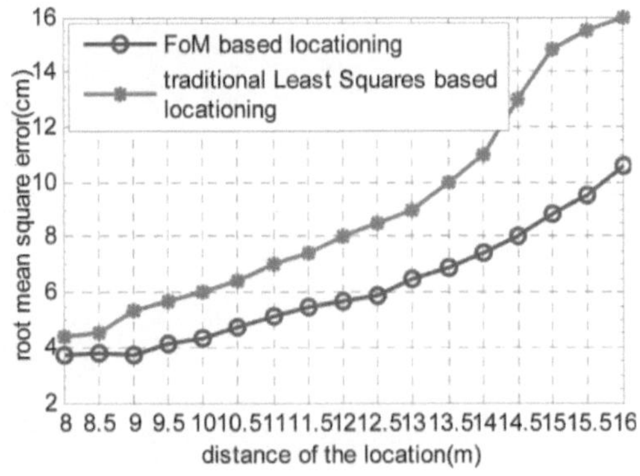

Figure 9. RMSE comparison of the two positioning methods in CM2 channel.

positioning and the traditional Least Squares based positioning.

Figures 8 and 9 find out that in the CM1, CM2 channel, the positioning accuracy can be improved about 30% with the help of FoM. Therefore, it is proved that the proposed FoM estimation method can be used to estimate the quality of communication and ranging or location, and then can improve the overall quality of UWB-based communication and ranging.

6 CONCLUSIONS

Although, IEEE802.15.4a gives the format of FoM, how to calculate or produce the FoM is not given. A low complexity FoM estimation algorithm has been provided for UWB positioning. Four different parameters were investigated, and from the obtained results, it is found out that no single parameter is a good measure of SNR change over a wide range of values but based-on Maximum Slope and Skewness, the joint metric is sensitive to both low and high SNRs, so it is developed for FoM estimation.

The simulation results show that the positioning accuracy can be improved using FoM to discard the worst ranging result.

ACKNOWLEDGMENTS

This work was supported by the Nature Science Foundation of China under grant No. 61301139, the Fundamental Research Funds for the Central Universities under grant No. 13CX02026A and 14CX02139A, and the Key Technology Development Projects in Qingdao Development Zone under grant No. 2013-1-52.

REFERENCES

[1] IEEE802.15.4a-2007. 2007. Wireless Medium Access Control (MAC) and Physical Layer (PHY) Specifications for Low-Rate Wireless Personal Area Networks (WPANs).

[2] Giorgetti Andrea, Chiani Marco. 2013. Time-of-arrival estimation based on information theoretic criteria. *IEEE Transactions on Signal Processing*, 61(8): 1869–1879.

[3] Fujiwara R., K. Mizugaki, T. Nakagawa, et al. 2011. TOA/TDOA hybrid relative positioning system based on UWB-IR technology. *IEICE Transactions on Communications*, E94B(4): 1016–1024.

[4] Lee. Y.U. 2011. Weighted-average based AOA parameter estimations for IR-UWB wireless positioning system. *IEICE Transactions on Communications*, E94B (12): 3599–3602.

[5] Xue-rong Cui, Hao Zhang, T. Aaron Gulliver. 2012. Threshold selection for ultra-wideband TOA estimation based on neural networks. *Journal of Networks*, 7(9): 1311–1318.

[6] Taponecco L., A.A. D'Amico, U. Mengali. 2011. Joint TOA and AOA estimation for UWB localization applications. *IEEE Transactions on Wireless Communications*, 10(7): 2207–2217.

[7] Fujii, Masahiro; Watanabe, Yu. 2013. Primary Signal to Noise Ratio Estimation based-on AIC for UWB Systems. *IEICE Transactions on Fundamentals of Electronics Communications and Computer Sciences*, E96 A (1): 264–273.

[8] Dissanayake, Tharaka; Tsselle, Karu P. Correlation based pattern stability analysis and a figure of merit for UWB antennas. 2006. *IEEE Transactions on Antennas and Propagation*, 54(11): 3184–3191.

E-Commerce, E-Business and E-Service – Lee (Ed)
© 2014 Taylor & Francis Group, London, ISBN 978-1-138-02646-9

An E-Service of a stand-alone hybrid renewable energy generation system

H.C. Chen & W.X. Bai
National Chin-Yi University of Technology, Taichung, Taiwan

ABSTRACT: This paper presents the dynamic modeling and the simulation results of a renewable energy based hybrid power generation system. In order to meet sustained load demands during varying natural conditions, different renewable energy sources need to be integrated with each other. The paper focuses on the combination of Solar Cell (SC), Wind Turbine (WT), Hydro Turbine (HT), Fuel Cell (FC) and Vanadium Redox Flow Battery (VRB) systems for power generation. Dynamic modeling of various components of this isolated system is presented. Transient responses of the system to step changes in the load, ambient temperature, radiation, wind speed and water speed in a number of possible situations are studied. The results show that the proposed hybrid power system can tolerate the rapid changes in natural conditions and suppress the effects of these fluctuations on the voltage within the acceptable range.

1 INTRODUCTION

The fossil fuel will produce pollution after burning, such as carbon dioxide, nitrogen oxide, sulfur oxide compound and the carbon hydrogen compound etc (Kobayashi, Matsuo & Sekine 2006). All of these poisonous airs are the cause of the air pollutions and green house effects. The fossil fuel is the non-renewable resources and less and less after massive consumptions. Comparing with the nuclear energy and thermal power, the renewable energy is inexhaustible and has non-pollution characteristics. The solar energy, wind power, hydro power and tide energy are natural resources of the interest to generate electrical sources. An unique hydro resources in Taiwan is irrigation water in agriculture canals and some countries like Japan and India also use the irrigation water in an agriculture canal (Adhaul, 2010; Lee, 2010). In Japan had been use the irrigation water in agriculture canals by water current speed and this form of generation don't need high water head (Lee 2010).

The wind, hydro and solar energy are welcome substitution for many other energy resources because it is natural, inexhaustible resource of sunlight to generate electricity (Pan, Chen and Chu 1999). The renewable energy generation had a significant problem is to store the energy generated for future usage when no wind is available but the user demand exists (Pan, Chen and Chu 1999). For this problem can be solving by energy storage system. The energy storage systems can be classified as Cr-Fe, Cd-Ni, lead-acid and vanadium redox flow batteries. The VRB is deemed as a promising green solution in practical applications (Wang and Wang 2006), due to the advantages of large scale, simple structure, long life cycle and high SOC. After many technological advances, proton exchange membrane fuel cell technology has now reached the test and demonstration phase. The recent commercial availability of small PEMFC units has created many new opportunities to design hybrid energy systems for remote applications with energy storage in hydrogen form (Gow and Manning 1999).

In this paper, a detailed dynamic model and simulation of a solar cell/wind turbine/hydro turbine/fuel cell/vanadium redox flow battery hybrid power system is developed using a novel topology to complement each other and to alleviate the effects of environmental variations. Modeling and simulations are conducted using MATLAB/Simulink software packages to verify the effectiveness of the proposed system. The results show that the proposed hybrid power system can tolerate the rapid changes in natural conditions and suppress the effects of these fluctuations on the voltage within the acceptable range.

2 DYNAMIC SYSTEM MODELS

2.1 *Solar cell*

A solar cell module consists of many jointly connected solar cells. A number of solar cell models have been developed, but the one diode electrical equivalent circuit shown in Figure 1(a) is commonly used for cell based or module based analysis. It consists of a diode, a current source, a series resistance and a parallel resistance. The current source generates the photo-current that is a function of

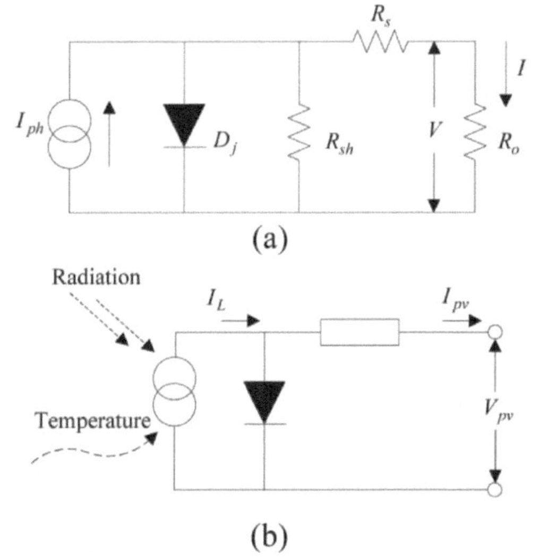

Figure 1. Solar cell (a) equivalent circuit (b) simplified circuit.

the incident solar cell radiation and temperature (Khan, 2005; Shaahid, 2007; Nelson, 2006). The diode represents the p-n junction of a solar cell. The temperature dependence of the diode saturation current and constant diode ideality factor are included in the modeling. At real solar cells, a voltage loss on the way to the external contacts is observed. This voltage loss is expressed by a series resistance (R_s). Furthermore leakage currents are described by a parallel resistance (R_{sh}). However, the series resistance is very small and the parallel resistance is very large (Nelson 2006). So we can ignore R_s and R_{sh}. The solar cell simplified circuit is then shown in Figure 1(b).

The solar cell current equation is

$$I_{pv}(t) = I_{sc}\left\{1 - C_1\left[\exp\left(\frac{V_{mp}}{C_2 V_{oc}}\right) - 1\right]\right\} + \left(\frac{E_{tt}(t)}{E_{st}}\right)[\alpha(T_a(t) + 0.002E_{tt}(t) + 1)] - I_{mp} \tag{1}$$

The solar cell voltage equation is

$$V_{pv}(t) = V_{mp}\left[1 + 0.0539\log\left(\frac{E_{tt}(t)}{E_{st}}\right)\right] + \beta(T_a(t) + 0.02E_{tt}(t)) \tag{2}$$

$$C_1 = \left(1 - \frac{I_{mp}}{I_{sc}}\right)\exp\left[\frac{-V_{mp}}{C_2 V_{oc}}\right] \tag{3}$$

$$C_2 = \frac{\frac{V_{mp}}{V_{oc}} - 1}{\ln\left(1 - \frac{I_{mp}}{I_{sc}}\right)} \tag{4}$$

2.2 *Wind turbine*

The power output of wind turbine is relating to wind speed with a cubic ratio. The power curve of the wind turbine WE-CXF-600 studied is nonlinear, which is digitized and the resulting table is used for simulation as Figure 2. The dynamics of the wind turbine due to its rotor inertia and generator are added by considering the wind turbine response as a second order slightly under-damped system (Khan, 2005; Kuo, 1995). Using this simple approach, small wind turbine dynamic is modeled as

$$P_g(s)/P_{wt}(s) = 0.25/(s^2 + 0.707s + 0.25) \tag{5}$$

2.3 *Hydro turbine*

The hydro turbine is similar to wind turbine (Ben, 2010), because that have same principle to extract kinetic energy to electricity. Thus, the power output of hydro turbine is relating to water current speed with a cubic ratio. The power curve of the river turbine EMC-005-F4 studied is nonlinear, which is digitized and the resulting table is used for simulation as Figure 3. The dynamics of the hydro turbine due to its rotor inertia and generator are used by considering the hydro turbine response as a second order slightly under-damped system (Khan 2010). Because it likes wind turbine, so we used the same approach, small hydro turbine dynamic is modeled as

$$P_g(s)/P_{ht}(s) = 90.23/(s^2 + 11.22s + 90.23) \tag{6}$$

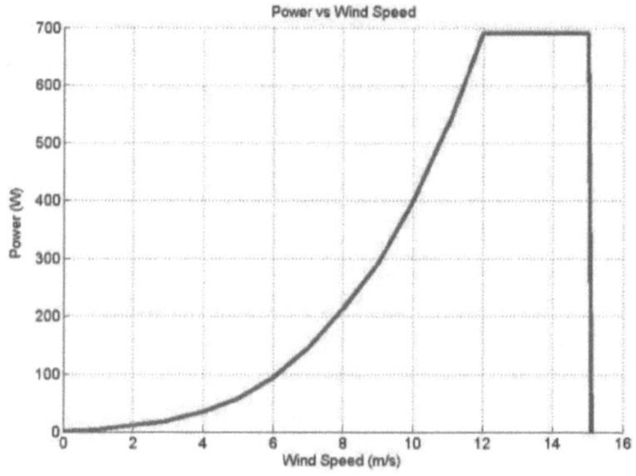

Figure 2. WE-CXF-600 power curve.

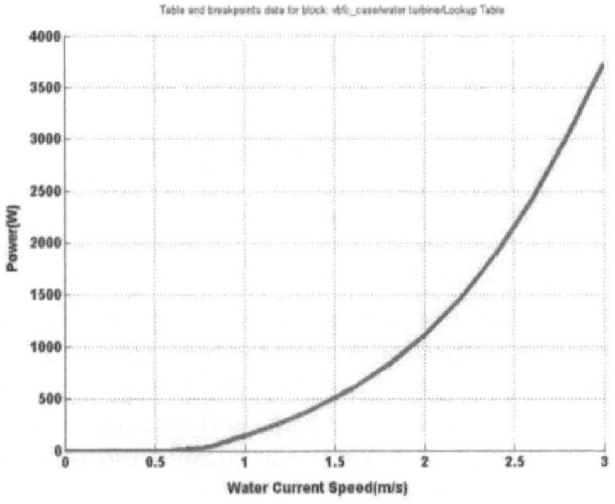

Figure 3. ENC-005-F4 power curve.

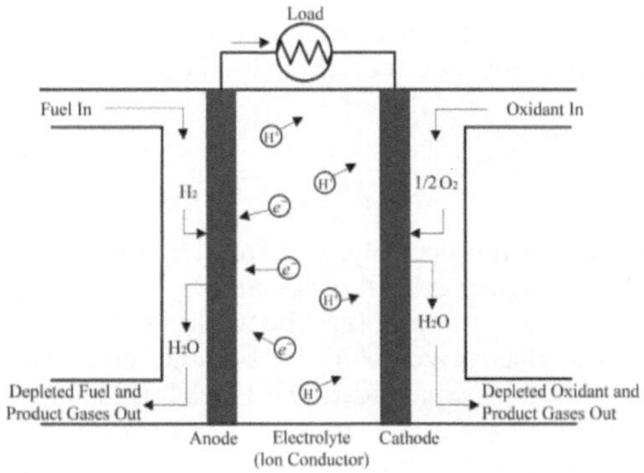

Figure 4. PEM fuel cell diagram (Obara 2007).

2.4 *Fuel cell*

A fuel cell is a stationary power generator. The fuel cell chosen for this type of application must provide competitive, reliable, and quality power without emitting pollutants such as oxides of nitrogen, carbon or sulphur. It must respond quickly to changes in load and have low maintenance requirements as well as a long cell life (Delfino 2003). The PEM fuel cell consists of porous carbon electrodes bonded to a very thin sulphonated polymer membrane. As illustrated in Figure 4, this Membrane Electrode Assembly (MEA) is sandwiched between two collector plates, which provide an electrical path from the electrodes to the external circuit (Obara, 2007; Amphlett, 1995; Rowe, 2001; Ulleberg, 1997). Flow channels cut into the collector plates distribute reactant gases over the surface of the electrodes. Individual cells consisting of collector plates and MEAs are assembled in series to form a fuel cell stack (Rowe 2001).

The thermodynamic potential E is defined via a Nernst equation in expanded form as (Shaahid 2007)

$$E = 1.229 - 0.85\times10^{-3}(T - 298.15) + 4.3085\times10^{-5}\cdot T\cdot(\ln P_{H_2} + 0.5\ln P_{O_2}) \quad (7)$$

The parametric equation for the over-voltage due to activation and internal resistance developed from the empirical analysis is given as

$$\eta_{act} = -0.9514 + 0.00312T - 0.000187T\ln(i) + 7.4\times10^{-5}T\ln(co_2) \quad (8)$$

$$co_2 = \frac{P_{O_2}}{5.08\times10^{6}e^{\frac{-498}{t}}} \quad (9)$$

$$R_{\text{int}} = 0.01605 - 3.5\times10^{-5}T + 8\times10^{-5}i \quad (10)$$

The combined effect of thermodynamics, mass transport, kinetics, and ohmic resistance determines the output voltage of the cell as defined by

$$V_{cell} = E - v_{act} - \eta_{ohmic} \quad (11)$$

The fuel cell system consists of a stack of 65 similar cells connected in series. Therefore, the total stack voltage is given by

$$V_{stack} = 65\,V_{cell} \quad (12)$$

Using the mole conservation principle, the gas pressure of the fuel cell anode is given as

$$\frac{V_a}{RT}\frac{dP_{H_2}}{dt} = \dot{m}_{H_2in} - (\rho_{H_2}UA)_{out} - \frac{i}{2F} \quad (13)$$

The gas pressure of the fuel cell cathode is given as

$$\frac{V_c}{RT}\frac{dP_{O_2}}{dt} = \dot{m}_{O_2in} - (\rho_{O_2}UA)_{out} - \frac{i}{4F} \quad (14)$$

2.5 *Electrolyzer*

Water can be decomposed into its elementary components by passing electric current between two electrodes separated by an aqueous electrolyte (Ulleberg, 1997; Sapru, 1997). The electrochemical reaction of water electrolysis is given by

$$H_2O(l) + \text{electrical energy} \Rightarrow H_2(g) + 1/2\ O_2(g) \quad (15)$$

According to Faraday's law, hydrogen production rate of an electrolyzer cell is directly proportional to the electrical current in the equivalent electrolyzer circuit (Shaahid 2007).

$$n_{H_2} = \frac{\eta_F n_c i_e}{2F} \tag{16}$$

where i_e is the electrolyzer current, n_c is the number of electrolyzer cells in series, and η_F is the Faraday efficiency. Assuming that the working temperature of the electrolyzer is 40°C, Faraday efficiency is expressed by (Shaahid, 2007; Ulleberg, 1997)

$$\eta_F = 96.5e^{(0.09/i_e - 75.5/i_e^2)} \tag{17}$$

2.6 *Vanadium redox flow battery*

A VRB is an electrochemical energy storage device that employs vanadium solution as activated materials on both anode and cathode. An open circuit voltage up to 1.5 V is achieved in a fully charge condition. Vanadium exhibits a number of oxidation states, i.e. V^{2+}, V^{3+}, V^{4+} and V^{5+}, but both V^{4+} and V^{5+} are essentially the vanadium oxides VO^{2+} and VO_2^+ respectively (Blanc 2009). Illustrated in Figure 5 is a redox schematic diagram of VRB charge/discharge processes.

In the discharge process, vanadium V^{5+} in the anolyte and V^{2+} in the catholyte turn into V^{4+} and V^{3+} respectively, while electrons e—migrate from the cathode toward the anode through an external load, and protons H^+ flow the other way around through an internal membrane. Yet, during the charge process, the same type of particles migrates in exactly the opposite direction to that in the discharge process. All the above chemical mechanism are formulated as

$$VO_2^+ + 2H^+ + e^- \underset{\text{charge}}{\overset{\text{discharge}}{\rightleftarrows}} VO^2 + H_2O \tag{18}$$

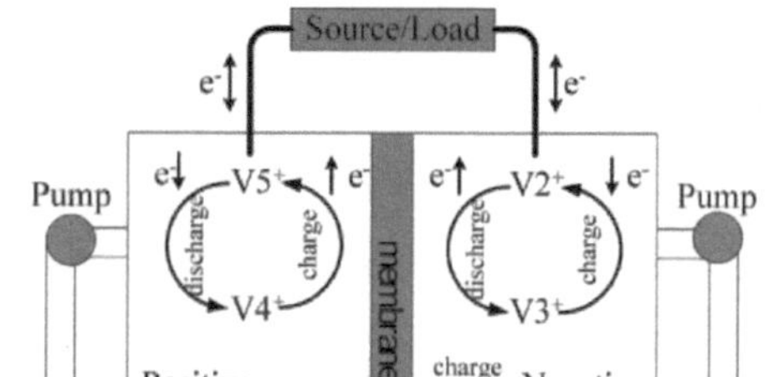

Figure 5. Redox schematic diagram of charge/discharge of VRB.

$$V^{2+} \underset{\text{charge}}{\overset{\text{discharge}}{\rightleftarrows}} V^{3+} + e^- \tag{19}$$

$$VO_2^+ + 2H^+ + V^{2+} \underset{\text{charge}}{\overset{\text{discharge}}{\rightleftarrows}} VO^2 + H_2O + V^{3+} \tag{20}$$

Through the Nernst equation, the equilibrium potential of a VRB is found subject to the electrolyte concentration (Li 2008). Given the concentrations of both the vanadium and hydrogen ions in the cell stack of a VRB, the equilibrium potential V_{eq} is evaluated as

$$E = E' + \frac{RT}{F}\ln\frac{[V^{2+}]_C[VO_2^+]_C[H^+]_C[H^+]_C}{[V^3]_C[VO^{2+}]_C} \tag{21}$$

where $E' = 1.255$ V denotes standard potential, a quantity derived form the reduction potential of two redox couples, R the ideal gas constant, and T the temperature.

Accordingly, $[VO_2^+]$, the concentrations of VO_2^+ is made equal to $[V^{2+}]$ in a fully charged state, following which $[VO_2^+] = [V^{2+}]$ and $[VO^{2+}] = [V^{3+}]$ at any given state of charge, and Eq. (27) is simplified into

$$E = E' + \frac{RT}{F}\ln\frac{[V^{2+}]_C^2[H^+]_C^2}{[V^{3+}]_C^2} \tag{22}$$

3 SIMULATION RESULTS

The system consists of a 250 W solar cell, a 600 W wind turbine, a 5HP hydro turbine, a 100 W proton exchange membrane fuel cell, a 1 kWh vanadium redox flow battery, an ultra-capacitors, an electrolyzer, and a power conditioner. The power conditioner includes a boost circuit and a SPWM inverter. It is used to step up ultra-capacitor voltage to DC 200 V and invert to 120 V_{rms}, 60 Hz AC. The flowchart of the energy dispatch strategy is depicted in Figure 6.

Simulation results with step changes in load demand, wind speed, water speed, radiation, and ambient temperature are analyzed and shown in Figures 7–11. Wind speed decreases at t = 20 s from 10 to 7 m/s at t = 30 s. The solar cell initially supplies power at the radiation 400 W/m² and temperature 25°. The radiation increases at 13 s from 400 W/m² to 600 W/m² at 17 s and then decreases at 28 s from 600 W/m² to 300 W/m² at 30 s. The temperature keeps unchanged at 25°. Water speed decrease at t = 10 s from 0.9 m/s to 0.8 m/s at

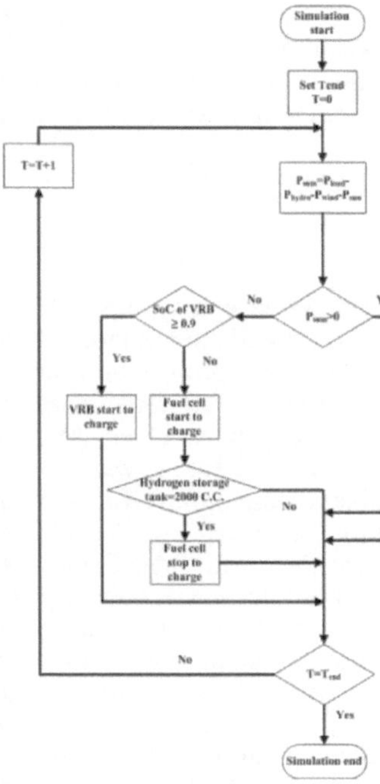

Figure 6. The flowchart of the energy dispatch strategy.

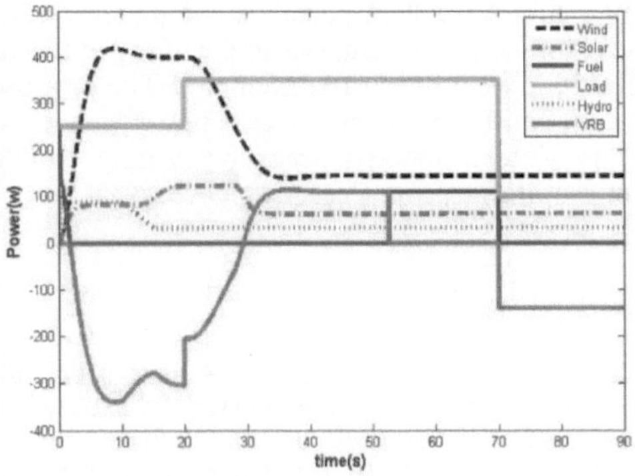

Figure 7. Power tracking performance of the hybrid power generation system.

t = 15 s. The initial SoC (state of charge) of VRB is 0.1. These step inputs cause changes in available power and load consumption. The power tracking performance of the hybrid topology with respect to load demand change and environmental variations is shown in Figure 7.

With variation in load, the power demand changes from 250 W to 350 W at 20 s and from 350 W to 100 W at 70 s as shown in Figure 7. At start-up transient stage, the fuel cell provides power for load requirement because of the output powers of the wind turbine, hydro turbine and solar cell are not sufficient enough to supply load demand at t = 0 s to 2 s when the SoC of VRB is 0.1 (reached the discharge limit). While the captured power increases, the contribution of the fuel cell decreases. Any excess power is diverted to the VRB during 2 s to 28 s. However, as the wind speed

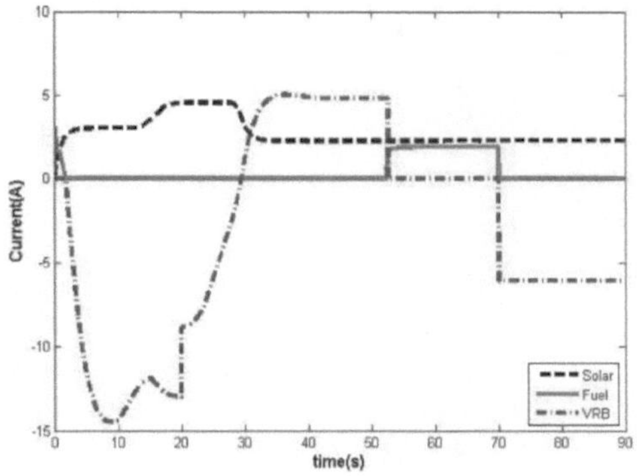

Figure 8. Current variations.

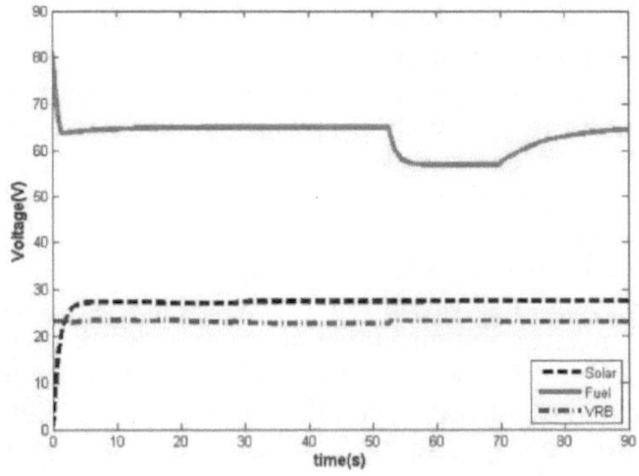

Figure 9. Voltage variations.

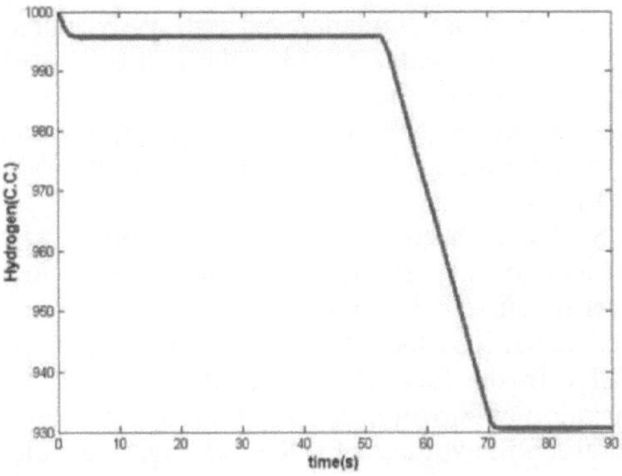

Figure 10. Hydrogen variation in storage tank.

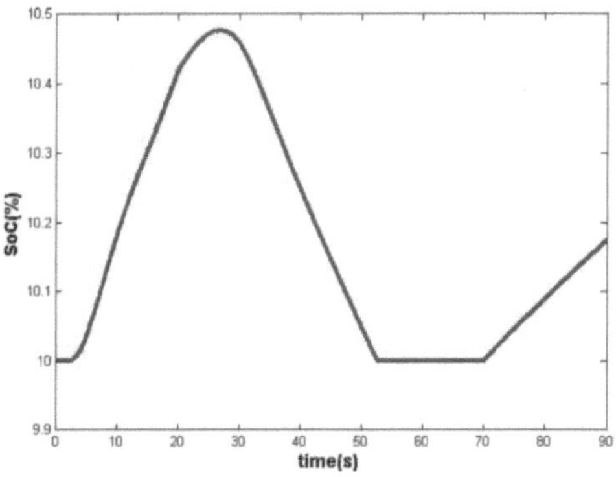

Figure 11. SoC variation of VRB.

and water speed decrease, the VRB begins to provide power to supply load demand at $t = 28$ s. The fuel cell provides power again for load requirement at 53 s while the VRB reaches its discharge limit. The load demand changes from 350 W to 100 W at 70 s and the excess power is again diverted to the VRB.

The current variations of the solar cell, the fuel cell, and the vanadium redox battery with load changes and environmental conditions are shown in Figure 8. These changes are reflected in the performance of the fuel cell system and vanadium redox battery. The stack current variation at $t = 0$ s to 2 s is due to start-up transient and load demand, as the solar cell's, hydro turbine's and wind turbine's contributions are limited and fixed. During $t = 2$ s to $t = 53$ s, the fuel cell current decreases to zero because the solar cell and VRB increase output power. After $t = 53$ s, the fuel cell current increases while the VRB reaches its discharge limit. After $t = 70$ s, the fuel cell current increases to zero because of the load changed from 350 W to 100 W.

Such changes in fuel cell current cause the stack voltage to vary significantly. Generally, a lower level of current implies higher stack voltage and vice versa. The use of an ultra-capacitor in parallel with the fuel cell reduces the stack's output variation as shown in Figure 9. However it can be observed the voltage of VRB has slight variations for the charge current changed.

The hydrogen is a fuel of fuel cell. At $t = 0$ s to $t = 2$ s, the fuel cell generated electricity by hydrogen in storage tank. The variation of hydrogen in storage tank is shown in Figure 10.

Shown in Figure 11 is the SoC variation of VRB. At $t = 2$ s, the VRB has charge current, but the SoC of VRB hadn't any variation. Because the VRB system had a pump that used for circulating the electrolyte, so it will consume power when charging and discharging. In this system, the pump power is 57 W. At $t = 2.3$ s, the VRB start to charge when the exceed power bigger than 57 W. Thus, the system can circulate supply load demand and renewable energy will not be wasted.

4 CONCLUSION

In this paper, a renewable energy based hybrid power generation systems and modeled for a stand-alone user with appropriate power controllers is proposed. The available power from the renewable energy sources is highly dependent on environmental conditions such as wind speed, water speed, radiation, and ambient temperature. To overcome this deficiency of the solar cell, hydro and wind system, we integrated them with the FC and VRB system using a novel topology. The voltage variation at the output is found to be within the acceptable range. The output fluctuations of the wind turbine varying with wind speed, hydro turbine with water speed and the solar cell varying with both environmental temperature and sun radiation are reduced using a fuel cell and vanadium redox flow battery. Therefore, this system can tolerate the rapid changes in load and environmental conditions, and suppress the effects of these fluctuations on the equipment side voltage. This hybrid topology exhibits excellent performance under variable radiation, ambient temperature, wind speed, water speed and load power requirements. The proposed system can be used for off-grid power generation in non-interconnected areas or remote isolated communities.

REFERENCES

Adhaul, S.P. Moharil, R.M. & Adhau, P.G. 2010. Reassessment of irrigation potential for micro Hydro Power Generation, *Proc. of IEEE International Conference on Sustainable Energy Technologies (ICSET)*, Kandy, Sri Lanka, pp. 1–5.

Amphlett, J.C. Baumert, R.M. Mann, R.F. Peppley, B.A. Roberge, P.R. & Harries, T.J. 1995. Performance modeling of the Ballard Mark-IV solid polymer electrolyte fuel cell, *Journal of the Electrochemical Society*, Vol. 1, pp. 9–15.

Blanc, C. & Rufer, A. 2009. Optimization of the operating point of a vanadium redox flow battery, *IEEE Energy Conversion Congress and Exposition (ECCE2009)*, pp. 2600–2605.

Delfino, B. & Fornari, F. 2003. Modeling and control of an integrated fuel cell-wind turbine system, *Proc. of IEEE Bologna PowerTech Conference*, pp. 23–26.

Elghalil, S.E.B. Benbouzidl, M.E.H. & Charpentier, J.F. 2010. Modelling and control of a marine current turbine-driven doubly fed induction generator, *IET Renewable Power Generation*, Vol. 4, pp. 1–11.

Gow, J.A. & Manning, C.D. 1999. Development of a Photovoltaic Array Model for Use in Power-electronics Simulation Studies, *IEE Proc. of Electric Power Application*, pp. 193–200.
Ismail, B. Taib, S. Saad, A.R. Isa, M. & Hadzer, C.M. 2006. Development of a single phase SPWM microcontroller-based inverter, *Proc. of 2006 IEEE Power and Energy Conference*, pp. 437–440.
Khan, M.J. & Iqbal, M.T. 2005. Dynamic modeling and simulation of a small wind-fuel cell hybrid energy system, *Renewable Energy*, pp. 421–439.
Khan, M.J. Iqbal, M.T. & Quaicoe, J.E. 2010. Dynamics of a vertical axis hydrokinetic energy conversion system with a rectifier coupled multi-pole permanent magnet generator, *IET Renewable Power Generation*, Vol. 4, pp. 116–127.
Kobayashi, K. Matsuo, H. & Sekine, Y. 2006. An excellent operating point tracker of the solar-cell power supply system, *IEEE Transactions on Industrial Electronics*, Vol. 53, pp. 495–499.
Kuo, B.C. 1995. *Automatic control systems*, 7th ed. New York: Prentice Hall Inc.
Lee, Y.Z. 2010. Agriculture water resources for effective renewable generation, *Department of Irrigation Management Technology Council of Agriculture Taiwan Joint Irrigation Association*, Harvest Vol. 60, No. 11.
Li, M.H. & Hikihara, T. 2008. A coupled dynamical model of redox flow battery base on chemical reaction, fluid flow, and electical circuit, *Ieice. Fundamentals*, Vol. E91-A, No. 7.
Lin, W.X. 2006. A new approach to the harmonic analysis of SPWM waves, *Proc. of IEEE Mechatronics and Automation Conference*, pp. 390–394.
Nelson, D.B. Nehrir, M.H. & Wang, C. 2006. Unit sizing and cost analysis of stand-alone hybrid wind/PV/ fuel cell power generation systems, *Renewable Energy*, pp. 1641–1656.
Obara, S. 2007. Analysis of a fuel cell micro-grid with a small-scale wind turbine generator, *International Journal of Hydrogen Energy*, Vol. 32, pp. 323–336.
Pan, C.T. Chen, J.Y. Chu, C.P. & Huang, Y.S. 1999. A fast maximum power point tracing for photovoltaic power systems, *Proc. of IEEE Industrial Electronics Society Conference*, Vol. 1, pp. 390–393.
Rowe, A. & Li, X. 2001. Mathematical modeling of proton exchange membrane fuel cells, *Journal of Power Sources*, Vol. 102, pp. 82–96.
Sapru, K. Stetson, N.T. & Ovshinsky, S.R. 1997. Development of a small scale hydrogen production storage system for hydrogen applications, *Proc. of the 32nd Intersociety Conference*, Vol. 3, pp. 1947–1952.
Shaahid, S.M. & Elhadidy, M.A. 2007. Technical and economic assessment of gridindependent hybrid photovoltaic-diesel-battery power systems for commercial loads in desert environments, *Renewable and Sustainable Energy Reviews*, pp. 1794–1810.
Ulleberg, O. 1997. Stand-alone power systems for the future: optimal design, operation and control of solar-hydrogen energy systems, *Ph.D. dissertation*, Norwegian University of Science and Technology.
Wang, W.H. & Wang, X.D. 2006. Analysis and measurement of SOC in the vanadium flow batter, *Journal of Zhejiang University of Technology*, Vol. 34.

E-Commerce, E-Business and E-Service – Lee (Ed)
 ISBN 978-1-138-02646-9

Employing an E-Service of BSMSTOL to the swimming learning

Peng-Chen Chieh
National Tou Cheng Home Economics and Commercial Vocational High School, Toucheng Township, Yi-Lan County, Taiwan

Huay Chang
Department of Information Management, Chihlee Institute of Technology, New Taipei City, Taiwan

ABSTRACT: This work employs an E-Service of Breast Stroke Movement Simulation Trainer On Land (BSMSTOL) to increase the Swimming Learning Efficiency. The machinery structure theory and ergonomics are combined together to submit the BSMSTOL. The advantages of this trainer are: 1. The learners can realize the whole movements of the breast stroke on land without the environment resistance of learning, 2. The learners can learn the whole details of the breast stroke, it includes (1) The kicking water of the breast stroke. The learners can realize the main points of kicking water by using the BSMSTOL. (2) The arm pulls of the breast stroke. The learner can realize the main points of arm pull by using the BSMSTOL. (3) The change breath of the breast stroke. The learners can realize the usage method of the back' power while changing breath by using the BSMSTOL. (4) The learners can stimulate the whole process of the breast stroke by using the BSMSTOL. This trainer modifies the traditional swimming teaching and improves the existed swimming training equipment that focuses on the lower limbs strength training only. The leg simulation movements are refined to become the breast stroke kicking water. The simulation machinery of the breast stroke arm pull movements is increased. This trainer helps the beginning swimmers learn the breast stroke and the advanced swimmers modify the movements. Numerous tests have been made to demonstrate the efficiency of the proposed BSMSTOL.

Keywords: E-Service; breast stroke; swimming; simulation; trainer

1 INTRODUCTION

Swimming is a physical activity that is good for good body and mind. In our country, the Department of Physics of the Ministry of Education promotes the Student Swimming Ability 121 Plan actively [1]. The student swimming ability is divided in to many levels that can upgrade the students' self-swimming ability and self-rescuing ability. Due to the constraints of humid and raining weather, the learning in the outdoor swimming pool is restricted. Therefore, how the beginning swimmers can improve their swimming ability under the constraints of time and space become a very important course. In this research, the BSMSTOL is designed on the basis of the existed 4th swimming simulation training equipment [2]. This trainer provides the beginners the fixed space track of the kicking water and arm pull of the breast stroke on land. As shown in Figure 1.

2 METHODOLOGY RESEARCH

The detailed methodology of building the Breast Stroke Movement Simulation Trainer On Land is introduced in this section.

2.1 *The importance of swimming gesture*

Although there are difference of different swimming types of body gesture, the streamline form body gesture are the same requirement. Correct body gesture is good for decreasing the resistance and bring the arm pull frequency of the limbs [3].

The two arms and two legs make the symmetrical movements of the breast stroke. The forward resistance comes from the body's upright vertical direction resistance while the swimmer is changing breath. Therefore, learning the correct body gesture is the only method to decrease the resistance.

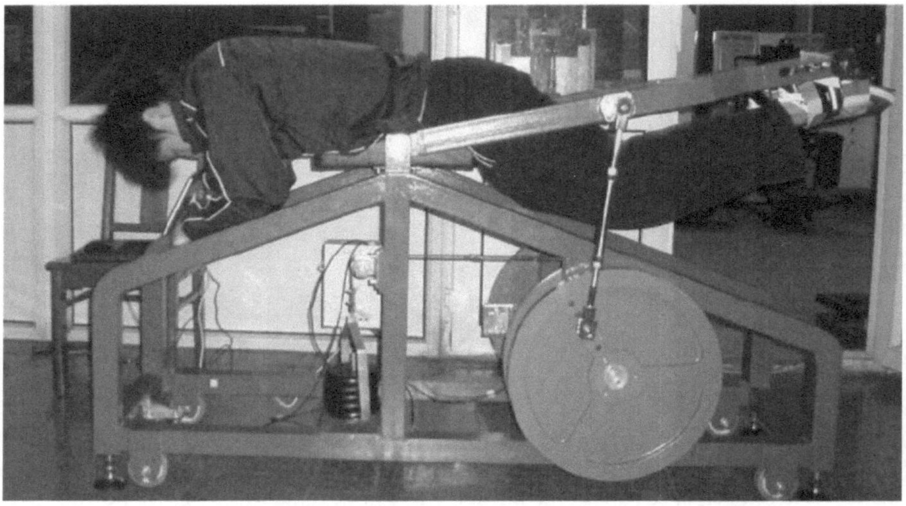

Figure 1. The 4th swimming simulation training equipment.
Source of data: The research and development of the YX-4 4th swimming simulation legs kicking water trainer [2].

2.2 *The leg movement process of the breast stroke*

The breast stroke is most welcomed by the teenagers. It has practical function and beautiful gesture. It can also provide the learners longer practicing time and physical training time [4]. The key points of the learning the breast stroke is on the legs' skills. The correctness of the leg movements appears a direct influence for the learners to keep correct gesture in the water and produce the propulsive power.

The succession movements of the breast stroke are:

1. Pack up the legs:
   Two knees must take apart when packing up the legs. The lower legs pack up ahead, the heel should be near to the bottom, the ankle should appear like ectropion and clip. The degree between the thigh and belly is about 120 degree. Just like the thigh, the lower legs should try to pack up tightly and just let the bottom to store the power [5]. As shown in Figures 2 and 3.
2. Legs ectropion
   Before the end of the legs' pack up movement, the two feet's thenars are ready to make the ectropion. This ectropion movement can make the legs' clipping movement more efficiency. As shown in Figure 4.
3. Tread and clip water
   This step is combined with two movements, the first is treading water, the second is clipping water. These two movements link as a continuing movement that pushes the body moving forward in the water. As shown in Figure 5.
4. Slide in the water
   After clipping the water, two lower legs straight up and tight together. This movement appears the minimum forward resistance and forbids the loss of the propulsive force after the tread and clip movement.

2.3 *The arm movement process of the breast stroke*

There are two types of the arm movement process:

1. Outward arm pull: Straight up two arms, the hands' thenars turns outward and slide ahead.
2. Inward arm pull: This is the main pushing stage of the inward arm pull process. The elbow is as an axle center, then appear the high elbow gesture to show the catching water movement inward with a great degree. The hands' thenars turn inward gradually. As shown in Figure 6.
3. Slide water: After clipping the water, two lower legs are combined and straight up that minimizes the forward resistance and decrease the loss of the propulsive force after the tread and clip.
4. Straight up the arsms: After the inward arm pull, the arms straight up forward and close to the body. This movement decreases the resistance shown in Figures 7 and 8.

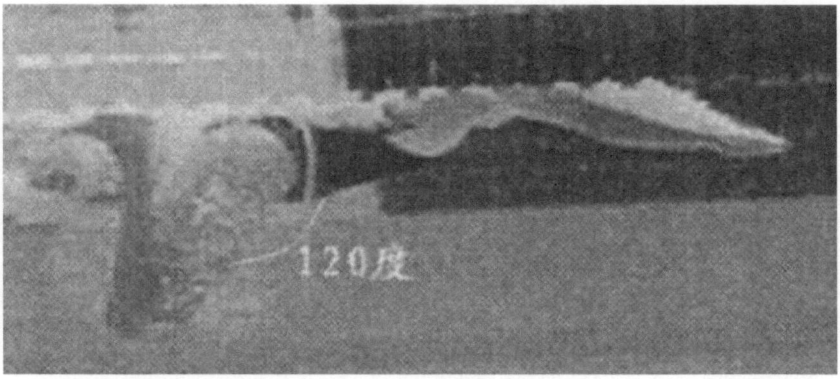

Figure 2. The degree between the thigh and belly is about 120 degree.

Figure 3. The degree between the thigh and belly is about 120 degree.
Source of data: The improvement process of the skills of the breast stroke of the excellent women sporters Swun Hwa and Ji Li Ping [5].

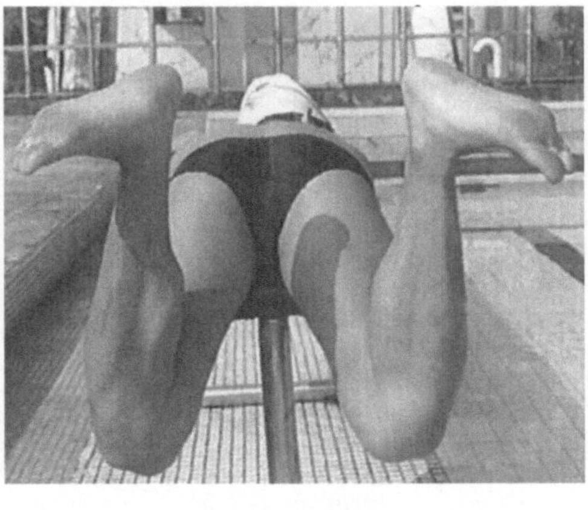

Figure 4. The ectropion movement of the breast and stroke.
Source of data: The Physical Education of the Institute of Technologies III [6].

3 PRACTICAL DESIGN

Mechanical Structure and Implementation Method of the BSMSTOL are introduced in this section.

3.1 *The mechanical structure of the breast stroke movement simulation trainer on land*

Mechanical Structure of the BSMSTOL is introduced in the following and as shown in Figures 9 and 10.

1. The support pad in the belly can simulate the changing breath of the body's raiding movement of the breast stroke. 2. The support stick of the elbow can simulate the hand movements of the breast stroke, such as, outward arm pull, inward arm pull and arm straight up. 3. The support sticks of the knees. 4. The support axle can fix the

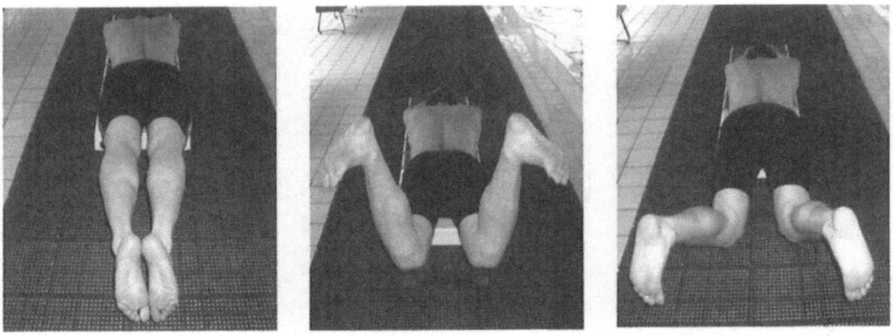

Figure 5. The continuing movements of feet's treading and clipping movement of the breast stroke.
Source of data: Physical Education III [6].

Figure 6. The catching water movement inward with a high elbow gesture.
Source of data: The Physical Education of the Institute of Technologies III [6].

Figure 7. The hands prepare movements --> outward arm pull --> inward arm pull (from left to right).
Source of data: Physical Education III [6].

learner's knee and forbid the knees over spreading while kicking water. It helps the learner realizes the standard movements of the breast stroke. 5. The wheels help the machine moves easily. 6. The fix support stick helps the machine fixed in the practicing field and keeps the dangerous away.

3.2 *The implementation method of the breast stroke movement simulation trainer on land*

The BSMSTOL includes the base. The base includes the following devices:

1. The support pad in the belly. 2. The arm support structure. 3. The legs support structure.

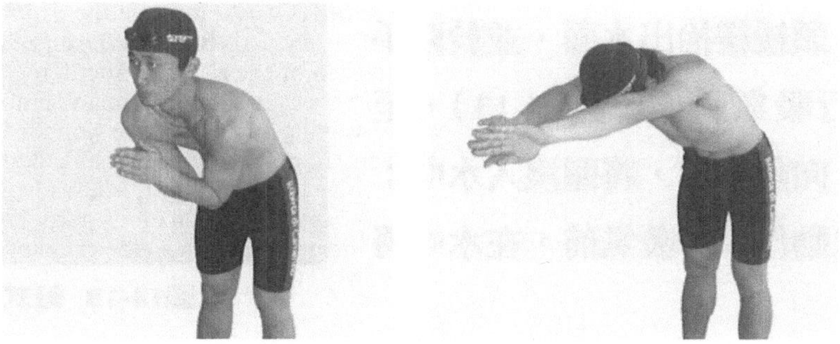

Figure 8. The movement of inward arm pull --> arms straight up (from left to right).
Source of data: Physical Education III [6].

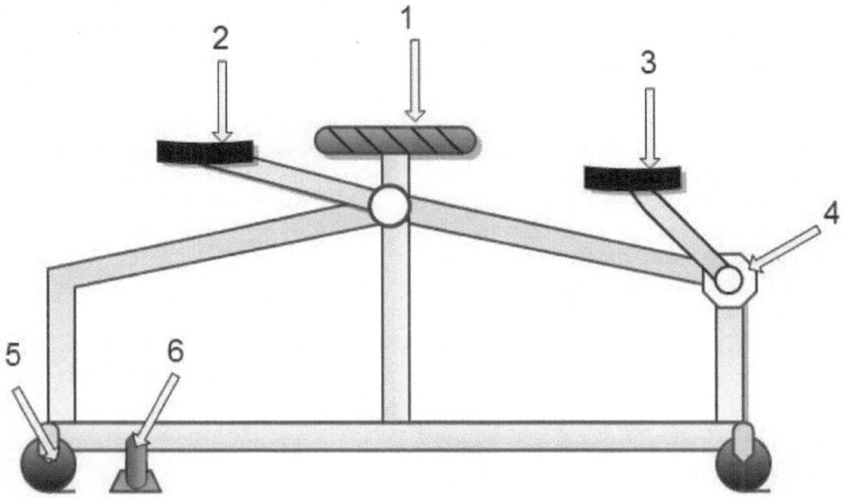

Figure 9. The side diagram of the breast stroke movement simulation trainer on land.

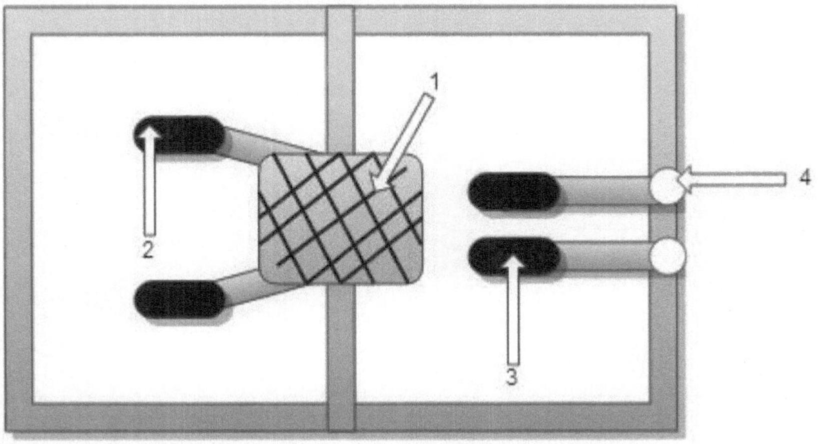

Figure 10. The vertical view of the breast stroke movement simulation trainer on land.

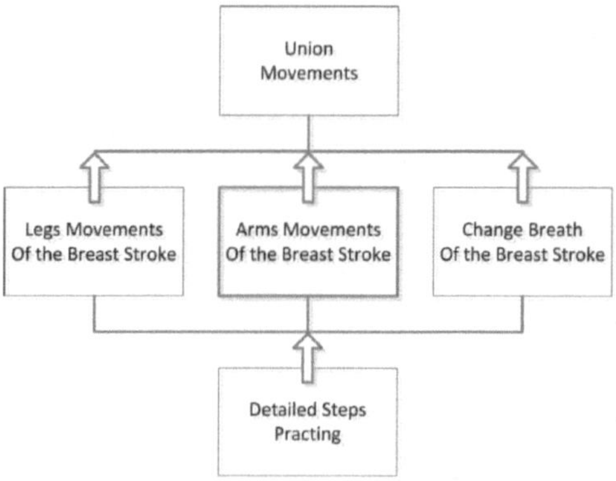

Figure 11. The implementation method of the practicing of the breast stroke.

4. The legs support stick. 5. The moving wheels. 6. The fix support stick.

In this research, the learner faces down and lies on the support pad of the belly. Two elbows put on the arms support structure, two knees put on the legs support structure. The learner can self-control the machine and simulates the detailed movements of the breast stroke. Divided movements can be practiced first and adopt the union practice. These learning processes are shown in Figure 11.

4 SUMMARY

This work employed an E-Service of Breast Stroke Movement Simulation Trainer On Land (BSMSTOL) to increase the Swimming Learning Efficiency. This BSMSTOL can be operated easily. Instead of the electric structu, it employs the support sticks to support human's limbs. It forbids the possibility of the getting wounded for the beginners, it provides the human's personal power to involve in practicing. It is a really good trainer for the beginners. Numerous tests have been made to demonstrate the efficiency of the proposed BSMSTOL.

REFERENCES

[1] Physical Dept. of the Ministration of Education (2012), Students Swimming Ability 121 Plan.

[2] Wei-Yan Gwo, The Research and Development of the YX-4 4th Swimming Simulation Legs Kicking Water Trainer, ShangTung Physical College, p. 9–11, (2010).

[3] Wu-Shan Chen, The Emphasis of Body Gesture of Swimming Training, Physical Teachers and Friends Journal, p. 19020, (2002).

[4] Ywe-Nan Lin, Swimming, Edition, Vol. 01, (2002).

[5] Yi Chang, The Improvement Process of the Skills of the Breast Stroke of the Excellent Women Sporters Swun Hwa and Ji Li Ping, ShangHai Physical College, p. 17, (2010).

[6] Chen-Chang Lin, et al. The Physical Education of the Institute of Technologies III, Long-Teng Culture Publisher, p. 97–100, (2009).

[7] De-Shou Hwung, et al. Physical Education III, OwShing Publisher, p. 133–136, (2001).

E-Commerce, E-Business and E-Service – Lee (Ed)
© 2014 Taylor & Francis Group, London, ISBN 978-1-138-02646-9

An E-Service of security in electric power systems

Ming-Chen Huang & Shieh-Shing Lin
Department of Electrical Engineering, Saint John's University, Tamsui, Taipei, Taiwan

ABSTRACT: This work proposes an E-Service of security detecting tool in electric power systems. This work also proposes a technique combining a method with the dual-type based method to solve the considered problem. This work has been demonstrated by the test to verify the efficiency of the proposed E-Service of security detecting tool and compares with some other existed tools on a typical electric power system.

Keywords: E-Service of security; block systems; dual-type method

1 INTRODUCTION

Most of the distributed power systems are constructed by the block systems interconnected with each other by tier lines; For example, the ground transportation system uses the freeway system to interconnect the local transportation systems; the communication network contains local networks and long distance network, which is used to connect different confined networks. The nonlinear big network optimization problems, for example, the state estimation problems and nonlinear multi- network problems are computationally hard since of their vast size and nonlinearity. There were many nonlinear encoding techniques urbanized for these problems [1]–[7]. This work adopted that the big systems is created by n block systems consistent by tier lines, and each block system corresponds to a different area. An instance with four block systems is given in Figure 1, where $x_{1_b}^2 = (x_{11}, x_{12}), x_{2_b}^1 = (x_{21}, x_{22})$.

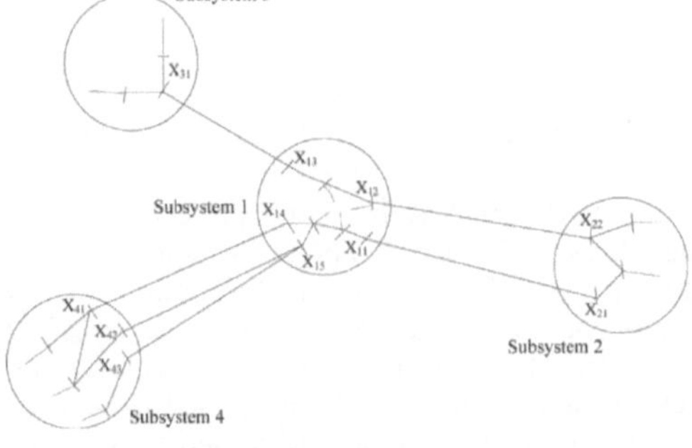

Figure 1. An instance block system.

2 THE E-SERVICE OF SECURITY DETECTING TOOL PROBLEM FORMULATION

The representations for the i th area ($i = 1, ..., n$):

N_i: Digit of bus of area i of the E-Service of security detecting tool.

z_i: Limited measured vector in the i th block of the E-Service of security detecting tool.

x_i: Limited vector collected of N_i magnitudes and N_i angles.

$x_{i_b}^j$: Status vector of the orientation boundary buses of area $i, j \in L(i)$.

h_i: Nonlinear measured vector function matching to z_i.

η_i: Gaussian random measured error vector matching to z_i of the E-Service of security detecting tool.

$v(x)$: The inequality constraint of the E-Service of security detecting tool.

The measured matching to Gaussian random fault vector in each block i is expressed as

$$z_i = \eta_i + h_i(x_i, x_{i_b}^j) \tag{1}$$

Based on (1), the measured fault vector of the whole system is uttered as

$$\eta = z - h(x) \tag{2}$$

The problem of the proposed E-Service of security detecting tool is to solve the following constrained problem in a distributed way:

$$\min f(x) \tag{3}$$

subject to

$$g(x) = 0 \tag{4}$$

$$v(x) \leq 0 \tag{5}$$

where $f(x) = \eta^T \eta$ denotes the objective function.

3 SOLUTION METHOD

3.1 *The dual-type method*

The dual problem can be stated as follows [7]:

$$\max_{\lambda} \phi(\lambda)(\phi(\lambda)$$

$$= \min_{dx \in \Gamma} \sum_{i=1}^{n} \left\{ \begin{array}{l} \frac{1}{2} dx_i^T D_i(x_i(k), x_{i_b}^j(k)) dx_i \\ + \nabla_{x_i} f_i^T(x_i(k), x_{i_b}^j(k)) dx_i \\ + \lambda_i^T [g_i(x_i(k), x_{i_b}^j(k)) \\ + \nabla_{x_i} g_i^T(x_i(k), x_{i_b}^j(k)) dx_i] \end{array} \right\} \tag{6}$$

Γ denotes the set of constraints, that is

$$\Gamma = \{dx |\ \underline{x}_i \leq x_i + dx_i \leq \overline{x}_i, i = 1, \ldots, n\}. \tag{7}$$

The dual-type method solve (8):

$$\lambda_i(t+1) - \lambda_i(t) = \beta_i(t) \Delta\lambda_i(t),\ i = 1, \ldots, n \tag{8}$$

where $\Delta\lambda^T(t) = [\Delta\lambda_1^T(t), \ldots, \Delta\lambda_n^T(t)]$ is the solution of the following approximate quadratic problem of (8) at $\lambda(t)$:

$$\max_{\Delta\lambda} \frac{1}{2} \Delta\lambda^T \begin{bmatrix} \Phi_1 & \cdots & 0 \\ \vdots & \ddots & \vdots \\ 0 & \ldots & \Phi_n \end{bmatrix} \Delta\lambda + \nabla_{\lambda} \phi^T \Delta\lambda. \tag{9}$$

where Φ_i is solved by

$$\Phi_i = -\nabla_{x_i} g_i^T \left(x_i(k), x_{i_b}^j(k)\right) D_i^{-1} \nabla_{x_i} g_i \left(x_i(k), x_{i_b}^j(k)\right) \tag{10}$$

The $\nabla \phi^T(\lambda(t))$ in (9) is of $\nabla_{\lambda} \phi\ (\lambda(t)) = [\nabla_{\lambda_1} \phi\ (\lambda(t)), \ldots, \nabla_{\lambda_n} \phi\ (\lambda(t))]$, and $\nabla_{\lambda_i} \phi(\lambda(t))$ can be computed by

$$\nabla_{\lambda_i} \phi(\lambda(t)) - g_i(x_i(k), x_{i_b}^j(k)) = \nabla_{x_i} g_i^T(x_i(k), x_{i_b}^j(k)) d\hat{x}_i(\lambda(t)) \tag{11}$$

where $d\hat{x}^T(\lambda(t)) = [d\hat{x}_1^T(\lambda(t)), \ldots, d\hat{x}_n^T(\lambda(t))]$ is the solution of the optimization problem on the RHS of (11) [7].

The solution of the approximate quadratic dual problem can be obtained by solving

$$\Delta\lambda(t) = -\Phi^{-1} \nabla_{\lambda} \phi(\lambda(t)) \tag{12}$$

which, based on the structure of Φ and $\nabla_{\lambda} \phi(\lambda(t))$, can be seperated into the following n set of linear equations

$$\Delta\lambda_i(t) = -\Phi_i^{-1} \nabla_{\lambda_i} \phi(\lambda(t)), i = 1, \ldots, n. \tag{13}$$

The n set of linear equations can be distributed solved in a computer network.

To solved $\nabla_{\lambda} \phi(\lambda(t))$, this work need to solve the minimization problem on the RHS of (13) to obtain $d\hat{x}\ (\lambda(t)) = [d\hat{x}_1\ (\lambda(t)), \ldots, d\hat{x}_n(\lambda(t))]$ and can be achieved using the two stages algorithm [8].

4 SIMULATION AND TEST RESULTS

This work employed the proposed E-Service of security detecting tool on the typical IEEE block system for the accuracy state problems. This work partitioned the IEEE block system into four areas (A1, A2, A3 and A4), as shown in Figure 2. To reveal the competence of the E-Service of security detecting tool in the power system, this work set the value δ in D_i to be small value and the area convergence criterion value set to be small in all cases. In this real devoted set-up this paper experienced the efficiency on the accuracy of the power systems of the E-Service of security detecting tool. The sizes of the IEEE block system of these four areas (A1, A2, A3 and A4) are shown in Table 1. The corresponding data of each block were stored in the corresponding computer. This work has tested many cases and obtained some successful results corresponding to the traditional detecting tools while the error data is increasing.

5 SUMMARY

This work proposed an E-Service of security detecting tool in electric power systems. This work also proposed a technique combining with the dual-type based method to solve the considered problem. This work has made some simulations to reveal the efficiency of the proposed E-Service of security detecting tool and obtained some successful results corresponding to the traditional detecting tools.

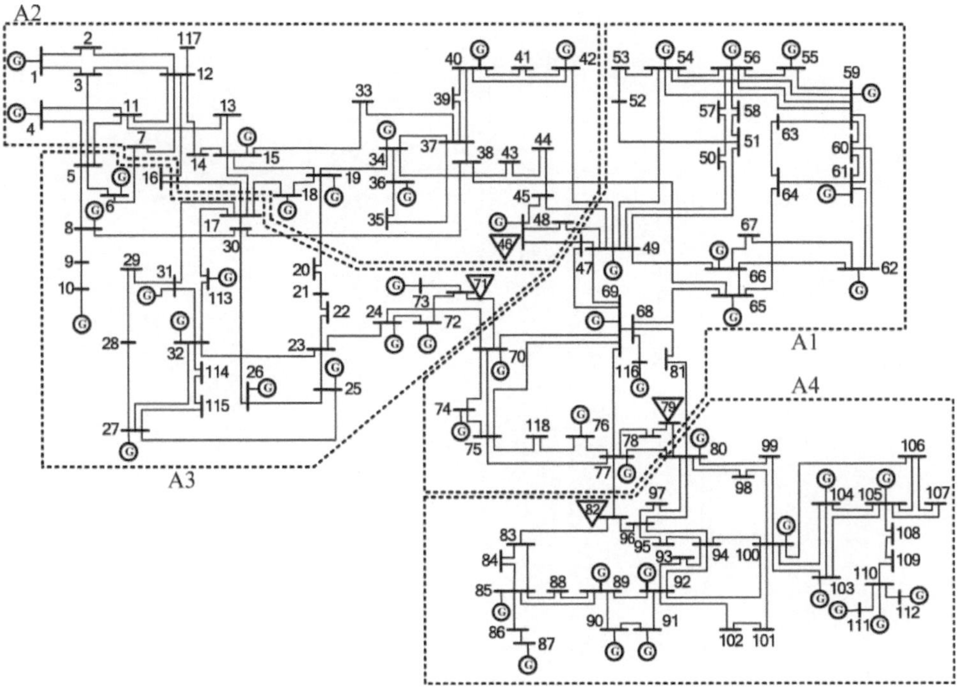

Figure 2. A typical IEEE block system.

Table 1. The cardinality of measured value in the IEEE block system.

Block	A1	A2	A3	A4
Cardinality	32	28	26	32
# of E-Service of security detecting tool				
Power injection	16	24	16	18
Voltage	6	6	7	8
Power line flow	84	54	55	74
# of no load and no generation buses	8	6	6	8

ACKNOWLEDGEMENT

This paper was partially supported by NSC in Taiwan under grand NSC101-2221-E-129-007-MY2.

REFERENCES

[1] Aschmoneit F.C., N.M. Peterson and E.C. Adrian, "State estimation with equality constraints," Tenth PIGA Conference Proceedings, May 1977, Toronto, pp. 427–430.

[2] Wu F.F., W.E. Lin and S.M. Lun, "Obscrvability analysis and bad data processing for state estimation with equality constraints", IEEE Trans. on Power Systems, vol. 3, no. 2, 1988, pp. 541–548.

[3] Clements K.A., G.W. Woodzell and R.C. Burchett, "A New Methods for Solving Equality-Constrained Power System Static-State Estimation", IEEE Trans. on Power System, vol. 5, no. 4, 1990, pp. 1260–1266.

[4] Nucera R.R. and M.L. Gilles, "A Blocked Sparse Matrix Formulation for the Solution of Equality-Constrained State Estimation" IEEE Trans. on Power Systems, vol. 6, no. 1, 1991, pp. 214–224.

[5] Lin S.-Y. and C.-H. Lin, "An implementable distributed state estimator and distributed bad data processing schemes for electric power systems", IEEE Trans. on Power System, vol. 9, no. 3, 1994, pp. 1277–1284.

[6] Bose A. and K. Clements, "Real-time modeling of power networks," Proceedings of the IEEE, vol. 75, no. 2, 1987, pp. 1607–1622.

[7] Luenberger D.G., Linear and nonlinear programming, 2nd ed., Addison Wesley, 1984.

E-Commerce, E-Business and E-Service – Lee (Ed)
 ISBN 978-1-138-02646-9

Employing an E-Service for establishing the improved hurdles of the hurdle race

Chi-Kung Ho
Yi-Lan Fu-Shing Junior High School, Yi-Lan City, Yi-Lan County, Taiwan

Huay Chang
Department of Information Management, Chihlee Institute of Technology, New Taipei City, Taiwan

ABSTRACT: This paper submitted an E-Service for the improved hurdles of the hurdle race of the track and field by using mechanical structure and safe quality. The new design teaching hurdles can be storage and help the beginners overcome the fear of the hurdle race activity while facing the standard hurdles. This design is suitable for the Junior High students training programs and can increase the competition grades. The hurdle race is a competition sport that sets fix height and numbers of hurdles at fixed distance in the race. The skills include begin run, begin run to the first hurdle, cross hurdle, running between the hurdles and reach line. There are lots of metal elements involved in the cross hurdle stage and become the most difficult training skill to overcome for the beginners of the hurdle race sport. Therefore, we submit a new design to solve this problem. The design includes the following advantages: 1. The cross bar uses the safety quality that reduces the fear of the beginners. 2. The height of the hurdle expands and contracts to 60 cm and 100 cm and reaches the training requirements of all learning stages. 3. The hurdle can be folded completely and easily storage and carry. 4. Decrease the hurting ratio of the players before the competition. This improved hurdles of the hurdle race of the track and field applied in the learning of the Junior High School players and gained a good practical result.

Keywords: E-Service; mechanical structure; cross hurdle; jump hurdle; fear; standard hurdle

1 INTRODUCTION

1.1 *Standard hurdle specifications*

The hurdle race was origin in 17 and 18 century. The English shepherd cross the fences those are used to control the sheep. This activity became the hurdle race gradually. The first hurdle race began in 1860. Up to now, there is a unified standard of the hurdles in the hurdle race of track and field. The structure of the hurdle should be metal or other suitable materials. The cross bar of the hurdle can be wood or other suitable materials. The hurdle is combined with the base and two sticks and becomes a rectangle holder. There is one or more cross bar are used to strengthen the fix of the holder's base. Once the upper cross bar of the hurdle is pushed by the 3.6 kg stuff, the hurdle may fall. The height of the hurdles can be adjusted for different competition purpose. Because the power to push down the hurdle is about 3.6 kg and no heavier than 4 kg, no matter the hurdle is high or low. Therefore, the design of the hurdle's center of gravity should be adjustable. The width of the hurdle should be 1.18 to 1.20 m. The maximum length of the hurdle's base is 70 cm. The width of the upper cross bar is 7 cm and the thick is 1–2.5 cm. The upper edge should be ground and both sides should be fixed. The weight of the hurdle should be no less than 10 kg. The upper cross bar of the hurdle should be painted black and white or other significant color. The color of the both sides of the hurdle could be lighter. The width of the upper cross bar should no less than 22.5 cm. As shown in Figure 1.

2 THE METAL FEAR RESULTS FROM THE STANDARD HURDLES FOR THE BEGINNERS IN HURDLE RACE TEACHING

The hurdle race is a short-distance barrier running competition. In order to cross a barrier in high speed, the runner needs long-time training. The high metal confidence is needed in the training. For a beginner, facing the hurdle that is 10 kg and 76 cm height, it is quite difficult to jump cross it not to run cross it. The paper [1] described that, in the hurdle race competition, the condition of the

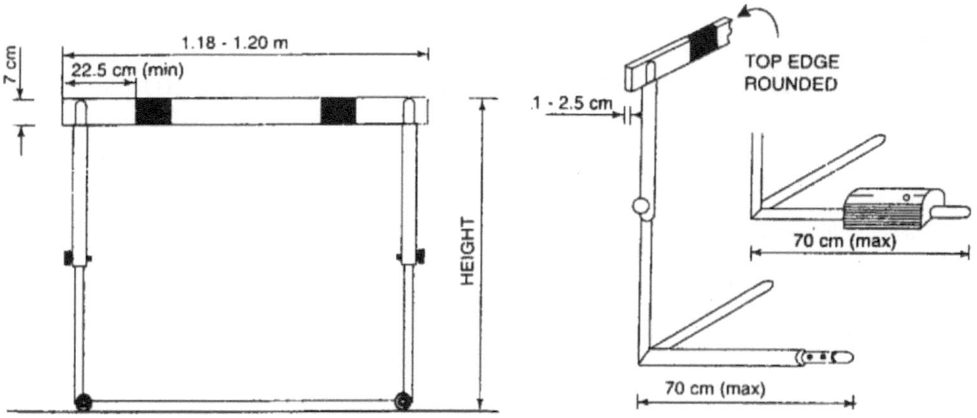

Figure 1. The standard hurdles diagram (2012–2013 track and field rules).

track, the height, weight, quality and looks of the hurdles affect the hurdle race teaching. The paper [2] described that, in the teaching of hurdle cross, besides the subject elements those make the students feel fear, the objective elements are one of the reason that cause students feel fear. The paper [3] described that the teenagers usually appear fear while learning hurdle cross and become the metal barriers that cause the student can't learn the hurdle cross skills. The paper [4] described that while teaching in the Middle School, in order to help the students complete the hurdle cross movements, the metal elements, the body elements and the sport skills should be combined together. In addition, the paper [5] described that the fear of hurdle create the students' negative feelings in the teaching process. Due to the influence of the metal element of the fear of hurdle, the students appear nervous, strange skill and can't complete the movements. Furthermore, the paper [6] described that, in the hurdle race competition, the human body is required to cross the standard height.

3 HOW TO HELP THE BEGINNER DECREASE THE FEAR WHILE FACING THE HURDLE

Improve the Teaching Equipment The design of the improved hurdles of the hurdle race of the track and field is busmitted is to help the beginners of the hurdle race to overcome their fear.

Using Assisted Teaching Tools During the training process, use the fold pads to increase the students' sport level gradually, increase the students' confidence, overcome the fear and cross the hurdle finally [7].

Detailed Explanation of the Cross Hurdle Movement Process Besides the "hurdle" is an element that makes the learners feel fear, another element is the lack of understanding of the hurdle race sport.

Table 1. Students' learning feedback of materials statistical analysis by using multimedia theory in teaching (n = 84).

Content	# of students	%
Good for understanding and control the theory knowledge of hurdles race.	77	88.1
New, practical, acceptable and wish to adopt from now on.	65	77.4
Adopt multimedia teaching method and bring interesting, image and easy understanding.	60	71.4
Good for establishing the correct movements and understanding the skills.	59	70.2
Easily to attract attention, increase learning interesting and become actively learning.	57	67.9

Using Multimedia Multimedia assisted teaching method appears the specific advange in learning hurdle race theory for the students. Comparing with the regular teaching, the multimedia assisted teaching is more active, imagible that increase students' learning interesting [8] [13] shown in Table 1.

Reasonable Teaching method The teaching of hurdle cross skills must be step by step. The movements begin from the simple to the difficult moement. The height of the hurdle increases from the lower to the standard height. The distance between hurdles can be shortened to appropriate distance. Once the learners are fluent, distance can be adjusted to the standard distance of the hurdles [9]–[11].

4 THE STRUCTURE ANALYSIS OF THE IMPROVED HURDLES OF THE HURDLE RACE OF THE TRACK AND FIELD

The purpose of the improved hurdles of the hurdle race of the track and field is to overcome the beginning learners' fear while facing the standard hurdles. The hurdles are easily storage and moving. Before the learning of the cross hurdle, the learner must move the hurdles to the track. Ten hurdles are needed in the hurdle race sport. Each hurdle is 10 kg. After settle down all the hurdles, the learner is tired. While the learner begins to cross the hurdle, the first sight is the heavy cross bar of the hurdle. After the settlement of the hurdles, the learner's learning decreases. The heavy hurdles stand in front of the learner and the learner is afraid to get hurt. The learner appears fear slightly. The practicing effect is worse while the learning willing decrease and the occurrence of fear. Not to say the refuse of learning this sport. Today, the quality of the corss bar of the hurdle is wood or hard plastic. Even the open safe cross bar is made of hard plastic. Those can't eliminate the fear of the learners. In this paper, we designed a connecting cross bar that is made of sponge. The appearance is just like the standard hurdle. This design decreases the incompatibility of transferring from the improved hurdles of the hurdle race of the track and field to the standard hurdle. The students' fear of the hurdle is the key point of the influence of the grades of the hurdle race competition [12]. Adoping the "soft hurdles" in assisting teaching not only overcome the students' fear of hurdles but also increase the students' learning interesting and efficiency. This increase the students' learning effect and teaching quality of the hurdle race education. The quality of the main structure of the hurdle uses the aluminum alloy that matches the concept of safety, easily storated and easily moving. There are 10 parts of the hurdle. They are 1. The Cross Holder, 2. The Connected Sponge, 3. The Velcro, 4. The Rotating Height Adjusting Stick, 5. The Fastener Fix Points, 6. The Strengthen Holding Fastener in the Base, 7. The Holding Fastener and the Base Bar Bearing, 8. The Moving Joints of the Base Bar 9. The Base Bar, 10, the Base. As shown in Figure 2.

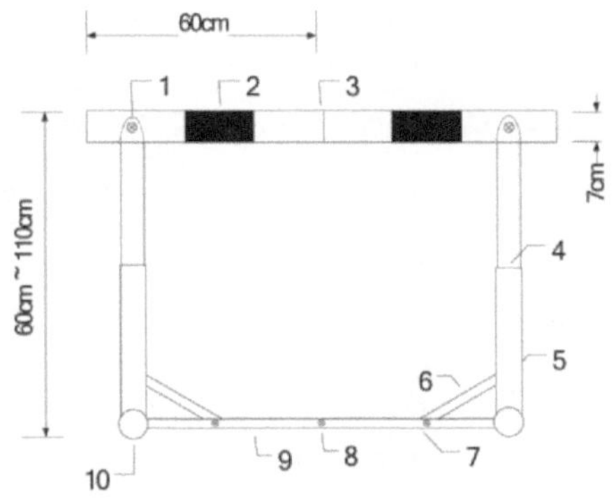

Figure 2. The hurdle front view of the improved hurdle race of track and field.

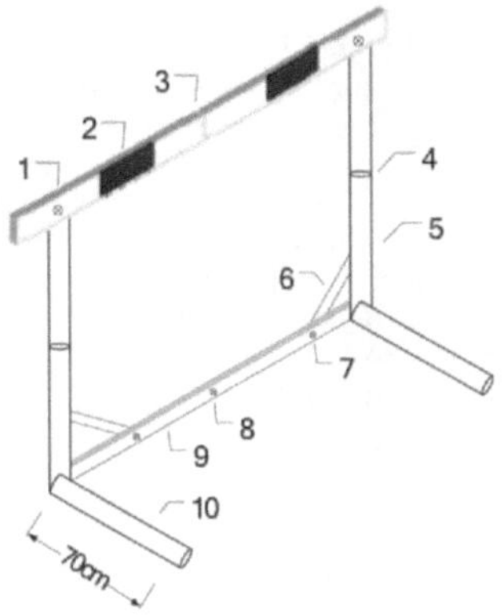

Figure 3. The hurdle of the improved hurdle race of track and field.

4.1 *Practical research*

In this session, we introduced our practical research of the improved hurdles of the hurdle race of the track and field toward the Junior High students.

4.2 *Practice design*

The practical design is toward the 1st grade students of the Junior High school. First, these students attended the hurdle race competition which used the tradition hurdles. The traditional competition grades are the research basis. Then, these students gained the training program of new running method between hurdles by using the the improved hurdles of the hurdle race of the track and field. Four male students and four female students joined this program. These students attended the 2013 Yi-Lan Middle School's Track and Field Hurdle Race Competition. Their grade is recorded as the basis of this practical research. Then these students attended the hurdle race training three times a week for two months by using the improved hurdles of the hurdle race of the track and field. The comparison record is listed in Tables 2 and 3.

4.3 *Analysis and discussion*

Tables 2 and 3 shows, after using the improved hurdles of the hurdle race of the track and field for two months, the number of steps decrease 50%

Table 2. The comparison table of the male 100 m hurdle race record—before learning v.s. after learning.

Attendant	Number of steps between two hurdles (steps)	The record before learning (second)	The new record after learning (second)
Before learning			
A	4	14.92	–
B	5	18.21	–
C	5	19.32	–
D	5	19.85	–
After learning			
A	3	14.70	–0.22
B	4	18.10	–0.11
C	4	18.56	–0.76
D	4	18.63	–1.22

Table 3. The comparison table of the female 100 m hurdle race record—before learning v.s. after learning.

Attendant	Number of steps between two hurdles (steps)	The record before learning (second)	The new record after learning (second)
Before learning			
A	4	18.5	–
B	5	18.76	–
C	5	19.14	–
D	5	21.85	–
After learning			
A	4	16.9	–1.6
B	4	18.2	–1.56
C	4	17.93	–1.21
D	4	18.91	–2.94

of the male team. Besides the A students of the female team, the rest three learners' number of steps decrease 50%. As to the grade, the average grade increase 0.58 seconds of the male team and 1.83 of the female team.

5 SUMMARY

The most difficulty thing in hurdle race activity is to overcome the feal of hurdles. When the learners can breakthrough the most barrier "fear" of the hurdle race activity, the hurdles in front of the runners wouldn't be the barrier in the runway. The learners of the hurdle race become braver to elaborate their speed. This work applied an E-Service for the improved hurdles of the hurdle race of the track and field in our research. In this research, eight attendants' number of steps decrease 50%, nine attendants average grades increase 1.20 seconds. These results definitely verify the the improved hurdles of the hurdle race of the track and field can decrease the fear of the learners.

REFERENCES

[1] Jyan-Chen Gung and Jane Wu, The Initial Researh of Overcoming the High School Students' Metal Fear of the Hurdle Cross Courses, Physical Teaching Monthly Journal, (2013).
[2] Jung Lee, The Research of the Fear Metal Problem of the Physical Professional Students Toward the Hurdle Cross, Technique Information, Vol. 26, (2012).
[3] Da-Ying Chen, The Beginning Research of the Hurdle Cross Beginners' Fear of the Teenagers, Fu-Jyan Physical Techniquie, Vol. 4, No. 32, (2013).
[4] Feng-Yin Liu, The Research of the Orientation of the Students' Mental Barrier and the Overcoming Methods in the High School Hurdle Cross Teaching Courses, Phy. Ed. Training, Vol. 7, No. 5, (2013).
[5] Wen Jiandong, Deng Lin, An Analysis of the Countermeasures for Overcoming the Fear of Hurdles in Hurdle Teaching, JIAOXUE, TANSUO, (2013).

[6] Jyan-Chr Liu, The Introduction of How to Eliminate the Cross Hurdle Phenomon in the Hurdle Cross Competition, Nei-Jyan Technique, Vol. 9, (2011).
[7] Bin Liu, The Application of Using the Folding Pads in Overcoming the Fear of the Elementary Students While Joining the Hurdle Cross Training, China Student Physical, (2011).
[8] Lin-Jung Ho and Shyan-Ru Lee, The Development and Experiments of the Multimedia Software of the Hurdle Cross Teaching, Wu-Han Physical College Journal, Vol. 3, No. 40, (2006).
[9] Ke-Jung Chen, The Influence of the Increasing Height of the Hurdles in Hurdle Cross Learning, The Physical Constitution Journal of the R.O.C., Vol. 17, (2011).
[10] Jya-Kwei Shu, The Research of How the Game Teaching Method Influence the Volleyball course in High School, Thesis of Graduate School of the Shang Tung Normal University, (2013).
[11] Shin-Yan Chu, The Research of How the Dean and the Teachers 'Physical Concept Influence the Students' Phy. Behavior in the Junior High of SuJou City, Thesis of Grad. of SuJou Univ., (2013).
[12] Jing-Shan Hwung and Shing Lin Wang, The Research of Using the Soft Hurdle in the Hurdle Cross Teaching, Beijing Physical University Journal, (2003).
[13] Shyang-Ru Lee, The Practical Research of Concerning the Modern Hurdle Cross's Teaching Method, SiAn Physical College Journal, (1991).

E-Commerce, E-Business and E-Service – Lee (Ed)
© 2014 Taylor & Francis Group, London, ISBN 978-1-138-02646-9

Relationships between the quality of online shopping service and customer loyalty in women online shopping website

Kai-Han Won
Department of Graphic Communication Arts, National Taiwan University of Arts, Taiwan

Sheng-Hao Tsao
Department of Management Information Systems, National Chengchi University, Taiwan

ABSTRACT: **Purpose:** The Market Intelligence & Consulting Institute of Taiwan (MIC), a division of the Institute for Information Industry (III), estimated that Taiwan's E-Commerce has the potential to reach NTD one trillion (approx. USD 33 billion) in 2015, especially in the women-based E-Commerce filed. The purpose of this research is to discover the relationships between the quality of online shopping service and brand loyalty. **Design/methodology/approach:** This study uses questionnaire simple sampling with online shopping user participants, 302 online and 40 paper surveys were collected, with 42 invalid and 300 valid surveys. **Findings:** Findings are as follows: 1. "quality of website service" is significantly differed by "age" in women-based E-Commerce, which "under 20" is greater than "over 51". 2. "quality of website service" is significantly differed by "degree of education" in women-based E-Commerce, which "college" is greater than "graduate school" and "high school". 3. "quality of website service" is significantly differed by "occupation" in women E-Commerce, which "IT industry" is greater than "manufacturing industry", "other", "student", "public servant", "health care industry", "service industry, "financial industry" and "freelancer". 4. "customer loyalty" is significantly differed by "gender" in women-based E-Commerce, which "female" is greater than "man". 5. "customer loyalty" is significantly differed by "degree of education" in women-based E-Commerce, which "college" is greater than "high school". 6. **The equation of quality of website service from regression analysis is:** customer loyalty = 0.319 efficiency + 0.178 reaction + 0.249 compensation + 0.154 connection. Moreover, the higher the efficiency, reaction, compensation and connection, the higher the customer loyalty will be. **Practical implications:** Research findings are expected to provide recommendations for improving E-Commerce website service for women and enhancing customer loyalty. **Originality/value:** This research will provide women-based shopping websites data regarding online service quality and customer loyalty, and further contribute an insight of customer utilization.

1 INTRODUCTION

1.1 *Background*

Taiwan's E-Commerce revenue in 2013 is estimated to reach NTD 7,673 billion (approx. USD 255 billion) with annual growth rate 16%, and it has the potential to reach NTD one trillion (approx. USD 33 billion) in 2015. Since the development of E-Commerce has considerable amount of potential, the Ministry of Economic Affairs has started the Asia-Pacific Chinese E-Commerce Development Program in August 2010. Until December 2012, it has counselled 406 Taiwanese companies to sell goods to China through E-Commerce; furthermore, the B2C and C2C E-Commerce revenues had reached NTD 6,605 billion (approx. USD 220 billion) (Ministry of Economic Affairs, 2013). In addition, with the Cross-Straits Agreement on Trade in Services, there can be more business opportunities for E-Commerce (The Bureau of Foreign Trade, MOEA, 2013). With supports form the government, although the E-Commerce industry grows rapidly over the years, there still are uncertainty regarding the quality of online shopping service, comparing with retail shops that have fully functioned services. This study discusses the relationship between website service and customer loyalty.

1.2 *Research motivation*

In recent years, as the women-based E-Commerce has risen, web-based shopping platforms have provided goods and services to satisfy different target audiences. A MIC E-Commerce study in 2012 presented that many online stores target 30 to 39 years old working women, and best-selling items are household related goods. Thus, the market for women has become one of the major groups in E-Commerce, and there are platforms specifically created for women.

Women have high buying power in E-Commerce, and there are still wide market for its market development. This research explores the relationships between women and E-Commerce, and further contribute findings to the industry.

1.3 *Research purpose*

- To explore the relationship between different demographic variables and research variables.
- To explore the relationship between different demographic variables and website service quality.
- To explore the relationship between different demographic variables and customer loyalty.
- Website service quality has significant prediction to customer loyalty.

2 LITERATURE REVIEW

2.1 *Relationship between women and E-Commerce*

Kalakota and Whinston (1997) proposed that E-Commerce is a modern way for business, it uses the Internet as a medium for purchases, sales, exchanges, in order to reduce costs, meet consumer demand and improve the quality of services.

E-Commerce is an innovative way of business. It uses the web as a medium, consumers can shop, make payments and decide ways of delivery with clicks of buttons, which not only saves time, but also eliminates the need for service personnel comparing to traditional retail shops, as well as rental costs.

A representative of CLUB VENTA, an online marketing firm, Mario Blanch mentioned that in Peru, 7% of 30 million population, 2.1 million people, have shopped online. In addition, women between the ages of 25 to 35 take 65% of the total number. (The Bureau of Foreign Trade, MOEA, 2013). According to this information, it indicates that not only in Taiwan, but women do take significant parts in E-Commerce around the world.

2.2 *Quality of website service*

The quality of website service affects consumers' loyalty toward companies both directly and indirectly. Studies and discussions have been done on this matter, including definition, measurement, aspects and opinions.

Zeithaml, Parasuraman and Malhotra (2000), first proposed the concept of E-Service Quality (E-SQ) and created 11 aspects of website service quality in 2000. In 2005, they had used Wal-Mart and Amazon.com as study objects, and developed E-Core Service Quality scale (E-S-QUAL) and E-Recovery Service Quality scale (E-RecS-QUAL).

2.3 *Customer loyalty*

Customer loyalty is defined as consumers stay with certain brands, products or services, and have the preference to choose the same brands, products or services when needed.

Kolter (1999) refers loyalty as customers who make regular purchases, often provide recommendations to particular companies, can tolerant the raise of prices, and the cost of goods and services are lower to the companies.

Thiel's (1999) indication of online shopping are as follows: numbers of visiting particular websites in a fixed period of time, the length of the time spending browsing websites, and the depth of browsing websites' information.

Lee et al. (2001) defines loyalty as making repurchases, welling to make recommendation, immune to competitors' goods, and welling to pay higher for similar goods for preferred brands.

Based on above researches, this study determines customer loyalty as by word of mouth, making repurchases, welling to pay higher for similar goods, and making particular companies to be ideal transaction objects.

2.4 *Relationship between quality of website service and customer loyalty*

PZB's research (1985, 1988) mentioned about service quality and customer loyalty. Frederick (1990) used zero defections concepts to guide that service quality can affect levels of customer loyalty. Thus, the quality of website service and customer loyalty do have degree of relations.

3 METHODOLOGY

3.1 *Research framework*

This research discusses different demographic variables, quality of website service, relationship between the Internet users and degree of loyalty, finding factors that influence degree of loyalty and relationship between quality of website service and customer loyalty in women-based E-Commerce. Moreover, it is to discover different demographic variables' Internet usage behaviors, quality of website service and loyalty.

3.2 *Hypotheses*

- H1: demographic variables have significant differences to the quality of women-based website service.
- H1.1 gender has significant differences to the quality of women-based website service.
- H1.2 age has significant differences to the quality of women-based website service.

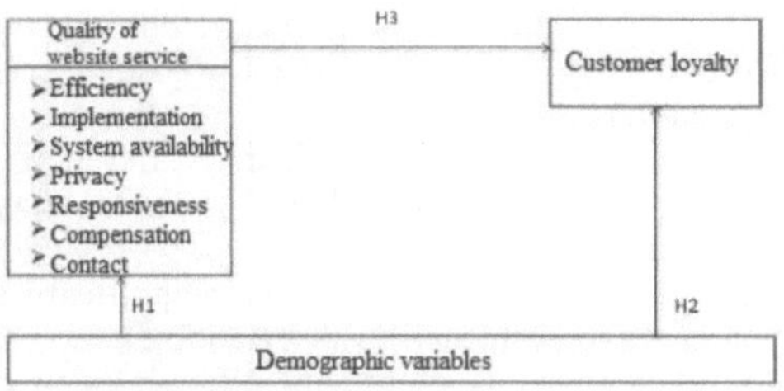

Figure 1. Framework.

- H1.3 education has significant differences to the quality of women-based website service.
- H1.4 occupation has significant differences to the quality of women-based website service.
- H1.5 revenue has significant differences to the quality of women-based website service.
- H2: demographic variables have significant differences to customer loyalty.
- H2.1 gender has significant differences to customer loyalty.
- H2.2 age has significant differences to customer loyalty.
- H2.3 education has significant differences to customer loyalty.
- H2.4 occupation has significant differences to customer loyalty.
- H2.5 revenue has significant differences to customer loyalty.
- H3: quality of women-based website service has significant forecast ability to customer loyalty.
- H3.1 efficiency of the quality of women-based website service has significant forecast ability to customer loyalty.
- H3.2 implementation of the quality of women-based website service has significant forecast ability to customer loyalty.
- H3.3 system availability of the quality of women-based website service has significant forecast ability to customer loyalty.
- H3.4 privacy of the quality of women-based website service has significant forecast ability to customer loyalty.
- H3.5 responsiveness of the quality of women-based website service has significant forecast ability to customer loyalty.
- H3.6 compensation of the quality of women-based website service has significant forecast ability to customer loyalty.
- H3.7 contact of the quality of women-based website service has significant forecast ability to customer loyalty.

3.3 *Sampling methods*

This study uses Taiwan Internet users as samples. Questionnaires were given online through my Survey with simple sample method. 302 online and 40 paper surveys were collected, with 42 invalid and 300 valid surveys.

In order to check survey's reliability, a pilot test was done with 30 test surveys. Irrelevant questions were deleted after collecting the samples, and Cronbach's α coefficient test (1951) were applied to check survey's internal unity.

4 ANALYSIS AND DISCUSSION

4.1 *Descriptive statistics analysis*

The samples have significant differences in gender. 214 females (71.3%) use women-based websites, and most of them are age 21 to 30, which consist 226 people (75.3%). For the degree of education, college has 239 people (79.7%), and it represents that people with higher education tend to shop through E-Commerce. Student has the most number in occupation, and there are 164 of them (54.7%). Lastly, results for revenue show that most of the samples have under NTD 10,000 per month (approx. USD 333), due to the fact that 125 participants (41.7%) are students.

Most of the Internet users spend 2 to 4 hours per day to be online, which consists 100 people (33.3%). In women-based websites, 85 people (28.3%) browse cosmetic section the most, 76 people (25.3%) choice apparel and bags section the most. 185 participants (61.7%) use the women-based websites under a year, and it means the importance of retaining those user follows. 238 people (79.3%) use the women-based websites 1 to 2 times per week, and there are 169 people (53.6%) use the women-based websites under half hour per week.

4.2 *Reliability analysis*

This study collected 300 valid surveys, and used Cronbach's α coefficient test, which needs to be greater than 0.7 to gain reliability. Furthermore, results show that efficiency = 0.840, implementation = 0.824, system availability = 0.822, privacy = 0.718, responsiveness = 0.866, compensation = 0.816, contact = 0.707, customer loyalty = 0.853. All aspects are above 0.7, and it indicates the survey is reliable.

4.3 *Factor analysis*

According to Kaiser's (1974) theory, KMO value within 0.5 to 1 is appropriate for factor analysis. The aspects for analyzing are: efficiency = 0.845, implementation = 0.803, system availability = 0.746, privacy = 0.500, responsiveness = 0.834, compensation = 0.500, contact = 0.500, loyalty and the Internet usage.

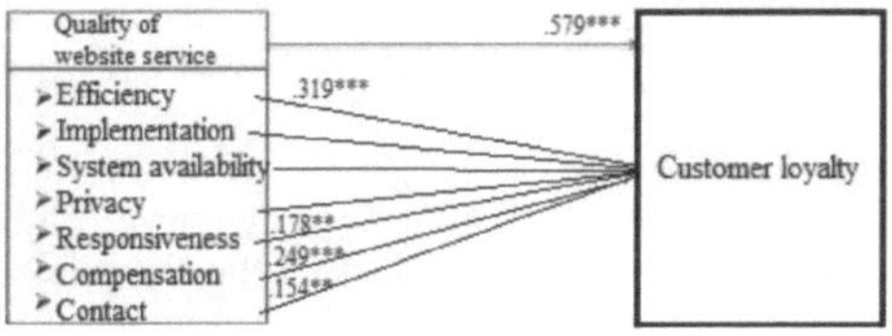

Figure 2. Regression analysis.

4.4 *Variance analysis*

In women-based E-Commerce, quality of website service is significantly differed by "age", "degree of education" and "occupation". Customer loyalty is significantly differed by "gender" and "degree of education".

4.5 *Regression analysis*

Based on different aspects, the results of regression analysis are illustrated in Figure 2.

Customer loyalty = 0.319 × efficiency + 0.178 reaction + 0.249 compensation + 0.154 connection.

5 CONCLUSION AND RECOMMENDATION

5.1 *Analysis on usage of women-based websites*

Based on sample data, there are 71.3% of female participants, which fits the research from the Bureau of Foreign Trade, MOEA that there are more female online shoppers. 75.3% of the users are age 21 to 30, 79.7% has College degree of education, 54.7% are student, and 41.7% have revenue under NTD 10,000 per month.

The Internet usage behavior shows that 33.3% users spend 2 to 4 hours per day to be online, 28.3% browse cosmetic section the most, 25.3% browse apparel and bags section the most, 61.7% use the websites under a year, 79.3% visit the websites 1 to 2 times per week, and there are 53.6% use the women-based websites under half hour per week.

5.2 *Results of hypotheses*

- H1: demographic variables have significant differences to the quality of women-based website service. (tenable)
- H1.1 gender has significant differences to the quality of women-based website service. (untenable)
- H1.2 age has significant differences to the quality of women-based website service. (tenable)
- H1.3 education has significant differences to the quality of women-based website service. (tenable)
- H1.4 occupation has significant differences to the quality of women-based website service. (tenable)
- H1.5 revenue has significant differences to the quality of women-based website service. (untenable)
- H2: demographic variables have significant differences to customer loyalty. (tenable)
- H2.1 gender has significant differences to customer loyalty. (tenable)
- H2.2 age has significant differences to customer loyalty. (untenable)
- H2.3 education has significant differences to customer loyalty. (tenable)
- H2.4 occupation has significant differences to customer loyalty. (untenable)
- H2.5 revenue has significant differences to customer loyalty. (untenable)
- H3: quality of women-based website service has significant forecast ability to customer loyalty. (partial tenable)
- H3.1 efficiency of the quality of women-based website service has significant forecast ability to customer loyalty. (tenable)
- H3.2 implementation of the quality of women-based website service has significant forecast ability to customer loyalty. (untenable)
- H3.3 system availability of the quality of women-based website service has significant forecast ability to customer loyalty. (untenable)
- H3.4 privacy of the quality of women-based website service has significant forecast ability to customer loyalty. (untenable)
- H3.5 responsiveness of the quality of women-based website service has significant forecast ability to customer loyalty. (tenable)
- H3.6 compensation of the quality of women-based website service has significant forecast ability to customer loyalty. (tenable)
- H3.7 contact of the quality of women-based website service has significant forecast ability to customer loyalty. (tenable)

5.3 *Relationships in different demographic*

The relationship between different demographic and quality of website service illustrate that there are significant differences in age "under 20" and "over 51". Education has significant differences, which "college" is greater than "graduate school" and "high school". The "quality of website service" is significantly differed by "occupation", which "IT industry" is greater than "manufacturing industry", "other", "student", "public servant", "health care industry", "service industry, "financial industry" and "freelancer".

Customer loyalty" is significantly differed by "gender", which "female" is greater than "men".

Moreover, "college" has higher loyalty than "high school" in the degree of education.

5.4 *Relationship between quality of website service and customer loyalty*

After analysis, the equation of quality of website service is: customer loyalty = 0.319 efficiency + 0.178 reaction + 0.249 compensation + 0.154 connection. The higher the efficiency, reaction, compensation and connection, the higher the higher customer loyalty will be.

REFERENCES

ACECI. July 2013. Shopping Mall for Female. *E-commerce News:* https://www.facebook.com/aceci99/posts/471284972963760.

Amazon.com. Jan 2013. Form 10-K, *Annual Report*: http://pdf.secdatabase.com/1562/0001193125-13-028520.pdf.

Bureau of Foreign Trade, MOEA. 2013. Business Revenues in December, *Economic News*: http://www.moea.gov.tw/Mns/dos/bulletin/Bulletin.aspx?kind=8&html=1&menu_id=6727&bull_id=745.

Bureau of Foreign Trade, MOEA. 2013. Cross-Straits Agreement on Trade in Services. *Economic News*: http://www.moea.gov.tw/Mns/populace/news/News.aspx?kind=1&menu_id=40&news_id=32017.

Bureau of Foreign Trade, MOEA. 2013. Results of Ministry of Economic Affair. *Economic News*: http://www.moea.gov.tw/Tapp/main/content/ContentImages.aspx?menu_id=3630.

Cronbach, L.J. 1951. Coefficient Alpha and the Internal Structure of Tests. *Psychometrika, 16 (3)*: 297–334.

Frederick F. Reichheld and W. Earl Sasser, Jr. 1990. Zero Defections: Quality Comes to Services. *Harvard Business Review. March–April 1996*: 57–69.

Institute for Information Industry. 2013. Operation of E-commerce. *Award News:* http://www.iii. org.tw/(S(sdt1ohz31ikjrkvtstv2qnri))/m/News-more.aspx?id=1261.

Jones, T.O. and W.E. Sasser, Jr. 1995. Why Satisfied Customer Defect. *Harvard Business Review, Vol. 73, No. 6*: 88–99.

Kaiser, H.F. 1974. An Index of Factorial Simplicity. *Psychometrika, 39 (1)*: 31–36.

Kalakota, R., and A. Whinston. 1997. Electronic Commerce. *A Managers Guide*. Addison: Wesley.

Kotler, P. 1999. Marketing Management: Analysis, Planning, Implementation and Control, 9th. *Englewood Cliffs, NJ*: Prentice-Hall Inc.

Lee, J., J. Lee, and F. Lawrence. 2001. The Impact of Switching Costs on the Customer Satisfaction-Loyalty. *Mobile Phone Service in France. Journal of Services Marketing, Vol. 15, No. 1*: 35–48.

Market Intelligence & Consulting Institute. 2012. Development of Internet Shopping. *MIC News*: http://www.iii.org.tw/m/ICT-more.aspx?id=596.

Oliver, R.L. 1997. Satisfaction: A Behavioral Perspective on the Consumer. *Boston*: McGraw-Hill.

Parasuraman, A., V.A. Zeithaml and L.L. Berry. 1988. "SERVQUAL: A Multiple-Item Scale for Measuring Consumer Perceptions of Service Quality. *Journal of Retailing, Vol. 64, No. 1*: 12–40.

Parasuraman, A., V.A. Zeithaml and L.L. Berry. 1994. Alternative Scales for Measuring Service Quality: A Comparative Assessment Based on Psychometric and Diagnostic Criteria. *Journal of Retailing, Vol. 70, No. 3*: 201–230.

Parasuraman, A., V.A. Zeithaml and L.L. Berry. 1994. Improving Service Quality in American: Lessons Learned. *Academy of Management Executive, Vol. 8, No. 2*: 32–52.

Parasuraman, A., Zeithaml, V.A., & Malhorta, A. 2005. E-S-QUAL: A multiple-item scale for Assessing Electronic Service Quality. *Journal of* Service Research, 7(3): 213–233.

Parasuraman, A., Zeithaml, V.A., and Berry, L.L. 1985. A Conceptual Model of Service Quality and its Implications for Future Research. *Journal of Marketing, Vol. 49, No. 4*: 41–50.

Peru, Taipei Economic and Culture Office. 2013. Peru and Online Shipping. *The Office News*: http://www.trade.gov.tw/Pages/Detail.aspx?nodeid=45&pid=437661.

Thiel, T. 1999. Reflections on Critical Incidents. *In Prospect, 14 (1)*: 44–52.

Zeithaml, V.A., Parasuraman, A. and Malhotra A. 2000. Services Marketing. *Integrating Customer Focus Across the Firm (2nd ed.)*: McGraw-Hill Inc.

Zeithaml, V.A., Parasuraman, A. and Malhotra A. 2002. Service Quality Delivery Through Web Sites: A Critical Review of Extant Knowledge. *Journal of Academy of Marketing Science, Vol. 34, No. 4*: 362–375.

Section 2: Internet computing and information services

E-Commerce, E-Business and E-Service – Lee (Ed)
© 2014 Taylor & Francis Group, London, ISBN 978-1-138-02646-9

An empirical analysis of the impact on economic growth from trade in services

Dejun Yu & Xiangjing Xu
Oxbridge College, Kunming University of Science and Technology, Kunming, P.R.China

ABSTRACT: Based on the total service trade and economic growth data of China during 1985–2012, using the statistical software Eview6.0, this paper makes the Co-integration test and granger causality test for the long-term equilibrium relations and the causal relationship between the service trade and economic growth of China. The empirical analysis results show that there is at least a long-term equilibrium relationship between the service trade import and export and economic growth of China; Import and export of service trade has promoted economic growth of China, in turn, economic growth of China does not necessarily promote the development of trade in service. Finally, this paper shows the strategy of the service trade competitiveness of China combined with the quantitative empirical analysis results.

Keywords: trade in services; import and export; economic growth; empirical analysis

1 INTRODUCTION

Since 21st century, service industry has been rapid development in the whole world, and the service trade has more and more influential on the economic development of countries around the world. Since 1978, China has been sustained and rapid development of economy, trade and services in all. In 2012, China's service trade has as much as $470.58 billion, more than an average increase of 10.3% of world trade in services, accounting for 5.6% of total world trade in services, China's service trade has become the world's third one, behind the United States and Germany. So many scholars in China pay attention to service trade and the relationship between China's economic growth rapidly, and this has become a hot topic.

About the study of relationship between service trade and economic growth mainly has two aspects, namely qualitative research and quantitative research. In qualitative research, in 2005 Wangyu Shao thought that the level of service industry development could directly measure a country's level of economic development, generally services are more developed, the level of economic development will be more high; in 2006 Kaibao Song thought that the comparative advantage of Chinese service trade export was in he traditional service industry, such as labor-intensive and resources endowment department, but the form of trade in services was very single. In quantitative research, in 2005 Ridong Hu studied the relationship between import and export of China's service trade and economic growth on the basis of data from 1985 to 2004 in China and had concluded that service trade import and export had inhibition on China's economic growth and service trade exports had promoted economic growth. Most of the existing research literature selected Gross Domestic Product (GDP) for the economic growth indicators, however, this article found that service trade import and export and GDP did not belong to the same order cointegration sequence, this had violated the Cointegration tests. Therefore, this article selects the per capita GDP as an index of economic growth, then study the long-term equilibrium relationship and granger causality between the service trade import and export and China's per capita GDP.

2 THE PRESENT SITUATION OF CHINA SERVICE TRADE

Since 1978, China's trade in services keeps high-speed growth. China's trade in services has become a force to be reckoned with in the international market in recent years, but the development of China's service trade does not match with the overall trade, there are still a variety of defects, and shows obvious lags. At the same time, and compared to developed countries and some developing countries of the development of trade in services, there exists a large gap.

Firstly, the development of service industry lags behind. The development of China's service industry not only lags the other industries, but also lags the other countries. From the point of service industry output value, services industry occupied 31.3% of GDP in 1990, and the proportion was only 33% in 2004, the added value of service industry growth was small. And the proportion of services industry in the developed countries is as high as 60%–80%, such as American service industry occupied 75% of GDP, Hong Kong was as high as 85%, and the service industry in developing countries was 45% to 55% of GDP. This shows that China's service industry not only lags the developed countries, but also is lower than the average level of developing countries for a long time.

Secondly, China's trade in services rise faster, but the volatility is bigger. In 1985 China's trade in services was only $5.186 billion, to $134.567 billion in 2004, increased by 24.95 times; the average annual growth is 19.52%. Although China's trade in services is rising very fast in recent years, but the volatility is bigger, and there has experienced three big fluctuations. The first time is during 1983 to 1985, the service trade from $43.06 billion in 1983 in 25.8% of the high speed increased to $54.17 billion in 1984, but during 1984 to 1985, there is a small amount of negative growth. The second time is in the early 90s, trade in services of import and export amounted to $98.61 billion in 1990, and was $225.55 billion in 1993. The last time is in 1995–1998, in 1995 China's trade in services is $430.65 billion, while in 1996 was down to $429.36 billion, it increased to $522.28 billion in 1997, but in 1998 it was a 3.6% negative growth again, and until now.

Finally, the service trade deficit situation is serious, and the trade structure is not reasonable. Since 1994, service trade was in deficit, and its scale was high. 1992 years ago, China's trade in services continues to surplus, the surplus is between $7.39 billion and $38.3 billion. And it was $6.46 billion for the first time in 1993, China's service trade deficit, from 1997 to 2001 there is five consecutive years deficit, which is almost r in more than $5 billion every yea; China's service trade surplus was $97.84 billion in 2002. From 2003 to 2004, China's service trade deficit, it reached $85.72 billion and $96.99 billion respectively. China's service trade balance reflects China's service trade import and export with the imbalance. On the other hand, China's service trade export structure is not reasonable.

3 EMPIRICAL ANALYSIS BETWEEN SERVICE TRADE AND ECONOMIC GROWTH

3.1 *Variable selection and data processing*

In this paper the World Bank database and data in China statistical yearbook 2012 are samples according to the period 1985–2012. China's trade in services imports (IM), trade in services exports (EX), China's Gross Domestic Product (GDP) and China's per capita gross domestic product (RGDP) are the research object, their associated index data are shown in Table 1. In order to eliminate heteroscedasticity problems, take natural logarithm processing data of IM, EX, GDP and RGDP respectively as LNIM, LNEX, LNGDP and LNRGDP.

3.2 *Stationary test*

As a result of the existence of "false return" problem, the analysis of the above sequence requires stationary test before. This article firstly gives a n unit root test for the trade in services imports (IM), trade in services imports (EX), Gross Domestic Product (GDP) and per capita Gross Domestic Product (RGDP), and the test results are shown in Table 2.

From Table 2, ADF test value of the original sequence in each target is not significant when the

Table 1. Data of China's service trade and economic.

Item/ years	IM (billion $)	EX (billion $)	GDP (billion yuan)	RGDP (yuan)
1985	25.24	30.55	9040.7	858
1986	22.76	38.27	10274.4	963
1987	24.85	44.37	12050.6	1112
1988	36.03	48.58	15036.8	1366
1989	39.10	46.03	17000.9	1519
1990	43.52	58.55	18718.3	1644
1991	41.21	69.79	21826.2	1893
1992	94.34	92.49	26937.3	2311
1993	120.36	111.93	35260.0	1998
1994	162.99	166.20	48108.5	4044
1995	252.23	191.30	59810.5	5046
1996	225.85	206.01	70142.5	5846
1997	279.67	245.69	78060.9	6420
1998	266.72	238.95	83024.3	6796
1999	315.89	262.48	88479.2	7159
2000	360.31	304.30	98000.5	7858
2001	392.67	333.34	108068.2	8622
2002	465.28	397.45	119095.7	9398
2003	553.06	467.34	134977.0	10542
2004	721.33	624.34	159453.6	12336
2005	837.95	744.04	183617.4	14185
2006	1008.33	919.99	215904.4	16500
2007	1301.11	1222.06	266422.0	20169
2008	1589.24	1471.12	316030.3	23708
2009	1588.56	1294.76	340320.0	25608
2010	1933.21	1621.65	399759.5	30015
2011	2380.68	1828.39	472115.0	35181
2012	2820.98	1909.39	519322.0	38354

Source: China statistical yearbook of 2013 and the World Bank database.

Table 2. Unit root test.

Var	ADF	(*c*,*t*,*p*)	5%	Prob
LNIM	–0.3144	(*c*,0,2)	–2.9640	0.9103
DLNIM	–6.0735	(*c*,0,2)	–2.9810	0.0000
LNEX	–0.7009	(*c*,0,2)	–2.9763	0.8301
DLNEX	–4.7585	(*c*,0,2)	–2.9810	0.0008
LNGDP	–2.3384	(*c*,0,2)	–2.9763	0.6559
DLNGDP	–4.8629	(*c*,0,2)	–2.9810	0.1682
LNRGDP	–0.6065	(*c*,0,2)	–2.9763	0.8533
DLNRGDP	–6.2898	(*c*,0,2)	–2.9810	0.0000

Table 3. Johansen cointegration test.

Date: 01/13/14 Time: 11:43
Sample (adjusted): 1987 2012
Included observations: 26 after adjustments
Trend assumption: Linear deterministic trend
Series: LNIM LNEX LNRGDP
Lags interval (in first differences): 1 to 1

Unrestricted Cointegration Rank Test (Trace)

Hypothesized No. of CE(s)	Eigenvalue	Trace Statistic	0.05 Critical Value	Prob.**
None *	0.743794	42.55179	29.79707	0.0010
At most 1	0.231596	7.145636	15.49471	0.5608
At most 2	0.011328	0.296213	3.841466	0.5863

Trace test indicates 1 cointegrating eqn(s) at the 0.05 level
* denotes rejection of the hypothesis at the 0.05 level
**MacKinnon-Haug-Michelis (1999) p-values

Unrestricted Cointegration Rank Test (Maximum Eigenvalue)

Hypothesized No. of CE(s)	Eigenvalue	Max-Eigen Statistic	0.05 Critical Value	Prob.**
None *	0.743794	35.40615	21.13162	0.0003
At most 1	0.231596	6.849423	14.26460	0.5071
At most 2	0.011328	0.296213	3.841466	0.5863

Max-eigenvalue test indicates 1 cointegrating eqn(s) at the 0.05 level
* denotes rejection of the hypothesis at the 0.05 level
**MacKinnon-Haug-Michelis (1999) p-values

Table 4. Granger causality test.

Pairwise Granger Causality Tests
Date: 01/13/14 Time: 11:47
Sample: 1985 2012
Lags: 1

Null Hypothesis:	Obs	F-Statistic	Prob.
LNEX does not Granger Cause LNIM	27	5.69082	0.0253
LNIM does not Granger Cause LNEX		0.18688	0.6694
LNRGDP does not Granger Cause LNIM	27	1.56202	0.2234
LNIM does not Granger Cause LNRGDP		8.54221	0.0075
LNRGDP does not Granger Cause LNEX	27	1.38224	0.2513
LNEX does not Granger Cause LNRGDP		13.1522	0.0013

significance level was 5%, so these sequences are unit root, namely non-stationary series; However IM, EX and RGDP first-order difference sequence at the significant level of 5% reject the null hypothesis of unit root, namely, the first order difference sequences of IM, EX and RGDP are stationary sequences, and first order difference sequence of GDP at the 5% significant level accepts the null hypothesis of unit root, so this article selects RGDP to cointegration test on behalf of the Chinese economic growth.

3.3 *Cointegration test*

Cointegration relationship represents a long-term equilibrium relationship, namely there is a linear combination of the smooth among a set of non-stationary time series with single whole order. This article uses the Johansen test for cointegration relationship for the empirical analysis, what can be seen from the test results is that under significance level of 5%, the index variable LNIM, LNEX and there are at least a cointegration relationship (as shown in Table 3).

3.4 *Granger causality test*

The cointegration test indicates that there is significant correlation between the three indicators, but it can't show whether there is a definite economic meaning between them, whether a causal relationship need further verification. This article adopt the method of granger causality test, this paper tests the causal relationship between the various indicators. Test results are shown in Table 4.

Granger causality test results show that:

1. The service trade import and export of all is the granger cause of China's per capita GDP, as a service trade import and export can promote China's per capita gross national product (RGDP).
2. China's per capita GDP is not a service trade import and export, the granger cause of the China's per capita GDP increase can't promote the increase of China's service trade import and export. This conclusion also verifies the characteristics of China's export driven economy.
3. The export of Chinese trade in services is a granger cause of service trade import and export growth as a service trade will drive the service trade in the process of the increase in imports.

4 THE STRATEGY OF SERVICE TRADE COMPETITIVENESS

The development level and international competetiveness of a country in service trade have been influenced by a variety of factors, such as its overall economic development, domestic and foreign economic elements, which is the result of the influence of the external performance. Based on the current situation of service trade and the empirical analysis results on trade in services international competitiveness promotion some strategies are put forward.

Firstly, strengthening the coordination service trade related industries and support. Industries of national economy are the chain structure, and are influenced by each other. The service industry is the basis of trade in services, in terms of services within industries, their interaction relationship is obvious. To improve the international competitive force of China's service trade, we must strengthen industrial chain effect, pay special attention to the development of service trade relevant and auxiliary industries. We will continue to improve the service infrastructure and basic industries, develop and strengthen the industrial cluster, at the same time, we also pay attention to use China strong impetus to the develop service industry.

Secondly, encourage the service enterprise innovation, and enhance the competition ability. China's economic system reform has been a long time, because the service industry development is slow, a number of industries have yet to achieve complete enterprise management, and therefore they must continue the service trade enterprise reform property rights and the operational mechanism in accordance with the requirements of the market economy, which makes it a real market main body and speeds up the service industry for the construction of modern enterprise system. Enterprise shall establish a governance structure characteristics in line with the service economy, as far as possible, set up a strict management specification and quality standards, and they must be strictly enforced; When the enterprise performance is evaluated, we must pay attention to enterprise long-term development potential, internal processes and dynamic evaluation of external competitiveness, innovation performance evaluation index system. By improving the services and creating a good working environment, prevent the brain drain and strengthen employee loyalty to the enterprise and job skills.

Finally, a proper discount and policy are inclined to the international trade in services. The United States, a most developed country, in order to encourage service export, the government gives a lot of preferential service enterprise for a long time, who attaches great importance to technology research, development and education at the same time. On the eve of the arrival of service economy, the industry policy of China should further tilted to the services and pay more attention to technology and knowledge intensive service enterprises and support the policy tilt, and the relevant tax, credit and other financial preferential policies should be combined with industry tilt policy in order to promote the rapid development of productive service industry and financial industry which are knowledge intensive services.

REFERENCES

Chen shuangxi. Emprical analysis on intra-industry trade in service industry in China. Journal of Shenyang University of Technology (Social Science Edition), 2010(03).

He Jun, Guo Lan. Study of the Upgrade of the Competitive Power of Service Trade in China—Based on the Empirical Analysis of Globe Main Service Trade Coutries Panel-Data. Journal of Shanxi Finance and Economics University, 2013 (03).

Hoekman, B, Karsenty, G. Economic Development and International Transaction in Services. Deveopnt Policy Review. 1992(05).

Hu Ridong, Su Zhifang. Empirical Analysis of the Relationship between International Service Trade and GDP of China. China Business and Market, 2005(12).

Juan Fu. An Empirical Analysis on Influence of FDI on China's Trade in Services. Proceedings of 2011 3rd IEEE International Conference on Information Management and Engineering. 2011.

Li Yang, Cai Chunlin. Positive Research on the Influencing Factors to Development of China's International Trade in Service. Journal of International Trade, 2008(05).

Lu Xiao-jin, Hu Hui-min. The analysis of the balanced development problem of China Tourism service trade. Proceedings of 2013 International Conference on Information, Business and Education Technology. 2013.

Ren Liang. Influential Factors Analysis and Index Calculation of Service Intra-industry Trade in China. Proceedings of 2011 International Conference on Public Administration (7th). 2011.

E-Commerce, E-Business and E-Service – Lee (Ed)
© 2014 Taylor & Francis Group, London, ISBN 978-1-138-02646-9

A study on model development for information management service system of health service industry

C.P. Ko
Ching Kuo Institute of Management and Health, Keelung, Taiwan

B. Song
Nanfang College of Sun Yat-Sen University, Guangzhou, P.R.China

ABSTRACT: This study intends to analyze the dimensions of health service industry and investigate the model development for information management service system of health service industry based on the studies conducted by experts and schools in service health industry. This study used Analytic Hierarchy Process (AHP) to perform analyses, and found that among the 10 models developed for information management service system of health service industry, the most significant one is the model developed for medical tourism industry, which includes medical tourism industry license obtainment (weight = 0.0811), health and leisure industry product R&D and service (weight = 0.0751), medical tourism industry standard operating procedure management (weight = 0.0636), health and leisure industry curriculum design and planning (weight = 0.0599), health and beauty industry service quality (weight = 0.0559), medical tourism industry risk management (weight = 0.0556), health and leisure industry project management (weight = 0.0550), health and beauty industry customer relationship (weight = 0.0405), medical tourism industry expertise classification (weight = 0.0392) and medical tourism industry function analysis (weight = 0.0389).

1 RESEARCH BACKGROUND AND MOTIVES

In Taiwan, at present, the role played by human resource management in health industry has gradually been transformed from passive administrative operating unit to management strategic role supporting health industry. In addition, hospitals have aggressively participated in the development of various strategies in health industry, and the operating aspect has also been changed from the past focus on personnel and administrative affairs to the assistance in co-development of health industry and employees, in order to effectively exert the influence of human resources within organizations. After being aware of the important influence of human resource management on health industry, health industry intends to use information technology to improve the effectiveness of human resource management and create competitive advantage of health industry. Therefore, this study on the model establishment intends investigate how to use the help of decision-making support system in a competitive environment to assist health industry operators and general public who intend to engage in health industry in using the information management service system of health service industry, as well as limited resources and capital to make the right judgments, develop a specific platform health service industry and achieve the objective of maximum mutual benefit.

2 RESEARCH PURPOSE

The main purpose of this study is to investigate the model development for information management service system of health service industry. The specific research purposes are as follows: (1) To understand the current status of information management service system of health service industry; (2) to compare how different background variables lead to difference in health service industry; (3) to understand the current status of the current health service industry; (4) to compare the difference in information management service system of health service industry; (5) to understand the application of information management service system to health service industry; (6) to analyze the status of development of information management service system of health service industry.

3 RESEARCH SCOPE

The purpose of this study is to develop the model of information management service system of health service industry. This study used in-depth interviews to conduct a questionnaire survey, in order to understand how different background variables lead to the difference in the level of information management service system of health service industry, as well as the predictive power of model development behavior of information management service system for health service industry. This study selected experts as the subjects, and used in-depth interviews to conduct a questionnaire survey on experts.

4 LITERATURE REVIEW

Enterprises use information technology to create competitive advantage: Bakos (1991) integrated the argument proposed by many scholars concerning application of information technology to creating competitive advantage. The following four chances for application of information technology: (1) to improve the operational efficiency and functional effectiveness; (2) to develop organizational synergy; (3) to use information technology to improve product innovation; (4) to obtain bargaining advantages over customers and suppliers. The former two can be generalized into "comparative efficiency." The last (4) belongs to bargaining power. Information technology certainly will affect organizational pattern and operation. "Organization" is also regarded as the combination of "system" with "knowledge." In addition, in the future enterprises, "system" and "knowledge" are used to complete all of the mechanisms (Gans, Kralewski, Hammons & Dowd, 2005). If high-level supervisors are imparted with sufficient information (i.e. information is input in knowledge-based system). Lower-level employees in the organization eventually will also share and learn sufficient information (Carrera, Percivil & Bridges, 2006). Inductive learning is comprehensively applied to the field of data mining, including decision tree classification, association data analysis and concept-oriented inductive learning.

1. Decision tree classification uses various property values to properly classify the data in the database and develop a decision tree. The future added data can also be classified using decision tree. In addition, the most appropriate analysis model can also be developed using decision tree, including ID3 (Quinlan, 1986), CDP (Agrawal, Imielinski & Swami, 1993) and IC (Smith & Offodile, 2008), which are all inductive learning methods.
2. Association data analysis aims to find out the association among various data, and intends to understand the mutual interactions according to the distribution of various data in the database, including the association rules of transaction data (Agrawal, Ghosh, Imielinski, & Swamik, 1992), KID3 (Piatesky-Shapiro, 1991), etc. Regardless of the method of data mining, the overall framework includes the following five major dimensions: (1) user communication interface; (2) database; (3) application field knowledge; (4) extracted knowledge; (5) data mining method (Skok, Kophamel, & Richardson, 2001). User communication interface: users usually cannot understand what data they can obtain from database. In addition to lacking the knowledge of database design, users usually do not possess the basic ability to use programming language, either (Anwar, Beck & Navathe, 1992). Computers are frequently applied to commerce, management and administration, and are inseparable from the fields of commerce and management (Wu, Chaudhry, Wang, Maglione, Mojica, Roth, Morton & Shekelle, 2006). Management information system involves all the issues of technology and management aspects, including system planning, system development, system integration, system management and introduction, strategic system, information human resource management, ethical and legal issues of information society, etc (Devaraj, & Kohli, 2000). Management information system is an integrated, computerized and human-machine interaction system aiming to support the operation and decision making of middle-level and basic level managers in organization (Hoffman & Ford, 2008).

5 RESEARCH DESIGN

According to the research motives, purposes and results obtained from studies on relevant theories, this study mainly investigated the model development for information management service system of health service industry, and took into account the dimensions of health and beauty industry, medical tourism industry, health and leisure industry and creative life industry (Chiasson & Elizabeth, 2005). Every dimension includes the important decision making of model development for information management service system of health service industry. Based on the AHP of the model, the hierarchical calculation of various factors had to be taken into account, in order to provide reference for future development.

5.1 *Setting of AHP framework*

To develop the research framework of this study, important factors concerning management proposed by various scholars in relevant studies were arranged, and the experts were invited to assess the importance of various dimensions of health service industry. Moreover, the four hierarchies were set up according to the objective—property of various important factors is different from "model development for information management service system of health service industry." The explanations about the hierarchies are given as follows.

5.1.1 *Hierarchy 1*

The "system model development" in this study is defined as "model development for information management service system of health service industry," based on which hierarchy 2, hierarchy 3 and hierarchy 4 were developed.

5.1.2 *Hierarchy 2*

According to the research framework and the comments from relevant studies, industrial analyses, experts and scholars, hierarchy 2 was developed under "hierarchy 1," and were divided into four dimensions, "health and beauty industry," "medical tourism industry," "health and leisure industry" and "creative life industry" for investigation.

5.2 *Expert questionnaire*

This study used the AHP to find out the model development for information management service system of health service industry. This study mainly designed the questionnaire based on literature review on model development for information management service system of health service industry and industry experts' comments. The AHP questionnaire was designed according to the amended hierarchical framework confirmed by the experts. The items were assessed using the bipolar scale 1~9. This study used pairwise comparison to compare the factors of each hierarch and calculate the geometric mean and weight to the priority.

6 DATA ANALYSIS

This study used Power Choice software to calculate the pairwise comparison matrix weight of each item chosen by experts in the returned questionnaires, and obtained the consistency index and consistency ratio of each hierarchy and overall hierarchy to test the consistency of decision making. Moreover, this study assessed the priority of model development for various information management service systems according to the weights to deduce the formation of model development for information management service system of health service industry.

The research purpose is to develop the model for information management service system of health service industry. Therefore, it is important to determine factors of priority, how to assess the model development for information management service system and whether it is applicable to the current health industry. Most importantly, the order of importance of model development for information management service system should be determined. As a result, this study used AHP, which can systematize complicated issues, to perform analyses.

7 CONCLUSION AND SUGGESTIONS

7.1 *Conclusion*

There are successful medical tourism examples in the neighboring countries, such as Thailand, Korea and Singapore. Moreover, it is necessary to aggressively assist various health service industries in promoting international marketing, which may significantly expand the market of health service industry and improve the economy in Taiwan.

According to the industrial characteristics and expert questionnaire, the population (subjects) of this study on model development for information management service system of health service industry was mainly personnel of health service industry-related background. A total of 100 questionnaires were distributed, and a total of 90 valid questionnaires were returned. The questionnaires were arranged and analyzed, and the content included two major parts: Part 1 was the extraction of important factors affecting the model development of information management service system for health service industry. Part 2 was the use of AHP as analysis tool to determine the priority of factors for model development of information management service system of health service industry.

According to the analyses of AHP, among the 10 models developed for information management service system of health service industry, the most significant one is the model developed for medical tourism industry, which includes medical tourism industry license obtainment (weight = 0.0811), health and leisure industry product R&D and service (weight = 0.0751), medical tourism industry standard operating procedure management (weight = 0.0636), health and leisure industry curriculum design and planning (weight = 0.0599), health and beauty industry service quality (weight = 0.0559), medical tourism industry risk management (weight = 0.0556), health and leisure industry project management (weight = 0.0550), health and beauty industry customer relationship (weight = 0.0405), medical tourism industry expertise classification (weight = 0.0392) and medical tourism industry function analysis (weight = 0.0389). Based on the results above, in the model development for information management service system of health service industry, more importance is attached to medical tourism industry, especially the medical tourism industry license obtainment, which is associated with the characteristics of health service industry. The professional requirement of medical tourism industry is high, and it is not easy to obtain a license. Moreover, to accentuate the uniqueness of profession and to improve the stable development of model development for information management service system of health service industry, the obtainment of medical tourism industry license is indeed worthy of attention.

7.2 *Suggestions*

In terms of policy aspect, the suggestions for the government agencies proposed according to the research results are as follows: (1) To improve the certification system of health service industry: there is no certification system in certain parts of health service industry. It is necessary to improve the certification system in order to accentuate the professional and uniqueness of health service industry; (2) to formulate health service industry protection act: to develop a complete legal system and effective review ability to protect the innovative achievements of health service industry; (3) to attract more venture capital to the model development for information management service system of health service industry: to enable industries and people who intend to engage in health service industry to obtain information through the professional platform, achieve the maximum mutual benefit and advance the development of health service industry in Taiwan. The suggestions for industries prosed according to the research results are as follows: (1) to strengthen the uniqueness and inheritance of techniques in health service industry: because it takes a lot of time to learn and inherit the techniques in health service industry, learners tend to give it up halfway through. Therefore, there is a lack of talents. It is integrate abilities and gradually train future talents to avoid brain drain; (2) to strengthen the mutual authorization and profit sharing among industries, academia and other health service industries: to cooperate with other health service industry companies in technology transfer, marketing and authorization contract, as well as to co-develop new technology and share profits with other companies or research units; (3) to use information management service system of health service industry to cultivate professional talents with professional licenses: in the era of certification, it is necessary to provide systemic education to cultivate basic-level talents and attract international professional human resources in heath service industry.

REFERENCES

Agrawal, R., Ghosh, S., Imielinski, T., Iyer, B., & Swami, A. 1992. An Interval Classifier for Database Mining Applications, *Proceeding of the 18th International Conference on Very Large Data Bases*, 560–573.

Agrawal, R., Imielinski, T., & Swami, A. 1993. Mining Association Rules Between Sets of Items in Large Databases, *Proceedings of the ACMSIGMOD conference*, 207–216.

Anwar, T.M., Beck, H.W. & Navathe, B. 1992. Knowledge Mining by Imprecise Querying: A Classification-Based Approach, *IEEE 8th Internationa Conference on Data Engineering*, Phoenix, Arizona: 622–630.

Bakos, J.Y. 1991. A Strategic Analysis of Electronic Marketplaces, *MIS Quarterly* 15(3): 295–310.

Carrera, Percivil M. & Bridges. F.P. 2006. Globalization and Healthcare: Understanding Health Care and Medical Tourism, *Expert Review of Pharmacy economics and Outcomes Research* 6(3): 447–454.

Chiasson, W., & Elizabeth. 2005. Taking Industry Seriously in Information Systems Davidson, *MIS Quarterly* 29(4): 591–605.

Devaraj, S., & Kohli, R. 2000. Information Technology Payoff in the Health-Care Industry: A Longitudinal Study, *Journal of Management Information Systems* 16(4): 41–67.

Gans, D., Kralewski, J., Hammons, T., & Dowd, B. 2005. Medical groups' adoption of electronic health records and information systems, *Health Affair* 24(5): 1323–1333.

Hoffman, M.F., Ford, J.J. 2008. The effect of information technology investment on firm-level performance in the health care industry, *Health Care Management Review* 33(1): 60–68.

Piatetsky-Shapiro, G. 1991. *Discovery, Analysis, and Presentation of Strong Rules, Knowledge Discovery in Databases*, AAAI/MIT Press, 229–238.

Quinlan, J.R. 1986. Introduction of Decision Tree, *Machine Learning*. 1: 81–106.

Skok, W., Kophamel, A., & Richardson, I. 2001. Diagnosing information systems success: importance–performance maps in the health club industry, *Information & Management* 38(7): 409–419.

Smith, A.D., & Offodile, O.F. 2008. Data Collection Automation and Total Quality Management: Case Studies in the Health-Service Industry, *Health Marketing Quarterly* 25(3): 217–240.

Wu, S.Y., Chaudhry, B., Wang, J., Maglione, M., Mojica, W., Roth, E., Morton, S.C., &. Shekelle, P.G. 2006. Systematic Review: Impact of Health Information Technology on Quality, Efficiency, and Costs of Medical Care. *Annals of Internal Medicine* 144(10): 742–752.

E-Commerce, E-Business and E-Service – Lee (Ed)
 ISBN 978-1-138-02646-9

Labeling schemes for XML dynamic updates: A survey and open discussions

Wei-Siang Chiew, Su-Cheng Haw, Samini Subramaniam & Fang-Fang Chua
Faculty of Computing and Informatics, Multimedia University, Cyberjaya, Malaysia

ABSTRACT: eXtensible Markup Language (XML) has become the de facto standard for data exchange over the Internet in many web applications especially in most business transactions. However, these application data are subject to frequent changes. As such, it is essential to support not only queries but dynamic updates (insert, update and delete operations) over XML content. Labeling schemes are one of the well-known query optimization techniques as they provide a quick way to determine the type of relationships which present among the nodes. In this paper, we discuss some selected labeling scheme based on some classification. We also highlight the current trends of XML labeling especially in terms of the support for dynamic updates.

1 INTRODUCTION

The speedy growth of XML as data representation and data communication is due to its adaptable and flexible structure to share and communicate information over the Internet. For example, a business can easily model complex structures such as invoice, orders, and inventory system in XML format. The structural information especially the relationship among the nodes is critical to be determine so that query processing could be expedited. The four basic relationships are the Ancestor-Descendant (A-D), Parent-Child (P-C), sibling and order.

Many existing approaches merely support static query processing whereby it is assumed that the structural information will not change over time. This situation is definitely impractical as the data exchanged over the Web is subject to very frequent updates. Due to this reason, towards the recent years, researchers propose some robust labeling schemes which governs that re-labeling is not needed or at very minimum whenever dynamic update occurs. Even a small update could cause the entire XML tree to be re-labeled, and hence, it becomes worse if the XML database is huge. As such, a persistent, robust and durable labeling scheme which avoids re-labeling is very much desirable.

The aim of this paper is to propose the classification for the labeling schemes. In addition, we will be looking into the characteristics of each classification including their pros and cons. Most important, this survey provide an insight for new researcher venturing into these research areas.

2 LABELING SCHEME CLASSIFICATION

Generally, the labeling scheme could be classified into Subtree labeling, Prefix-based labeling, Multiplicative labeling and Hybrid labeling.

2.1 *Subtree labeling*

Subtree labeling is the earliest and simplest; where the label of a given document node v in document tree D encodes the position and the extent of the subtree D_v of D that is rooted in v, by means of offsets in sequence of nodes. Under this classification, there is two types of labeling, i.e., interval encoding and region encoding.

Dietz (1982) introduced an interval encoding labeling scheme called tree traversal order, whereby each node is labeled based on preorder and postorder traversal. Using this scheme, the A-D relationship could be determined easily based on the range (for two given nodes v and w of a tree D, v is an ancestor of w if and only if v occurs before w in the preorder traversal of D and after w in the postorder traversal). Yet, the P-C relationship could not be determined directly as the information stored in the label is insufficient. Also, this labeling scheme is inefficient for dynamic update as whenever new node is inserted or deleted, the preorder and postorder values need to be recomputed.

Li and Moon (2001) proposed extended preorder traversal whereby each node in the XML tree is labeled with a pair of numbers <order, size>. To enable insertions, *size* (v) can be an arbitrary integer large. Yet, a global reordering is necessary when the reserved spaces have been consumed.

Min et al. (2007) suggested Efficient XML Encoding and Labeling (EXEL), which is based on the concept of region numbering scheme (Zhang et al., 2001). Yet, they proposed using bit strings instead of decimal values. By using the bit strings instead, this completely eliminates re-labeling if updates occur. The start and end values are ordinal bit strings generated by the binary encoding of EXEL. In contrast to the region numbering, EXEL uses the parent information (i.e., the start value of the parent's label) instead of the level information to determine the P-C relationships.

Tekli and Chbeir (2011) proposed a framework which consists of four modules: discovering the commonalities between structural sub-trees, identifying the sub-tree semantic resemblances, computing tree-based edit costs, and computing tree edit distance to address the atomic and complex update operations.

2.2 *Prefix-based labeling*

Prefix-based labeling scheme encodes the nodes on the path from the document root down to node as a sequence. Thus, given a node v and its ancestor, u, their relationship could be determined precisely, i.e., u is an ancestor of v iff *label*(u) is a prefix of *label*(v). Nevertheless, this group of labeling scheme usually suffers from huge label size.

Tatarinov et al. (2002) proposed Dewey ID labeling scheme where by, for a non-root element v, label(v) = label(w).x, where x is the x-th child of w. Element u is an ancestor of element v if and only if label(u) is a prefix of label(v). As for the P-C relationship, the number of integer in label(u) is one more than that of label(v).

Lu et al. (2005) proposed *extended Dewey*, which uses the modulo function on a vector of integers to create mapping from an integer to an element name. The computation method in extended Dewey is rather expensive.

O'Neil et al. (2004) proposed ORDPATH which is the first scheme to allow for arbitrary updates without changing any existing label. ORDPATH encodes the P-C relationship by extending the parent's ORDPATH label with a component for the child. In contrast to Dewey ID, ORDPATH reserves even numbers for any further node insertions in ORDPATH.

Liu et al. (2014) proposed DEC (Dynamic Encoding scheme), which is tailored for both static and dynamic possibilistic XML documents. For any node, they used a 3-tuple (ID, PN, Sequence) to express related DES information, where ID is the identifier of the document, PN represents the position of a node in an XML document and Sequence is an ordered set which stores the information of possibility edges.

2.3 *Multiplicative labeling*

In this class of labeling scheme, relationships between nodes can be computed based on some arithmetic properties. This idea usually uniquely idenfied the A-D and P-C relationships, however, it is computationly very expensive.

Weigel et al. (2005) propose BIRD, which is an integer-based node identification scheme to answer structural queries (the dataGuide) based on weight assigned. When enumerating the nodes of the database, they reserve some interval size for all subtrees rooted at any of the nodes. Some numbers remain unused in the enumeration so that one could further find room to insert other nodes.

Some of more recent works include Improved Binary String Labeling (IBSL) (Ko and Lee, 2010) and Variable Length Endless Insertable (VLEI) code (Kobayashi, 2005). IBSL is computed based on the lexicographical order rather than numerical order. The labeling size is large (D, F, and N denote the maximal depth, maximal fan out, and number of nodes in an XML tree) -> $N * D * (F/2 + 1/2)$. It neither relabels any existing nodes, nor recalculates any values. As such, the cost of updating ordered XML is much lower than that of the existing schemes as IBSL reuses the deleted labels in updates.

On the other hand, the VLEI Code use a bit sequence code and apply this code to XML labeling to reduce the cost of insertions. The QED (Li and Ling, 2005) and CDBS (Li et al., 2008) include novel encoding methods to support code (a binary or quaternary string) insertion into a sequence of existing codes without disturbing the order between the nodes or having to relabel them. Nevertheless, CDBS cannot completely avoid relabeling. As for QED, it can avoid relabeling; however, its label size is large and its update and query performance is not as good as that of CDBS.

2.4 *Hybrid labeling*

Towards the recent years, most researchers focus on proposing some hybrid labeling to speed up query processing and yet robust and persistent enough to sustain updates.

Yun and Chung (2008) combined the interval-based labeling scheme and prefix-based labeling. For any inserted sub-tree, it is labeled as one leaf node using the interval-based scheme. As for the nodes in the subtree, they are labeled with a new prefix-based numbering scheme. With this scheme, the relationship between one node in the subtree and another node not in the subtree can be determined by comparing the label of the subtree and the label of the latter node.

Xu et al. (2012) introduced using vector order as the foundation of the dynamic labeling schemes instead of based on natural order or lexicographical order. They demonstrated that vector order can be gracefully applied to both sub-tree and prefix-based labeling schemes with little overhead.

O'Connor and Roantree (2012) proposed SCOOTER (the name encapsulates the Scalable, Compact, Ordered, Orthogonal, Trinary Encoded Reusable properties). SCOOTER has compact adaptive growth mechanism which ensures that label sizes growth rate is controlled under various insertions scenarios. The order is maintained between nodes at all times by way of lexicographical comparison. In addition, SCOOTER can be implemented using prefix-based labeling, or multiplicative labeling (prime numbering).

3 SUMMARY AND DISCUSSIONS

Table 1 concludes the advantages and disadvantages of the labeling schemes while Table 2 depicts the support for structural relationship, i.e., (A-D) and (P-C) as well as its support for dynamic update.

Table 1. The advantages and disadvantages of some reviewed labeling schemes.

Scheme	Advantages	Disadvantages
Dietz	Node consisting of preorder and postorder is able determine the A–D relationship.	Could not determine P-C relationship and requires relabeling when update occurs.
Extended preorder traversal	Determination of A-D and P-C is easy.	Nodes require re-labeling when the reserved numbers in *size* being used up.
EXEL	Encodes the node using bit strings, this completely eliminates re-labeling if updates occur.	Rather computational expensive to calculate to determine the P-C relationship.
DeweyID	A vector of numbers that represents the node-ID path from the root to the node. Efficient in determining A–D and P–C relationships.	Need to re-label the sibling nodes. Label's length scales up quadratically as number of fan-out and level increases. It is not practical to support XML tree with huge fan-out.
Extended Dewey	Ability to determine structural relationship	The construction of finite state transducer requires the scan of the whole XML data if the Data Type Definition (DTD) of XML data is not available.
ORDPATH	Encodes the P–C relationship relationship by extending the parent's ORDPATH label with a component for the child. Reserves even and negative integers for later insertions ancestor–descendant or parent–child structural relationship is determined almost no node re-labeling occurred.	The labeling size gets larger as the tree goes deeper.
DEC	Tailored for both static and dynamic posXML documents.	Huge label size
BIRD	Supports structural relationship. It does not require re-labeling as long as the assigned weight for each node is not overflowed.	Label generation is computationally expensive.
IBSL	Support structural relationship and reuse the deleted labels during dynamic update.	Label generation is computationally expensive.
VLEI	Uses a bit sequence code and apply this code to XML labeling to reduce the cost of insertions.	Reduction of update time by using floating point numbers.
QED	Offers compact labeling scheme and supports infinite number of insertions without the need of re-labeling.	Size of QED code rises for skewed insertions and query performance can be complex.
CDBS	Compact labeling and labeling cost is small.	Frequent update may cause overflow problem.
Nested tree and nested inverted list	Ability to determine structural relationship among nodes.	Labeling size for new nodes increases if there is mass insertion during dynamic update.
Vector order	Highly scalable for skewed insertions.	The labeling size gets larger as the tree goes deeper.
SCOOTER	Support an subjective number of node insertions and deletions while absolutely avoiding the need to relabel nodes.	Labeling size grows speedily when a large number of nodes are inserted at a fixed point.

Table 2. Support for structural relationship and dynamic update by each labeling scheme.

Scheme	Supported axes			Support for dynamic update
	A-D	P-C	Sibling	
Dietz	√	√	√	X
Extended preorder traversal	√	X	√	X
EXEL	√	X	√	√
DeweyID	√	√	√	X
Extended Dewey	√	√	√	X
ORDPATH	√	√	X	X
DEC	√	√	√	√
BIRD	√	√	√	X
IBSL	√	√	√	√
VLEI	√	√	√	√
QED	√	√	√	√
CDBS	√	√	√	√
Nested tree and nested inverted list	√	√	√	√
Vector order	√	√	√	√
SCOOTER	√	√	√	√

We have seen the advantages and drawbacks of each labeling scheme. Subtree group has a fixed-size and smallest labeling size. Nevertheless, labeling schemes in this group does not support dynamic update efficiently. Therefore, to cater for dynamic update, some approaches reserve some space for future insertion. Yet, in most cases, this scheme is weak in data updates as large numbers of nodes need to be relabeled. Prefix-based labeling scheme is a good approach for structural query processing. However, due to its label generation nature, the labeling size increases exponentially as the tree goes deeper. Thus, it consumes larger space and maybe inefficient for twig query processing. On the other hand, in thr multiplicative labelling, the label computations are complex and expensive and rather impractical for real applications. The hybrid labeling schemes supports larger class of queries and supporting dynamic updates efficiently. Conversely, label generation are computationally expensive and complex.

4 CONCLUSIONS

Efficient XML query processing technique is certainly vital to support flawless information retrieval. Thus, the adaptation of an appropriate labeling scheme which support structural relationship and generation of immutable labels are crucial to promise fast query processing with less resources consumption. In this paper, we have scrutinized and summarized the characteristics of labeling schemes in terms of the complexity of label generation, support for structural relationship and dynamic update. We expect further experimentations with hybrid approaches in the near future especially those capitalizing on the semantics inherent in data structures, supporting complex queries and dynamic updates.

ACKNOWLEDGEMENTS

This work is supported by funding of Fundamental Research Grant Scheme, from the Ministry of Higher Learning Education (MOHE).

REFERENCES

Dietz, P.F. 1982. Maintaining Order in a Linked List. In *Proceedings of the ACM Symposium on Theory of Computing*. 122–127.

Ko, H.K., and Lee, S.K. 2010. A Binary String Approach for Updates in Dynamic Ordered XML Data. IEEE Transaction on Knowledge and Data Engineering. 22(4): 602–607.

Kobayashi, K. 2005. VLEI Code: An Efficient Labeling Method for Handling XML Documents in an RDB. In Proceedings of International Coference on Data Engineering. 386–387.

Li, C. and Ling, T.W. 2005. QED: A novel quaternary encoding to completely avoid re-labeling in XML updates. In Proceedings of International Conference on Information and Knowledge Management. 501–508.

Li., C., Ling, T.W., and Hu, M. 2008. Efficient updates in dynamic XML data: from binary string to quaternary string. The VLDB Journal. 17(3): 573–601.

Li, Q. and Moon, B. 2001. Indexing and Querying XML Data for Regular Path Expressions. In *Proceedings of the VLDB*, 361–370.

Liu, J., and Ma, Z.M. 2014. Dynamically querying possibilistic XML data. Information Sciences 261: 70–88.

Lu, J., Ling, T.W., Chang, C.Y., and Chen, T. 2005. From Region Encoding To Extended Dewey: On Efficient Processing of XML Twig Pattern Matching. In Proceedings of the VLDB. 193–204.

Min, J.K., Lee, J., and Chung, C.W. 2007. An Efficient Encoding and Labeling for Dynamic XML Data. In Lecture Notes Computer Sciences 4443: 715–726.

O'Connor, M.F., and Roantree, M. 2012. SCOOTER: A Compact and Scalable Dynamic Labeling Scheme for XML Updates. In Lecture Notes in Computer Science, 7446: 26–40.

O'Neil, P. O'Neil, E., Pal, S., Cseri, L., Schaller, G., and Westbury, N. 2004. ORDPATHS: Insert-Friendly XML Node Labels. In *Proceedings of the ACM SIGMOD.* 903–908.

Tatarinov, I., Viglas, S., Beyer, K., Shanmugasundaram, J., Shekita, E.J., and Zhang, C. 2002. Storing and Querying Ordered XML using a Relational Database System. In *Proceedings of the ACM SIGMOD*. 204–215.

Tekli, J. and Chbeir, R. 2011. A novel XML document structure comparison framework based-on sub-tree commonalities and label semantics. Web Semantics: Science, Services and Agents on the World Wide Web 11. 14–40.

Weigel, F., Schulz, K.U., and Meuss, H. 2005. The BIRD Numbering Scheme for XML and Tree Databases—Deciding and Reconstructing Tree Relations Using Efficient Arithmetic Operations. In the *Lecture Notes In Computer Science* 3671: 49–67.

Xu, L., Ling, T.W., and Wu, H. 2012. Labeling Dynamic XML Documents: An Order-Centric Approach. IEEE Transaction on Knowledge and Data Engineering. 24(1): 100–113.

Yun, J.H., and Chung, C.W. 2008. Dynamic Interval-based Labeling Scheme for Efficient XML Query and Update Processing. Journal of Systems and Software, 81: 56–70.

Zhang, C., Naughton, J., DeWitty, D., Luo, Q., and Lohman, G. On Supporting Containment Queries in Relational Database Management Systems. In *Proceedings of the ACM SIGMOD*. 425–436.

E-Commerce, E-Business and E-Service – Lee (Ed)
© 2014 Taylor & Francis Group, London, ISBN 978-1-138-02646-9

Formal analysis and verification of a new improved NSSK protocol

Guangtian Shi & Huili Chen
Lanzhou Jiaotong University, Lanzhou, P.R. China

ABSTRACT: Authentication protocols are the basis of network security. Needham-Schroeder Shared-Key (NSSK) protocol is considered as one of the earliest and classical authentication protocol. This paper introduces NSSK protocol, analyzes its security, and proposes an improved NSSK protocol. Also modeling and verification the security of protocol within the framework of CSP and its model-checking tool FDR. The results of the analysis show that the proposed protocol can achieves the expectative authentication goals, and achieve the purpose of network security.

1 INTRODUCTION

Security protocol is also known cryptographic protocols. It is messaging protocol based on Cryptography, and it achieved the purposes of assign passwords and authentication by means of cryptographic algorithms. Authentication protocol is also an important part of the security protocol, and its main functions include authentication, key distribution, and non-repudiation of messages. The most famous authentication protocol is the Needham-Schroeder protocol and it is mainly divided into NSSK protocols and NSPK protocols. NSSK agreements have a significant impact on the design and verification of later authentication protocol. The most famous attack to NSSK agreement is Denning-Sacco. Denning and Sacco think the main security problem of NSSK is that the agreement responder B could not determine the freshness of messages 3. Many scholars have made some crack programs and improved protocol, for example, reference (Liao et al. 2009) proposed an improved NSSK protocol. This article use formal verification tools to verify the feasibility of the protocol in reference (Liao et al. 2009), and proposed another improved NSSK protocol.

2 NSSK PROTOCOL

2.1 *Security analysis of NSSK protocol*

The main objective of NSSK protocol is to assign session key to the communicating parties with the help of a trusted third-party agents, thereby establish a secure communications channel. The protocol includes communications agency A, B, and the authentication server S. It authenticates the identity of the main part of the communication at first, so as to obtain a session key of the secure communication process.

Steps of the protocol:

1. A→S: A, B, Na
2. S→A: {Na, B, Kab, {Kab, A}kbs}kas
3. A→B: {Kab, A}kbs
4. B→A:{Nb} kab
5. A→B: {Nb-1} kab

When A and B want to communicate with each other in secret, A request session key from the server S to ensure the secrecy of communication content, that is, user A must inform the server S that user A and user B want to communicate with each other, and a fresh random number Na is send to server S at first. After receiving the message (1), server S generates a session key Kab for user A and B, using the key of B to encrypt the identity of B and Kab to generate certificate, and in the next using the key of A to encrypt the certificate, Na, identity of B and Kab, then the message is returned to A. After decrypting the message A gets the key Kab, meanwhile cipher text of the certificate is send to B. User B decrypts the certificate to obtain the session key Kab, and using Kab to encrypt the fresh random number Nb, and then send Nb to A. After decrypting the message (4), A will get the Nb and then send the Nb-1 to B.

In the protocol, A confirms the identity of B by Nb, and B confirms the identity of A by Nb-1, so the Kab is the session key between A and B.

When the classic NSSK protocol is verified with Casper + FDR, some security vulnerabilities were found.

The first is that B can not determine freshness of the message (3), so the protocol is vulnerable to replay attacks.

The second is that the message (4) does not contain any information which can be verified by A, so

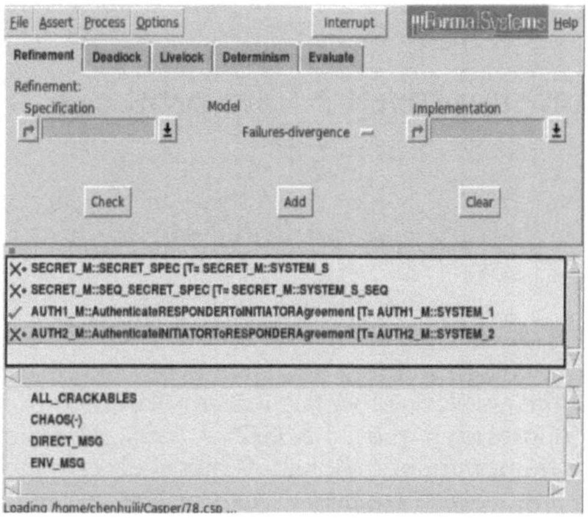

Figure 1. Verification of classical NSSK protocol.

A can not determine whether the message (4) was encrypted or not with the session key Kab, so the protocol is susceptible to false base station attack.

The Figure 1 is the verification results of classical NSSK protocol.

As it shows in Figure 1 the classical NSSK protocol is under replay attacks (Abadi et al. 1994) because it is not conform to the principle of identification of the main part. Participant identifier is very important, so participants must be clearly identified in the message. Either A, B or S receives the message, their own identity is not included in the response messages, therefore, they are unable to check the identity of the sender. If the message is intercepted and transmitted, the intruder can cheat and attack on it.

2.2 *Verify improved NSSK protocol in reference (Liao et al. 2009) by Casper +FDR*

An improved NSSK protocol is put forward in reference (Liao et al. 2009), which is proofed by BNA logic. The protocol steps are as follows:

1. A->S: A, Na
2. B->S: A, B, Na, Nb
3. S->B: {Na, B, Kab} Kas, {Nb, A, Kab} Kbs
4. B->A: {Na, B, Kab} Kas, {Na, Nb} Kab
5. A->B: {Nb}kab

Message (3) contains a random value Nb generated by B, and B can determine the freshness of message (3) by comparing the random values Nb in message (3) and message (2). B will believe message (3) is fresh if the two values are the same in the two messages. The freshness of message (4) is verified by comparing of Na in the message (1) and (4). It can be verified that {Na, Nb} are encrypted or not by Kab through the way that A

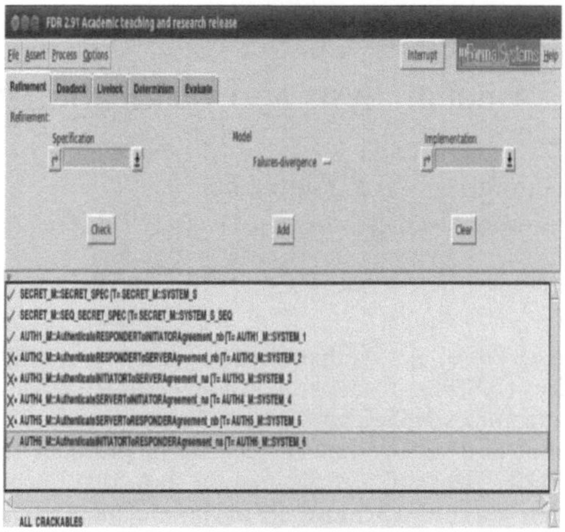

Figure 2. Validation of NSSK protocol, (Liao et al. 2009).

compares Na is the same value or not in the message (1) and (4). If the result is the same value of Na, A thinks that {Na, Nb} are encrypted or not by Kab. Similarly, that {Nb} is encrypted or not by Kab can be done by B with compare of the value of Nb in the message (1) and (5). The verification of the improved NSSK protocol by Casper + FDR has security problem. The Figure 2 is the validation result of the improved NSSK protocol.

The improved NSSK protocol realizes the encryption of messages and identification verification of sender and receiver. Though the improved NSSK protocol proposed in (Liao et al. 2009) is done with the help of trusted proxy server of third party, the validation result showed in the Figure 2 would occur when the server is unable to verify the identification of sender and receiver and fake base station attacks.

3 MODELING AND VERIFYING ANOTHER NEW IMPROVED NSSK PROTOCOL

3.1 *A new improved NSSK protocol*

This paper proposes a new improved NSSK protocol, its specific steps are as follows:

1. A→S: {A, B, Na} kas
2. S→B: {A} kbs
3. B→S: {Nb}kbs
4. S→A: {Na, B, Kab} kas
5. S→B: {Nb, A, Kab} kbs
6. B→A: {Nb'} kab
7. A→B: {Nb'-1} kab

When a plans to connect with B, the server will not immediately send session key to A. However, after it get the confirmation of B, the server will send the session key to A and B at the same time, and the message will contain the session on both sides of the identity. This improved NSSK protocol uses identification technology to complete the identity authentication, so it not only identifies the integrity of the information, but also its sources, which prevents attacks of fake base station and realizes the identity authentication of sender and receiver.

3.2 *Using the CSP to model the new improved NSSK protocol*

Protocol participant is represented as process and the message as the event, and the protocol is expressed as a set of communicating sequential process. Figure 3 is the CSP model (Hoare. 1985).

As show in Figure 3 the CSP model of NSSK protocol is composed of trusted process of participants and intruders, and the processes operate in parallel and interactively through channels. The CSP model of NSSK protocol includes the initiator a and responder b and server s which can perform key generation, transmission or authentication, so there are four processes of initial, responder, servers, and an intruder. Each process has one channel of *send* and one channel of *receive*. Here "send.a.b.m" means that a sends the message m to b, and receive.a.b.m means that b receives the message m from a, and a presents sender and b receiver. All of the messages are handled by the *send* and *receive* channels respectively.

When the protocol is modeled, sending message is thought as to communication media and receiving message from communication media. The operation of protocol in each process is just the visible procedure of sending and receiving messages. Intruder model is here based on Delov-Yao (Abadi et al. 1994) model.

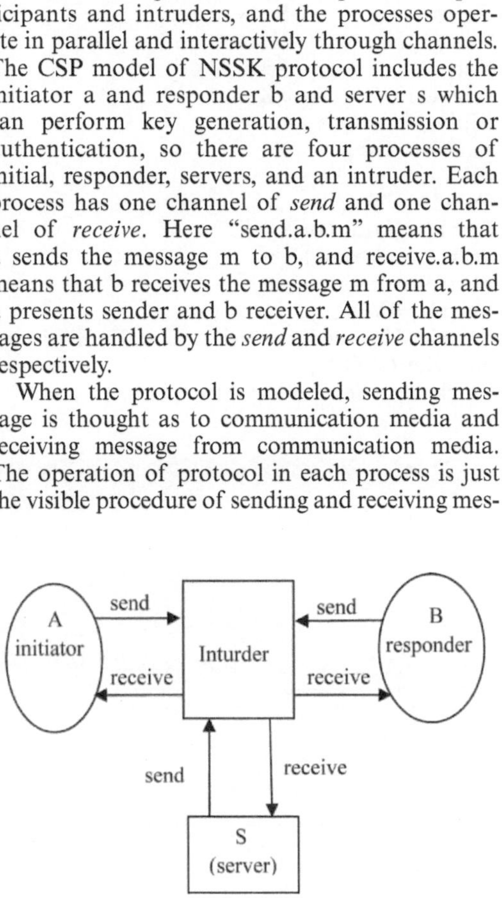

Figure 3. CSP protocol model.

The CSP process of initiator a is described as follows:

$$
\begin{array}{l}
\text{Initiator (a, na)} = \\
\text{Env ? s: Agent} \rightarrow \text{send.a.s.\{a.b.na\} Skeya} \rightarrow \\
\begin{array}{l} \text{kab} \in \text{Key} \\ \text{nb} \in \text{Nonce} \\ \text{b} \in \text{Agent} \\ \\ \end{array}
\left(\begin{array}{l}
\text{receive s.a.\{na,b,kab\}Skeya} \rightarrow \\
\text{receive b.a. \{nb'\}kab} \rightarrow \\
\text{send a.b.\{nb'-1\}kab} \\
\rightarrow \text{session (a, b, kab, na, nb)}
\end{array} \right)
\end{array}
$$

The CSP process of B is described as follows:

$$
\begin{array}{l}
\text{Responder (b, nb)} = \\
\begin{array}{l} \text{kab} \in \text{Key} \\ \text{nb} \in \text{Nonce} \end{array}
\left(\begin{array}{l}
\text{receive s.b.\{a\}Skeyb} \rightarrow \\
\text{send b.s.\{nb\}Skeyb} \rightarrow
\end{array} \right)
\end{array}
$$

$$
\begin{array}{ll}
\text{a} \in \text{Agent} & \text{receive s.b.\{nb,a,kab\}Skeyb} \\
 & \left(\begin{array}{l}
\text{send b.a.\{nb'\}kab} \rightarrow \\
\text{receive.a.b.\{nb'-1\}kab} \rightarrow \\
\text{session (a, b, kab, na, nb)}
\end{array} \right)
\end{array}
$$

The CSP process of server s is described as follows:

Server(s) = |||Serv(s, kab) kab $\in$ KEYS server
Serv(s, kab) =

$$
\begin{array}{l} \\ \\ \text{a,b} \in \text{Agent} \\ \text{na,nb} \in \text{Nonce} \\ \\ \\ \end{array}
\left(\begin{array}{l}
\text{receive a.s. \{a,b,na\}skeya} \rightarrow \\
\text{send s.b. \{a\} skeyb} \rightarrow \\
\text{receive b.s. \{nb\}skeyb} \rightarrow \\
\text{send s.a. \{na,b,kab\} skeya} \rightarrow \\
\text{send s.b.\{nb,a,kab\}skeyb} \\
\rightarrow \text{Server(S, Ks)}
\end{array} \right)
$$

The Intruder process is described as follows:

Intruder(X) = learn? M : messages$\rightarrow$
Intruder (close(X $\cup$ {m})
  say?m: X $\cap$ messages$\rightarrow$
Intruder(X)

The CSP protocol model merges the processes of above trusted proxy, sever and intruder into network through the way of rename, and thinks the intruder as unbelievable communication media in fact. The rename is to add the events of say.m and learn.m of intruder on the events of receive.a.b.m and send.a.b.m. And it renames the channels of *send* and *learn* as one *take* channel, and the channels of *receive* and *say* as one *fake* channel. Agent, server and intruder process join into a network

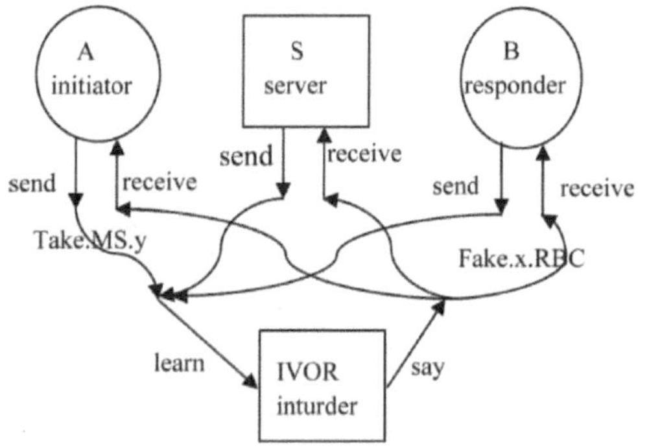

Figure 4. CSP model of new improved NSSK protocol after merge.

by the renaming way. The entire network of CSP model is described as follows:

Network=Initiator (a,na)[[fake,take/receive, send]]
  |||
Responder (b, nb) [[fake, take/receive, send]]
  |||
Server (sk) [[fake, take/receive, send]]
System = Network||
Intruder[[take.x.y,fake.x.y/learn,say|x,y∈ {Initiator} ∪{Responder}∪{Server}]]

After merged to network the new improved NSSK protocol can be described as follows and in the Figure 4.

Its CSP descriptions are:

(Agent (A) [[fake,take/receive,send]]
  |||
Agent (B)[[fake,take/receive,send]]
  |||
Server(S) [[fake,take/receive,send]]
  |||
Intruder (IVOR)[[take.x.y,fake.x.y/learn,say|x,y∈ Agents∪{Server}]]

4 THE VERIFICATION OF NEW IMPROVED NSSK PROTOCOL BY CASPER + FDR

Convert the new improved NSSK protocol to Casper (Lowe. 1998) specification.

#Free variables
A, B : Agent
S : Server
na, nb : Nonce
SKey : Agent->Server Key
Kab : Session Key
Inverse Keys=(SKey, SKey), (kab,kab)
#Processes
INITIATOR (A,S,na) knows SKey(A)
RESPONDER (B,nb) knows SKey(B)
SERVER(S,kab) knows SKey
#Protocol description
0.->A:B
[A!=B]
1. A->S:{A, B, na}{SKey(A)}
2. S->B:{A}{SKey(B)}
3. B->S:{nb}{SKey(B)}
4. S->A:{na,B,kab}{SKey(A)}
5. S->B:{nb,A,kab}{SKey(B)}
6. B->A:{nb}{kab}
7. A->B:{nb, nb}{kab}
#Specification
Agreement (A, S, [na, B])
Agreement (B, S, [nb])
Agreement (S, A, [na])
Agreement(S, B, [nb])
Agreement (B, A, [nb, nb])
Agreement (A, B, [nb, nb])
Secret (A, kab, [B])
Secret (B, kab, [A])
#Actual variables
Alice, Bob, IVOR : Agent
Jeeves : Server
Na, Nb, Ni : Nonce
Kab : Session Key
Inverse Keys = (Kab, Kab)
#Inline functions
symbolic SKey
#System
INITIATOR (Alice, Jeeves, Na)
RESPONDER (Bob, Nb)
SERVER (Jeeves, Kab)
#Intruder Information
Intruder=IVOR
Intruder Knowledge = {Alice, Bob, IVOR, Jeeves, Ni, SKey (IVOR)}
Crackable = Session Key(3)

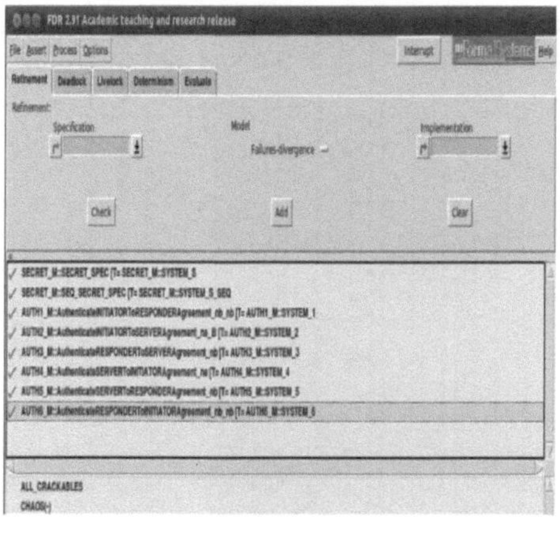

Figure 5. Verification of new improved NSSK protocol.

The verification results of the new improved NSSK protocol by Casper + FDR is described as in the Figure 5, and experimental results show that the new improved NSSK protocol did not find any security vulnerability, this protocol is secure.

5 CONCLUSION

This paper use formal validation tool CSP + FDR to verify the protocol which was put forward in (Liao et al. 2009), and the security vulnerabilities were found, so this paper proposes a new and improved NSSK protocol. After the test and verification of validation tools it is found that the security vulnerabilities existed in the old NSSK protocol are thoroughly prevented and it achieves mutual authentication, and avoids the man-in-the-middle attack, replay attack and attack of fake base station.

REFERENCES

Abadi M., Needham R. 1994. Prudent engineering practices for crypto-graphic protocols. *In:Proceedings of the 1994 IEEE Symposium on Security and Privacy*: 122–136. Los Alamitos: IEEE Computer Society Press.

Gavin Lowe. 1998. Casper-A Compiler for the Analysis of Security Protocols.

Hoare C.A.R. 1985. Communicating Sequential Processes. *Prentice-Hall.*

Liao Junguo, Zhu Bin, He Yong. 2009. Security Analysis of NSSK Protocol and Its Improvement. *2009 Eighth IEEE International Conference on Dependable, Autonomic and Secure Computing*:115–118.

E-Commerce, E-Business and E-Service – Lee (Ed)

Storing and retrieval of hybrid XML databases: A performance evaluation

Wei-Siang Chiew, Wan-Yee Yeow, Su-Cheng Haw, Samini Subramaniam & Fang-Fang Chua
Faculty of Computing and Informatics, Multimedia University, Cyberjaya, Malaysia

ABSTRACT: As XML has become well-known of its features and benefits, more applications and internet websites has increase their usage of XML. Nevertheless, it has not been fully utilized as the sole database due to the fact that relational databases are still widely use as the back-end repository for most organizations. Towards the recent years, hybrid XML databases have been introduced to enable the existence of XML technology in the traditional databases such as relational databases. In this paper, we conducted the performance evaluation on storing and retrieving XML data using (1) relational DTD approach, and (2) XML Column approach in IBM DB2. The experimental results indicate that the relational DTD approach performs better in most cases.

1 INTRODUCTION

There are two types of XML databases—XML Enabled Databases (XED) and Native XML Databases (NXD). NXD uses XML as the fundamental unit of storage. As for XED, it maps XML data to any traditional database such as relational database, object-oriented database, and object-relational database (Papamarkos et al., 2008; Amer & Hirzallah, 2013). However, the market still prefers relational database over XML database. This is because most enterprises have long secured the use of relational databases for high-end transaction processing system, they could not risk replacing relational database with a pure XML database. Thus, instead of replacing relational database, many researchers propose some mapping techniques to resolve the conflict between the hierarchical nature of the XML data model and the two dimensional structure of the relational data model.

Among one of the mapping approach is the Relational Data Type Definition (DTD) (Tian et al., 2002) approach which involves a shared-inlining method. Using this approach, the DTD provides the information necessary to determine where inlining is possible. The shared-inlining method ensures that any single element node is represented in exactly one relation so that the search space could be reduced.

Recently, commercial Relational Database Management Systems (RDBMS) vendors have shown significant interest in providing support for XML data management (Haw & Lee, 2011; Hall & Stromback, 2010). These hybrid systems have the added advantage of being capable of storing both relational (structured) and XML (semi-structured) data, thus, allowing applications to access a single data repository. For example, in IBM DB2, this is implemented as one of the data types named XML Column.

In this paper, we evaluate the performance of storing and retrieving XML database using the two approaches, i.e., the relational DTD approach and the XML Column approach on a real DBLP dataset (DBLP, 2014).

The outline of the paper is as follows. Section 2 discusses the related work. Section 3 presents our simulation engine. Section 4 presents the experimental setup, findings and benchmarking, and lastly, Section 5 concludes the paper.

2 RELATED WORK

Various studies have been conducted which compare on the performance of different method of storing and retrieving XML. Tian et al. (2002) conducted the experimental evaluation on storing XML documents on two real datasets. Yao et al. (2004) proposed XBench which could generate various types of XML for benchmarking purpose. They also proposed a simulator which could insert and read data using XQuery. Runapongsa et al. (2006) proposed the Michigan Benchmark, which is a widely used for performance evaluation. Their tools are able to compare 45 different queries for operation such as loading, insertion, updation and deletion. Nevertheless, all these approaches focus on performance comparison for either the native storage or using the relational repository.

The most directly related work to this paper is the work done by Nicola and Rodrigues (2006).

They utilized the IBM DB2 as the storage model and compare the pure XML storage against shredding and clobbing effect. Their study indicated that clobbing performs the best if XML documents are treated as the whole document; shredding suites well in the case if the structured of XML is fixed and data-oriented, and pure storage is the best choice if the XML is a document-centric type.

In the next sub-section, we review the two approaches of storing XML in the relational database. The Relational DTD approach is chosen due to the fact that most existing mapping technique is based on this as the reference unit. However, the XML Column is a feature provided by IBM DB2, to support the hybrid storage.

2.1 *The relational DTD approach*

This approach uses the strategy shared-inlining method which requires DTD of the XML document. The data element's occurrences define the DTD. A table is used to store and show relationship between an element and a set of children elements which have the same tag. The characteristics of the hierarchical structure include (Subramaniam et al., 2012):

- Every node consist of an ID and a ParentID
- If there were any one to many relationships, these elements will be stored in a separate table.
- Nodes with a degree zero is unreachable, thus is stored in separate relations.
- An element that can be shown only once in its parent is inlined as a column of the table which that table represents its parents

The following shows an example of a DTD of an XML note:

```
<?xml version="1.0" encoding="ISO-8859-1"?>
<!DOCTYPE note SYSTEM "Note.dtd">
<note>
<to>Tove</to>
<from>Jani</from>
<heading>Reminder</heading>
<body>Don't forget me this weekend!</body>
</note>
```

XML note

```
<!DOCTYPE note
[
<!ELEMENT note (to,from,heading,body)>
<!ELEMENT to (#PCDATA)>
<!ELEMENT from (#PCDATA)>
<!ELEMENT heading (#PCDATA)>
<!ELEMENT body (#PCDATA)>
]>
```

DTD

There are three types of transformation strategies that simplify the details of DTD. The transformation is described as follows. Flattering transformation is to convert recursive definition into a flat representation. Simplification transformation simplifies the unary operators into a single one. Lastly, grouping transformation groups all the sub-elements that has the same tag.

2.2 *The XML column*

XML Column is a feature of the IBM DB2 database. It is a column with an XML type which enables it to store XML document in a hierarchical form. By using XML Column, the XML data will be stored into a separate table spaces from the base table. The based table is linked to this separated table using each row of the XML Column.

For example, the following SQL is used to create the table for our program. It creates a new table with an XML column, which can be used to store our DBLP dataset which is in XML format:

```
CREATE TABLE XMLC
(ID BIGINT NOT NULL,
XMLINFO XML)
```

When a table with an XML Column is created, DB2 also generates a hidden column named DOC ID. DOC ID contains the unique identifier for the XML column. It links the base table to the separated XML table. The XML table is where it displays each XML document in a hierarchical form. When an XML document is being stored into the XML Column, it would be divided into multiple XML document and stored to each row of the XML Column.

3 OUR SIMULATION ENGINE

This system works as an XML-based database. To prove that our engine is working, we have implemented it on the real DBLP dataset (DBLP, 2014).

In the first part of evaluation, we compare the performance of storing XML in DB2 using XML Column and using DTD approach. Next, we perform some query retrieval based on some set of queries. User could choose up to 12 queries to evaluate the performance of query retrieval. Figure 1 depicts the graphical interface of our engine.

4 PERFORMANCE EVALUATION

4.1 *Experimental setup*

The goal of this paper is to evaluate the performance for two approaches which are the relational DTD approach and the XML Column approach. This experiment evaluates these approaches in

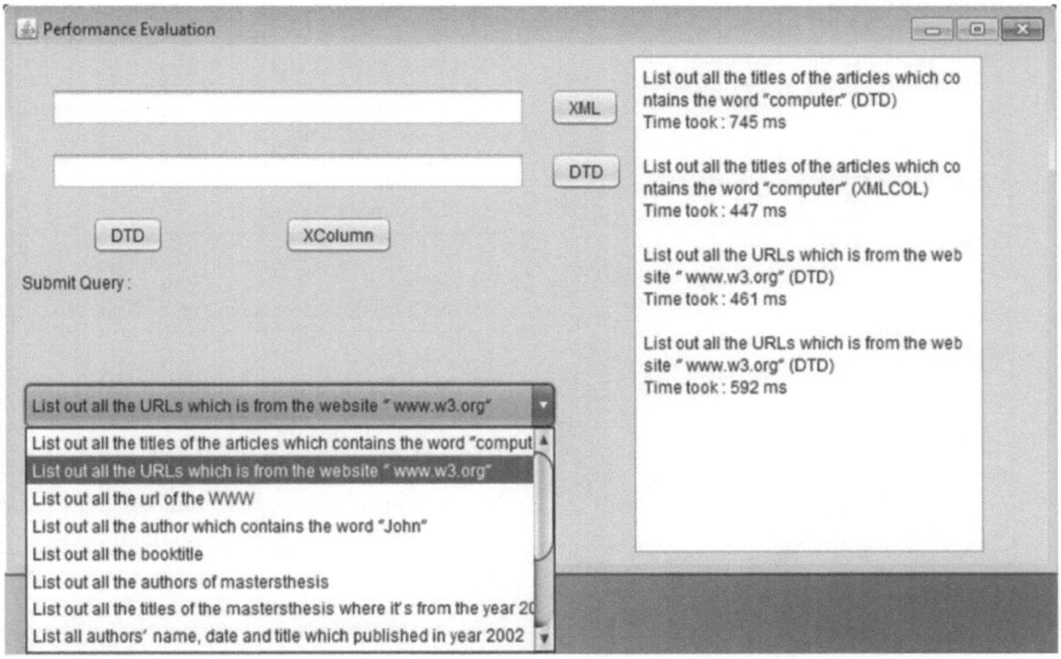

Figure 1. The user interface of our simulation engine.

terms of the performance of storing XML data into RDBMS and retrieving XML data from RDBMS. We have built an engine to receive XML file and inserts them into the database. This engine will then return the time consumed for these operations. Besides, this engine will also evaluate retrieving data by calculating the time for the query to execute. The details of the evaluation will be discussed later in this section.

To achieve a better and accurate result, we will use various sizes of data sets and several queries to be executed. We have prepared three datasets from small size to large size (See Table 1), and six queries to do the testing.

There are two types of query being used in this retrieval evaluation, Simple Query and Complex Query. Simple Query means query that searches using only one parameter. It does not search from one node to the other, no matter if it is a sibling node or a child node. From DTD database, this means that it would not use more than one table in the query. As for Complex Query, it searches for more than one parameter. It involves sibling nodes or child nodes in the search. It may also involve aggregate functions such as Max, Min, Count, and so on. In short, it is a query that would involve more than one table. We use second to measure the time performance for storing evaluation and millisecond for retrieving evaluation.

Table 1. Various dataset sizes.

DBLP dataset	Document	File size
D1	Dblp.xml	313 kb
D2	Dblpmid.xml	11 MB
D3	Dblpbig.xml	25 MB

All our experiments are performed on a Dual Core processor running on Windows 7. We will test each case for three times in order to achieve better accuracy in the evaluation. Thus, the result presented in the paper is an average of the three consecutive times.

4.2 *Storing evaluation*

For dataset that has a smaller size, relational DTD approach tends to be faster than XML Column. However, for dataset that has a larger size, relational DTD approach is slower than XML Column as depicted in Figure 2(a) and 2(b) respectively. This may be due to the process of shredding in relational DTD approach. For larger dataset, more shredding process is needed to divide the dataset into categories and stored into various tables. Thus, it uses more time to store the data into the database. As for XML Column, it always has been faster since the database only generates one table

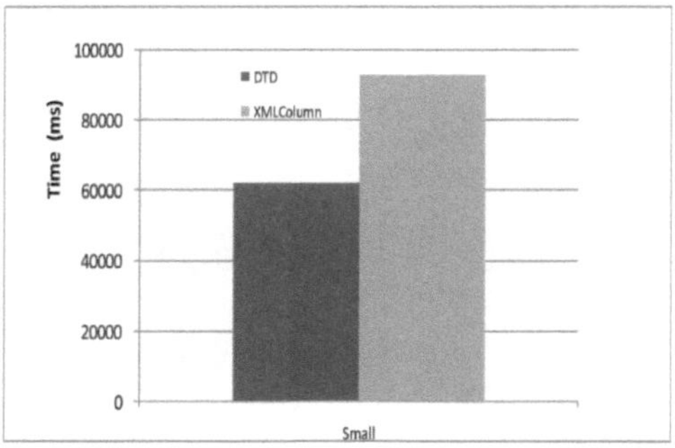

(a) Small dataset

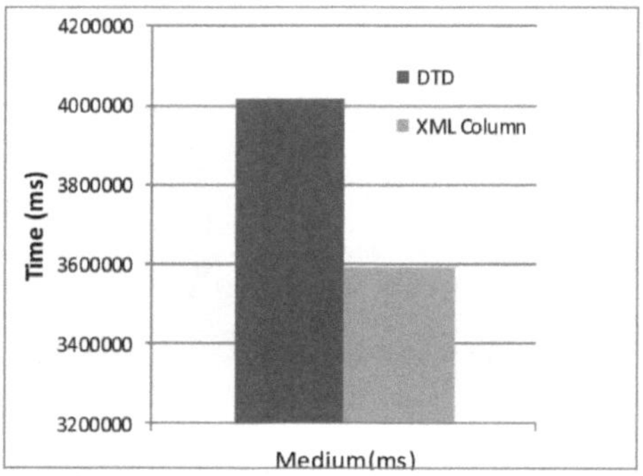

(b) Medium dataset

Figure 2. Storing evaluation on DBLP dataset.

and each XML document is stored into each row of the XML Column. Less shredding process is needed for XML Column approach.

4.3 *Retrieval evaluation*

For the retrieve evaluation, we will retrieve certain data to evaluate the time performance for both methods. We have prepared a list of description which defines which query to execute. After executing that particular query, we calculate the time used for the query to operate and determine the difference between relational DTD approach and XML Column approach. From the evaluation, we can get the best method that suits our engine. We used SELECT DB2 command to test the relational DTD approach and XQuery for XML Column database.

There are two types of queries used for this experiment. These two queries are Simply Query and Complex Query. We will be using these queries to test the three different sizes of datasets.

Tables 2 and 3 show the list of description of the selected queries to use in this test evaluation.

Result shows that relational DTD perform simple queries faster than XML Column during retrieval of data. As the dataset used becomes larger, XML Column uses more time to retrieve each data. Table 4 depicts the average time taken to execute the simple query for all sizes of dataset (as shown in Table 1). As can be seen from Table 4, the performance of Relational DTD approach is much faster in all cases. XML Column circulars the one column that stores all the data to retrieve the query result. For most of the queries that has a condition, XML Column would use extra time to retrieve the data. As for relational DTD, since simple query involve only one parameter, relational DTD would only search for one table which contains only a part of the dataset and not everything in it. Thus it uses lesser time to retrieve the query result as all the data are categorized and easy of being searched.

Figure 3 shows the average time taken to execute the simple query on various sizes of DBLP dataset.

As for complex queries, result shows that most of the complex queries favor the relational DTD

Table 2. Description for simple type of queries.

No.	Description
1	List out all the titles of the articles which contains the word "computer"
2	List out all the author which contains the word "John"
3	List out all the booktitle

Table 3. Description for complex type of queries.

No.	Description
1	List out all the titles of the mastersthesis where it's from the year 2006
2	List the number of article that its author is Tim Berners-Lee
3	List all the title of article except the article that is published in year 1990

Table 4. Query time on various DBLP dataset for simple query.

Approaches	No.	D1 (ms)	D2 (ms)	D3 (ms)
Relational DTD	Q1	91.6	358.33	665.33
	Q2	94.3	275.33	403.33
	Q3	61.6	150.33	176.67
XML column	Q1	93	940.67	1205.33
	Q2	115	951.67	1301.00
	Q3	92.6	1554.33	2656.67

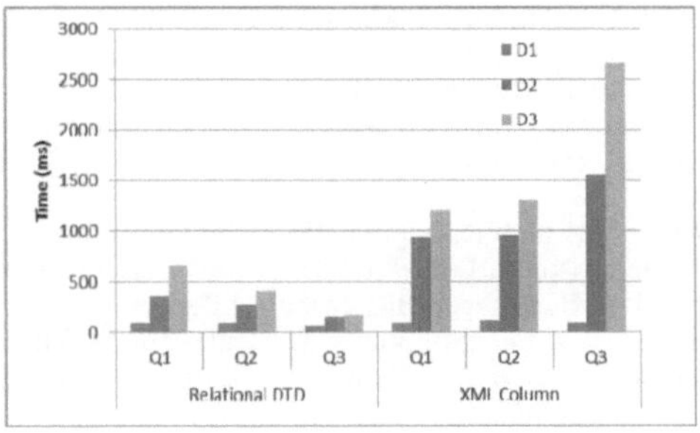

Figure 3. Time taken to execute simple query.

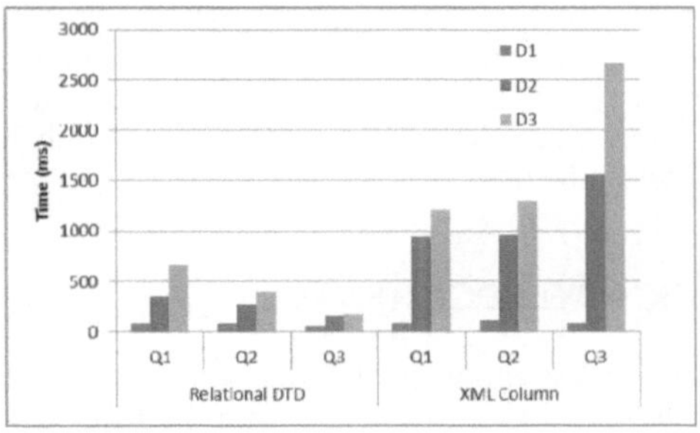

Figure 4. Time taken to execute complex query.

Table 5. Query time on various DBLP dataset for complex query.

Approaches	No.	D1 (ms)	D2 (ms)	D3 (ms)
Relational DTD	Q1	102.33	151	228.67
	Q2	532	140.67	171.33
	Q3	514.33	1643	3943.67
XML column	Q1	87.33	967	1320
	Q2	226.33	5740.67	8056.67
	Q3	242	868.67	1356.33

approach in terms of time performance compared to XML Column (see Table 5). For the smallest size dataset, relational DTD uses more time to retrieve data in contrast with XML Column. This is because with smaller size of data, XML Column could easily search for the result since it only has a column to store the data. Each row of the XML Column stores an XML document, allows it to read through each document one by one and look for those that fulfill the condition. However, relational DTD approach may use more time because it consist of many tables. It may need to match the data of one table to another to retrieve data.

However for the larger size dataset, it would be more troublesome to look through each XML document. It would be better to match a data from one table to another to retrieve the data faster. Thus, relational DTD approach has better time performance compared to XML Column.

For queries that have an aggregate function such as COUNT, it takes longer time for XML Column to retrieve the data as well. However, it does not affect much for relational DTD approach. This shows that relational DTD approach could cope with various techniques of queries better than XML Column.

Figure 4 shows the average time taken to execute the simple query on various sizes of DBLP dataset respectively.

4.4 *Overall discussion*

From the evaluation for both relational DTD approach and XML Column, it would seem that relational DTD approach best suits our engine. There are several points to support that statement.

First of all, the main function for our program is to retrieve data and display the retrieved data according to their tags. Most users would use our program to view the details of an input they entered. We would not want users to spend more time for waiting the result to be displayed. Thus, time efficient is our main basis. Since relational DTD approach has better time performance in retrieving query's result, it makes this the most important point for using it in our system.

Second, most of the retrieval query used in our program is complex query. Relational DTD approach has proven to have faster retrieval time for complex query compared to XML Column. Besides, some complex queries come with aggregate functions. XML Column would work longer for queries that have an aggregate function.

Thirdly would be the language. Compared to SQL, XQuery is a newer language. For further improvement, we or other developers may improve the program to have better performance or more features. It would be easier for us to use SQL instead of XQuery. Besides, for retrieving many tags of data, XQuery would be longer and it would be more complexity. For example, for the third simple query, to able to retrieve every booktitle from each bibliographic entry such as mastersthesis, article, www, and others, we need to write for each and every bibliographic entry to get the full result. If there are more than ten kinds of entry, then the XQuery would be longer and more complex.

5 CONCLUSIONS

In this paper, we have created a simulation engine to measure the performance of storing and retrieving XML using the relational DTD approach and

XML Column. The experimental results indicate that: (1) in terms of storing, XML Column approach has better performance, and (2) in terms of retrieving, the relational DTD approach has better performance.

ACKNOWLEDGEMENTS

This work is supported by funding of Fundamental Research Grant Scheme, from the Ministry of Higher Learning Education (MOHE).

REFERENCES

Amer, Y.Y., and Hirzallah, N. (2013). An Approach to Convert Object Oriented Database into XML. In Proceedings of the IEEE Computer Science and Information Technology, 238–242.

DBLP. (2014). DBLP XML records. http://dblp.uni-trier.de/xml/.

Hall, D., and Stromback, L. (2010). Generation of Synthetic XML for Evaluation of Hybrid XML Systems. Lecture Notes in Computer Science 6193, 191–202.

Haw, S.C., and Lee, C.S. (2011). Data storage practices and query processing in XML databases: A survey. Knowledge-Based Systems, 24(8), 1317–1340.

Hopfner, H., Schad, J., Mansour, E. (2009). On Analyzing the Database Performance for Different Classes of XML Documents based on the used Storage Approach. In: Proceedings of the International Conference on Software and Data Technologies, pp. 243–248.

Nicola, M. and Rodrigues, V. (2006). A performance comparison of DB2 9 pureXML and CLOB or shredded XML storage. IBM Technical Report.

Papamarkos, G., Zamboulis, L., and Poulovassilis, A. (2008). XML Databases. http://www.dcs.bbk.ac.uk/~sven/adm08/xmlDBs.pdf.

Runapongsa, K., Patel, J.M., Jagadish, H.V., Chen, Y., and Al-Khalifa, S. (2006). The Michigan benchmark: towards XML query performance diagnostics. Information Systems, 31(2),73–97.

Subramaniam, S., Haw, S.C., and Poo, K.H. (2012). Bridging XML and Relational Databases: An Effective Mapping Scheme based on Persistent Labeling. International Journal of Electrical and Computer Enginerring, 2(2), 239–246.

Tian, F., DeWitt, D.J., Chen, J., and Zhang, C. (2002). The Design and Performance Evaluation of Alternative XML Storage Strategies. Proceedings of the ACM SIGMOD, 5–10.

Yao, B.B., Ozsu, M.T., and Khandelwal, N. (2004). XBench Benchmark and Performance Testing of XML DBMSs. In Proceedings of ICDE 2004, 621–632.

E-Commerce, E-Business and E-Service – Lee (Ed)
 ISBN 978-1-138-02646-9

An evaluation model for embedded knowledge services using fuzzy AHP

Yijie Zhu
Shanghai Second Polytechnic University Library, Shanghai, China

ABSTRACT: This article is to study an evaluation problem for embedded knowledge services in academic libraries. It analyzes the main factors which impacting embedded knowledge services and construct an index system which includes effects of service, facilities, service skill knowledge resource configuration. It uses fuzzy analytic hierarchy process to evaluate embedded knowledge services in academic libraries. Finally, an example is given to illustrate the application of the proposed method.

Keywords: librarian; embedded knowledge services fuzzy library

1 ANALYSIS OF EMBEDDED KNOWLEDGE SERVICE BACKGROUND

Libraries contribute to users success through the resources and services they provide. [1] The resources involve a wide variety of forms, such as books, periodicals, patent document, database, multimedia resources, hypermedia resources, cyber source and experimental data. Librarians must be comprehensive collect and collate of the various types of information resources for effective organization, dynamic link and system integration.

In its first iteration, the term "embedded librarian" was adopted to describe librarians who were physically located within a discipline or faculty rather than in the library building. They were in situ liaison librarians who provided support for researchers and teachers, participated in faculty meetings, and provided discipline specific support for students. [2]

Library services could now be embedded virtually within faculties or courses. As a result, "embeddedness" evolved to encompass the embedding of library resources within online courses and the embedding of library services through librarians directly participating in online discussions. [3]

October 10, 2013 "Tsinghua University school online" open online courses (MOOC) platform officially released. Users watch the online courses at home, learning related knowledge. Teachers can answer users' questions, correction of users work online. For users who study online, these outcomes can be facilitated by participation of librarians in online discussion forums.

Open access, which provides free access to the information content, is widely expanding its domain because of enormous benefits accrued from it. [4]

2 EMBEDDED KNOWLEDGE SERVICES IMPLEMENTATION STEPS

2.1 *Build associated knowledge services team*

A diverse team is made up of librarians, experts and teachers. The expert cooperation is the essential condition of embedded knowledge services.

2.2 *User needs research*

Users focus on solve their complicated problems. This requirement calls for embedded knowledge services. Some users undertake teaching and scientific research, they need to focus on professional development trend and grasp the frontier. They have higher information literacy with a clear professional direction to search the materials.

2.3 *Embedded in the process of user research*

This is the first stage of librarians embedded in the process of user research. Librarians provide the project analysis, thematic literature, technology information, to discuss and formulate retrieval keywords, search strategies and to understand the main content of the research. [5]

In the research phase of the implementation adjust the retrieval keywords and search strategies based on dynamic project progress according to the project needs situation analysis, competitor analysis, and provide the project team with a comprehensive report.

In the research project completion stage librarians to use their professional background together with the project team to write articles, such as participation in the part of literature review; according to circumstances make recommendations on choosing publish in the relevant journals.

2.4 *Establishment of user feedback mechanism*

In the beginning stage of projects, librarians need to analyze project framework. In the course of research implementation phase, librarians need to consider the characteristics of different knowledge discovery retrieval system, and constantly adjust the angle of analysis report, and comprehensive constant communication. [6] In the concluding stage of research, librarians need to consult on the analysis results to users, constantly revised analysis results, in order to ensure user satisfaction.

3 FUZZY EVALUATION OF EMBEDDED KNOWLEDGE SERVICES

3.1 *Evaluation framework*

Embedded knowledge services evaluation is a process based on certain evaluation criteria. However, factors affecting the embedded knowledge services is multifaceted, which makes the evaluation of embedded knowledge services has become very complex and must weigh a variety of factors when conducting a comprehensive evaluation.

Considering the characteristics of the library under the circumstances, we choose effect of services, facilities, service skills, knowledge resources configuration, four evaluation factors, and established a secondary evaluation system.

Fuzzy comprehensive evaluation method is a comprehensive evaluation method based on fuzzy mathematics. [7] The comprehensive evaluation method turn the qualitative evaluation into quantitative evaluation, which uses fuzzy mathematics make an overall assessment on objects which affected by many factors. It has a clear result, systematic and strong features, can solve vague, and to quantify the problem, suitable for solving the problem of non-deterministic. For embedded knowledge services evaluation index system of indicators, fuzzy comprehensive evaluation method can better measure the indicators. In the following example evaluation of embedded knowledge services using fuzzy comprehensive method in a library. The quantified indicators of every affecting factors are determined. [8]

Table 1. Indicators for embedded knowledge services evaluation.

Embedded knowledge services evaluation index system	
First factors	Second factors
Effect of the services	Services means Services benefit User satisfactory
Facilities	Learning commons Online devices, network and width
Service skills	Knowledge acquisition ability, knowledge mining ability, knowledge innovation ability
Knowledge resource configuration	Database resource, institutional repositories OA resources

3.2 *Data analysis*

In factors $U = \{u_1, u_2, u_3, u_4\}$, u_1 represents effect the service, u_2 represents facilities, u_3 represents service skills and u_4 represents knowledge resource allocation.

In Sub-factors of the factors set $u_1 = \{u_1^1, u_1^2, u_1^3\}$, u_1^1 represents service means, u_1^2 represents service benefit and u_1^3 represents user satisfactory.

In parameter of $u_2 = \{u_2^1, u_2^2, u_2^3\}$, u_2^1 represents learning commons, u_2^2 represents the number of online devices and u_2^3 represents network bandwidth.

In parameter of $u_3 = \{u_3^1, u_3^2, u_3^3\}$, u_3^1 represents knowledge acquisition ability, u_3^2 represents knowledge mining ability and u_3^3 represents knowledge innovation ability.

In parameter of $u_4 = \{u_4^1, u_4^2, u_4^3\}$, u_4^1 represents database resource, u_4^2 represents the Institutional Repositories (IR) and u_4^3 represents OA resources.

In $V = \{v_1, v_2, v_3, v_4\}$, v_1 represents good, v_2 on behalf of better, v_3 represent the general and v_4 on behalf of poor.

Create sub-factor weight set and evaluation matrix. For sub-factors weight set into "important", "slightly important" "general" three levels. Invite a number of experts in related fields, university teachers and students (total 10), to be evaluated respectively on the basis of statistics. For example, there are four people think u_1^1, is important that there are three people in slightly important, three people in general; five people consider u_1^2, important, three people in slightly important, three people in general; four people consider u_1^3, important, four people in slightly important, two people in general; five people consider u_1^4, important, three people in slightly important, three people in general.

3.3 *Establishing fuzzy AHP weights*

Then the weight of u_1,

$$A(u_1) = ((4+5+4)/30, (3+3+4)/30, (3+2+2)/30) = (0.43, 0.33, 0.23).$$

Similarly can be calculated,

$A(u_2) = (0.35, 0.45, 0.19),$
$A(u_3) = (0.33, 0.48, 0.19),$
$A(u_4) = (0.41, 0.28, 0.30).$

Then sub-factor evaluation matrix is determined. For a sub-factor, based on past ratings (including the evaluation of the higher authorities, the school self-evaluation and students evaluation) the results were evaluated. Evaluation of sub-factors u_1^1 such as $P(u_1^1) = (0.1, 0.4, 0.4, 0.1)$, which means that a comprehensive evaluation of the past, 10% of people believe it good, 40% of people think it better, 40% of people think that in general, 10% of people considered it is slightly poor. Each sub-factors evaluation matrix obtained by factors evaluation matrix.

$$R(u_1) = \begin{bmatrix} P(u_1^1) \\ P(u_1^2) \\ P(u_1^3) \end{bmatrix} = \begin{bmatrix} 0.1 & 0.4 & 0.4 & 0.1 \\ 0.2 & 0.3 & 0.2 & 0.3 \\ 0.3 & 0.2 & 0.2 & 0.3 \end{bmatrix}$$

$$R(u_2) = \begin{bmatrix} P(u_2^1) \\ P(u_2^2) \\ P(u_2^3) \end{bmatrix} = \begin{bmatrix} 0.2 & 0.1 & 0.2 & 0.5 \\ 0.3 & 0.1 & 0.4 & 0.2 \\ 0.2 & 0.1 & 0.4 & 0.3 \end{bmatrix}$$

$$R(u_3) = \begin{bmatrix} P(u_3^1) \\ P(u_3^2) \\ P(u_3^3) \end{bmatrix} = \begin{bmatrix} 0.2 & 0.3 & 0.4 & 0.1 \\ 0.2 & 0.2 & 0.2 & 0.4 \\ 0.2 & 0.3 & 0.4 & 0.1 \end{bmatrix}$$

$$R(u_4) = \begin{bmatrix} P(u_4^1) \\ P(u_4^2) \\ P(u_4^3) \end{bmatrix} = \begin{bmatrix} 0.3 & 0.2 & 0.3 & 0.2 \\ 0.2 & 0.2 & 0.4 & 0.2 \\ 0.2 & 0.3 & 0.2 & 0.3 \end{bmatrix}$$

For sub-factor fuzzy comprehensive evaluation. The weights $A(u_i)(i = 1, \ldots, 4)$ and the single factor evaluation matrix $R(u_i)(i = 1, \ldots, 4)$, using the model of $M(\wedge, \vee)$ make comprehensive evaluation. $B(u_i) = A(u_i)R(u_i)\ (i = 1, \ldots, 4)$.

Calculated as follows:

$B(u_1)$	(0.23, 0.4, 0.4, 0.3)
$B(u_2)$	(0.3, 0.1, 0.4, 0.35)
$B(u_3)$	(0.2, 0.3, 0.33, 0.4)
$B(u_4)$	(0.3, 0.3, 0.3, 0.3)

The evaluation of the object as a comprehensive evaluation is considered based on the factors of $U = \{u_1, \ldots, u_4\}$. The weight $P(U)$ and single factor judge matrix R and $M(\wedge, \vee)$ are adopted for comprehensive evaluation.

The results are as follows:

$$R = \begin{bmatrix} B(u_1) \\ B(u_2) \\ B(u_3) \\ B(u_4) \end{bmatrix} = \begin{bmatrix} 0.23 & 0.4 & 0.4 & 0.3 \\ 0.3 & 0.1 & 0.4 & 0.35 \\ 0.2 & 0.3 & 0.33 & 0.4 \\ 0.3 & 0.3 & 0.3 & 0.3 \end{bmatrix}$$

$P(u_i)$ Similar to the determination of the weight $P(U)$. $P(U)$ is divided into "extremely important", "important", "general", "unimportant" four levels. Invited a number of experts in related fields, university teachers and students representatives (total 10), to be evaluated respectively on the basis of statistics. For example, u_1, consider an extremely important four people, three people think it is important that in general there are two people, one person considers unimportant; u_2, considered vital for five people, three people think it is important that one person in general, there is one person considered unimportant; u_3, considered extremely important for four people, four people think it is important that there is one person think in general, there is one person considered unimportant; u_4, considered extremely important for four people, two people think is important that there are two people in general, there are two people considered unimportant. Then

$$P(U) = (0.425, 0.3, 0.15, 0.125)$$
$$B = P(U)R = (0.4, 0.3, 0.4, 0.3)$$

4 CONCLUSION

Based on the evaluation results, knowledge resource allocation and facilities are both more satisfactory. For users more concerned of these two aspects, the library function is relatively conform an cetheir requirements. The level of the service skills, and effects of services are slightly poor. Analysis the model of fuzzy mathematics embedded knowledge services this model practical significance. Users can understand their own library embedded knowledge services level, try to improve the effect of several contingency factor embedded knowledge services, such as knowledge resource utilization rate, so as to enhance the level of embedded knowledge services.

User feedback mechanism is established to carry out the most important part of embedded knowledge services, throughout the whole process of knowledge services.

Under ubiquitous knowledge environment, embedding into user environment, user research team and research process changed direction of library services.

REFERENCES

[1] Becker, Bernd W. 2010. "Embedded Librarianship: A Point-of-Need Services." Behavioral Social Sciences Librarian 2 29(3):237–240 doi:10.1080/01639269.2010.498763.

[2] Drewes, Kathy, and Nadine Hoffman. 2010. "Academic Embedded Librarianship: An Introduction." Public Servicess Quarterly 6 (2–3):75–82. doi:10.1080/15228959.2010.498773.

[3] Mezick, E.M. 2007. "Return on Investment: Libraries and Student Retention." The Journal of Academic Librarianship 33(5):561–566. doi:10.1016/j.acalib.2007.05.002.

[4] Falk, H. (2004). Open access gains momentum. The Electronic Library, 22(6), 527–530. Directory of Open Access Journals. (2013). Retrieved from http://www.doaj.org/doi: 10.1108/02640470410570848.

[5] Liu, H. and Guo, S. (2007), "Research on fuzzy number ranking methods in management decision making analysis", Advances in Systems Science and Applications, Vol. 7 No. 1, pp. 110–6.

[6] Smith S. S, Sutton L. Embedded librarians: On the road in the Deep South[J]. College & Research Libraries News, 2008, 69 (2):71–74, 85.

[7] Lee, K.H. (2005), First Course on Fuzzy Theory and Application, Springer, New York, NY.

[8] Shumaker D, Tvler L.A. Embedded library services: An initial inquiry into practices for their development, management, and delivery[EB/OL].http://www.sla.org/pdfs/s1a2007/ShumakerD,EmbeddedLibSvcs.pdf.

E-Commerce, E-Business and E-Service – Lee (Ed)
© 2014 Taylor & Francis Group, London, ISBN 978-1-138-02646-9

Cloud-grid-based algorithm for data distribution management

Hong-Biao Mei
Applied Science College, Jiangxi University of Science and Technology, Jiangxi, China

ABSTRACT: Data Distribution Management (DDM) is an important service of the DoD High Level Architecture (HLA). For providing efficient, scalable mechanisms for distributing state updates and interaction information in large scale distributed simulations, DDM mechanisms (also known as filtering) in these interest management systems is the use of real time training simulations. This paper presents cloud-based algorithms for DDM. In this method, the initiation of the regions is made with the cloud model and an interest tree of pan-concepts is formed. The results of instance imply that this cloud DDM algorithm can be applied under the most situations and is self-adaptive, can reach the goal of the data distributed management, and greatly improves the effect and efficiency of the high complexity simulation system based on high-level architecture.

Keywords: grid; data distribution management; cloud; high level architecture

1 INTRODUCTION

Data Distribution Management (DDM) is one management of HLA. And it is also important for the parallel and distributed simulation community.[1] DDM is to limit and control the volume of the data exchanged during the simulation and reduce the processing requirements of simulation hosts by relaying events and state in formation only to those applications that require them.[2,3] For the goal, numbers of researches have been carried out, especially DDM, but there is a great space between the effect and the demand, and all the proposed algorithms have certain limitations as the system complexity advances.[4,5] The basic observation leading to this strategy is that real-world objects might be interested in only a fraction of the objects surrounding them.[6] For example, a spy plane is the most interested only in other planes surrounding it in Figure 1, but not interested in any other planes, some of which might be far away from its' interesting area. In Systems without DDM, however, the interest of the spy planes cannot be taken into account in the realization. The host simulating the spy plane cannot elect to receive only information about certain objects, namely the ones that are within the radar range of the spy plane, which in this example are the planes in Squadron A. Instead, it gets an update message anytime any plane in Squadron A or Squadron B, or any other aircraft, changes its position. All hosts will receive these state updates, just in case one of them should need it.[5]

The DDM techniques can be classified into two groups: region-based[7] and grid-based.[8] The region-based approach of the RTI 1.3 match the subscription and publication regions directly through comparing them. As is well known, in the worst case, every subscription need to be compared with every publication, and this leads to $O(N^2)$ scaling characteristics. The grid-based is applied for deducing the computational and communication overheads of the region-based method. While all publications necessitates to be matched with all subscriptions for the region-based approach, these extensive and costly operations are avoided for the grid-based technique. An RTI using a grid-based style maps each interest region onto a multidimensional grid, which represents the routing space.[5]

The grid-based includes two types: fixed grid-based and dynamic grid-based. Fixed grid-based allocation of multicast groups was the initial

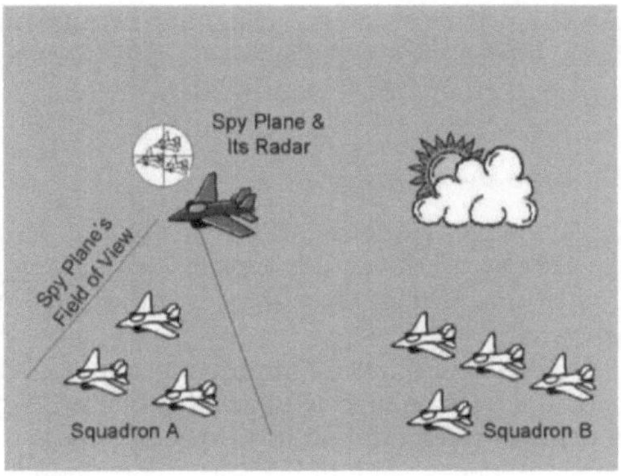

Figure 1. Domain region of interest.

implementation of the HLA-RTI, also known as the RTI prototype. It is a straightforward scheme.[8] But it is regretted that the fixed grid-based scheme does not directly determine intersection regions. The dynamic grid-based scheme is a hybrid of the fixed grid-based and region based methods since it reforms the fixed grid-based approach by giving it the intersection detection and the triggering abilities of the region-based scheme.[5] In this method, the most difficulty is to determine the grid sizes. The sizes are the key of the grid-based costs. In this paper, the cloud method-one probability method—is used to make the sizes adaptive.

2 CLOUD-BASED GRID ALGORITHM

In this section, we present a detailed description of our cloud-based grid algorithm. Firstly, we define the clouds by the normal rate to compute the overlaying terrain within a grid. Then the multicast clouds of the current regions in the system are dynamically allocated and then are distributed hosts trigger hosts. The proceeding is the same as in the region method, but different from the fixed grid-based scheme, which statically assigns multicast cloud groups to all of the terrain in the grid. The method of this paper is novel because only those cloudgrids includes at least one overlaying terrain of publishing federate and subscribing federate. These clouds are assigned to a multicast group, and when the messages sent by the cloud grid system are triggered, the federates will join or leave the appropriate clouds. The appropriate clouds are defined with cloud algorithm. Only when at least one overlaying cloud grid is created, Publishers will join and transmit on one cloud and the subscribers will join and listen on one group if there is at least one publisher transmitting on that group. There are dual advantages in this technique: one is reducing the data which senders transmitting, and the other is reducing the amount of multicast cloud groups t a federate joining, because only when at least one overlaying cloud grid is created, the cloud groups are joined and data are sent.

2.1 *The algorithm model*

In this paper, cloud algorithm is used to define the cells by overlapping the terrain within a grid. Firstly, the cloud algorithm is introduced as following.

Definition 1 (cloud) Suppose that the V is the fixed quantity domain with accurate values, X⊆T is the quality conception in U space. If there is a steady tendency random C(x)∈[0,1] corresponding to an element x (x∈X), x is called as the membership of T, namely:

$$C(x): V \to [0,1], \forall x \in X (X \subseteq V), x \to C(x) \quad (1)$$

Cloud model is an uncertain interconvert model with language values between quality conception and quantity conception. The digital characters of clouds are expressed with three values: expectation value Ex, entropy En, hyper entropy He, and noted as *C*(*Ex*,*En*,*H*), so on-dimension normal cloud operator *Arforward*(*C*(*Ex, En, He*)) is a mapping π: C→Π, which change the whole character of quality conception into quantity conception, and which satisfies the followings:

1. $\Theta = \{t_i | t_i$ is one implement of *Norm* (*En, He*), $i = 1 \ldots N\}$;
2. $X = \{x_i | x$ is one implement of *Norm* (*Ex*, t_i), $t_i \in \Theta, i = 1 \ldots N\}$;
3. $\Pi = \{(x_i, y_i) | x_i \in X, t_i \in \Theta, y_i = e^{-(x_i - ex)^2/(2t_i^2)}\}$;

$Norm(\mu, \delta)$ is normal random variable with the expectation μ and variance δ, and the N is the number of clouds. The normal cloud operator can convert the quality conceptions *C* (*Ex*, *En*, *He*) into clouds sets with values, and it is implemented that the conception space convert into value space. In cloud models, after the mapping is done, the value of a quality language is uncertain, and always has some little change, but the change doesn't affect the whole character of clouds. The cloud is flexible, non-edge and paddy. The distributed characters reflect the fuzziness and randomness, and its' whole shape is the most important. The cloud will be owned to node (or processor).

In this paper, our algorithm basic steps are described when a federate (F) is interested in a region containing cloud C_i. Recall that if there exits an intersection in the cloud C_i published and subscribed previously then that cloud will be distributed a multicast group. Consequently, F simply joins that cloud and is only recorded. Otherwise, if there is no intersection of publication or subscription in the cloud, a multicast group is then allocated to that cloud. But this multicast group will not only be allocated to F, but also to all of the other entities published or subscribed in that cloud C_i. Our cloud-based grid approach to DDM implement by finishing the following tasks.

1. Constructing a distributed cloud-based grid. Generally, during the simulation, the clouds are evenly allocated among the processors. The processor (or node) distributing to the cloud is named as the owner of the cloud. We utilize the cloud model to map clouds to nodes.

   We determine the *EX* in *C*(*Ex*, *En*, *He*) as the owner of a cloud. In our system, multiple independent grids are used to represent different routing spaces. A cloud is defined with the cloud number and the sub-cloud id. The sub-cloud is a conception of the dimensions and their sizes of a grid being used. And the sub-clouds are stored in each processor. Subscription and publication information for a particular cloud C_i is memorized in the processor that owns that cloud. For example, the publishing or subscribing list of processors is memorized.

2. Memorizing the actions of federates which are publishing or subscribing to each cloud. In this paper, we make use of two binary matrixes to identify that whether each cloud is owned by a processing (owner) or not, and if processing *i* is owned to that cloud, the *i*th binary in the matrix is set as 1, otherwise, it is as 0. When a node publishes or subscribes to a cloud owned by another n processing, the owner processing will receive a message to record the publishing (or subscribing) processing into the corresponding binary matrixes. Unsubscribing and unpublishing are done in a similar way.
3. Fix the owning groups of clouds in the intersection between publishing and subscribing. When a cloud owner manager distributes one action to a cloud where there are both publisher and subscriber at the same time, the MULTICAST library allocates one multicast group to the cloud. Otherwise, during the cloud owner manager distributing actions, the If the both publisher and subscriber do not present in the cloud at the same time, the cloud will be stored in an MULTICAST library routine. the cloud owner manager gets a new multicast group and becomes the group manager. The membership list of the MULTICAST library gets administered.
4. Federates' joining or leaving the groups. When the processing owner manager checks an intersection in a cloud and deal with the multicast group assigned to that cloud, all of the cloud processes must be notified to join the group. The FM (Federate management) communications library gets a join-group message from the cloud owner manager. In the other hand, there must be a leave-group message to be dispatched.
5. Finishing federates' joining or leaving the groups. When a process gets messages from the task TD, it sends a request to an MULTICAST library routine. The routine then gives an answer for the request with an update message to the cloud after it ensures to subscribe or unsubscribe that cloud to the group. For example, there is an update message about the manager of the multicast group. And the group manager updates its' membership table by giving a request message to the MULTICAST library according to the message.

2.2 *The algorithm programmer*

The basic steps of our algorithm are given the flowing when a federate F is triggered to join or leave to a group allocated to a terrain containing cloud $C_{i.}$

Step 1 Definition and initialization. The scale reference of the routing space is re-defined and the grid is initialized with the cloud theory;
Step 2 Initializing the terrain of publishing/subscribing with the cloud-grid;
Step 3 Detect the intersection of publisher and subscriber and match them if the intersection exists;
Step 4 Changing the scale of the cloud-grid and re-matching the terrain of publisher and subscriber, and then getting the max scale of the intersection;
Step 5 Distributing one multi-cast group owner to the cloud subscribed by other federates;
Step 6 Checking the cloud identification of the subscribing federate in the multicast group, and then send a message of federate's joining the groups;
Step 7 The MULTICAST library routine establishes the communication connection to get the new global time *tmin* and save;
Step 8 Programming the procedure of the simulation;
Step 9 End.

3 EXPERIMENT AND RESULTS

This experiment is applied for the development of MAKRTI1.3, and one city traffic simulation is done on this platform. In this simulation, one node is as one federate, and there are the same amount of vehicles in every federate. The vehicles are randomly set on the different roads. The references of this simulation are as Table 1.

We perform this simulation on WinXP system, the 10 computers are all Pentium II and they are linked with 100 MB network. And the results of simulation are as Table 2.

As Table 2 shows, the cloud-grid algorithm is superior to the grid algorithm in sending messages, DDM time and multicast groups. And it is superior to the region algorithm in multicast groups and DDM time, but it sends more messages than the region algorithm.

Table 1. The simulation references.

Reference	Value	Reference	Value
Federate	10	Interest level	5
Area	30 Km^2	Road length	50 Km
Step	100		

Table 2. Cloud vs region and grid schemes.

Algorithm	Objects	DDM-time (s)	Messages	Multicast groups
Region	200	73	26	176
	300	112	32	236
	500	406	109	423
	800	810	203	800
	1000	1030	287	983
Grid	200	92	154	30
	300	116	219	40
	500	230	444	105
	800	390	772	211
	1000	500	995	238
Cloud	200	60	132	42
	300	88	196	63
	500	200	357	76
	800	260	415	100
	1000	430	554	110

4 CONCLUSION

During the distributed simulation under the rules of the high level architecture of DOD, the DDM is essential. And a good DDM algorithm makes the volume of simulation network to reduce rapidly and makes the simulation the be high efficiency. In this paper, we have presented a new DDM algorithm-cloud-grid based algorithm. The grids of the region are modified with the cloud theory, and the simulation data is re-filtrated under this strategy. This method reduces the volume of the simulation considerably. Our DDM scheme combines the scalability of a grid-based method with the low multicast group usage provided by the intersection detection and triggering mechanisms of the region based strategy. Our simulation experiments indicate clearly that the cloud-grid based algorithm is a valid approach to data distributed management within the HLA/RTI. The cloud-grid based algorithm updates the rule of DDM, and enhances the RTI efficiency.

ACKNOWLEDGEMENT

Supported by the item (No. GJJ12346) (the nature foundation of Jiangxi provincial department of Education).

REFERENCES

[1] Azzedine Boukerche, Caron Dzermajko, Lu Kaiyuan. An enhancement towards dynamic grid-based DDM protocol for distributed simulation using multiple levels of data filtering[J]. parallel Computing. 2006,32:902–919.

[2] Boukerche A. and A. Roy, In search of data distribution management in large scale distributed simulations[C], in "Proc. of the 2000 Summer Computer Simulation Conference, Canada", pp. 21–28.

[3] Feng Chen, Yuhong Zhang, Yi-Ping Phoebe Chen, Subdividing globally important zones based on data distribution across multiple genome fragments[J], Computers in Biology and Medicine, 48 (2014) 109–118.

[4] Morse K.L, Bic L., Dillencourt M. Interest management in large scale virtual environments[J]. Presence, 2001, 4: 145–151.

[5] Yi Jiao, Yinghui Wang, et al. A cloud approach to unified lifecycle data management in architecture, engineering, construction and facilities management: Integrating BIMs and SNS[J]. Advanced Engineering Informatics 27 (2013) 173–188.

[6] Jian Wang, Ting Zheng, A hybrid multicast–unicast assignment approach for data distribution management in HLA [J]. Simulation Modeling Practice and Theory 40 (2014) 39–63.

[7] Morse K.L. and J.S. Steinman, Data distribution management in the HLA: Multidimensional regions and physically correct filtering, in "Proc. SIW, Spring, 1997."

[8] Ismael Herrera, Alberto A. Rosas-Medina, The derived-vector space framework and four general purposes massively parallel DDM algorithms, Engineering Analysis with Boundary Elements, 37 (2013) 646–657.

E-Commerce, E-Business and E-Service – Lee (Ed)
© 2014 Taylor & Francis Group, London, ISBN 978-1-138-02646-9

Trustworthiness measurements of real-time Web services

X.X. Wu & G.S. Hu
Department of Computer, Shanghai Technical Institute of Electronics and Information, Shanghai, China

ABSTRACT: This paper firstly modifies the existing service-based language introduced by He in UTP semantics for Web services into a dynamic Web service language. Then, it considers the quantification of real-time service trustworthiness, and discusses some characteristics of the operators in this language. Next, it studies the trustworthiness expectation and develops the tool TMRWS based on our theory. Finally, it summarizes this paper.

Keywords: dynamic Web services; trustworthiness; trustworthiness expectation; reputation degree; trustworthiness ratio; TMRWS

1 INTRODUCTION

Recently, Web services and Web-based applications become much more important in information systems. Many researchers have been engaged in this research, and kinds of business modelling languages have been introduced [1]–[7], such as WSFL [1], BPEL4 WS (BPEL) [4], StAC [2] and XLANG [3]. In order to accommodate the new program features, He [8] adds coordination and compensation combinators to synchronise the activity of programs, then extends the Guarded Command Languages [9]. He uses the new extended language to describe the transaction-based services. What's more, he shows the language's UTP [10] semantics with the operator H in [11].

However, in practice what we much more care about is, how to describe the trustworthiness of Web services. One of the best methods is based on software metrics theory. As well known, software metrics is a term that embraces a wide range of activities concerned with measurement in software engineering [12,13]. And the software metrics's history is almost as old as software engineering's history. And software trustworthiness Measurement is a much newer kind of software metrics [14]. Some work on QoS and trustworthiness of composite Web services have been introduced [15–18]. As Golbeck [19] tells that … trust is a good metric for identifying useful content … and it may provide a metric for establishing trust in user-generated content through the analysis of online social networks. And as Golbeck says, we think that the trustworthiness metrics of the atomic Web services are available when required, and the question we care is how to compute the trustworthiness of complex services being available. Chen et al in [20] have improved the language introduced in [8] by He by replacing the atomic statement $x := e$ with any atomic proposition P and present a service-based language, which describes the Web services's features, such as compensations, coordinations and some classic combinators of sequential composition, parallel and probabilistic choice. Their work focus on the Web services' evaluation, and map every Web service to a real number in [0,1] (the real number stands for the degree of success). After giving the evaluation of the atomic services, Chen et al give the complex services' evaluation model through the operators' compositions. However, this model doesn't discuss the timeliness. To stimulate the practical development of the service-based system, the language's timeliness is very important. Furthermore, the personalization of Web service is an important Internet's trend today. For these reasons, we will consider and discuss the timeliness and personalization.

In this paper, based on the previous work [20–23], we further improve the service-based language introduced by He in UTP semantics for Web services [11] into a dynamic Web service language (we think that the Web services provided by the website will change with the clients and time), which can be bused to describe the personalization and timeliness of Web services. And we analysis the quantification of real-time service trustworthinesses and some interesting characteristics of the operators (for example, or, sequential composition, parl and so on), and also discuss the trustworthiness expectation. Moreover, a tool based on our theory is developed.

We organize this paper as follows. In section 2, we improve the language proposed by Chen et al in [20] to a dynamic real-time Web service language. In section 3, we consider the quantification of real-time service trustworthinesses and discuss the operators' characteristics. In section 4, we analysis the Web services' trustworthiness expectation, for example, trustworthiness ratio and reputation degree. In section 5, we develop a tool TMRWS to show how to apply our theory in practice. The summarization follows in section 6.

Assumption. In rest of this paper, we assume that the service S at the website ω wouldn't be cancelled once ω provides S, and the atomic services always operate earlier than the composition services. For example, $t_\omega(S_1) \le t_\omega(S_1;S_2)$ and $t_\omega(S_2) \le t_\omega(S_1;S_2)$.

2 ADD DYNAMICS TO WEB SERVICE LANGUAGE

In this section, we give a example: a person takes a trip to Shanghai to book a train ticket or a flight ticket or a steamer ticket or a bus ticket, etc., and book a table and book a room and so on on websites. Clearly, it is necessary to book a ticket and book a table and book a room and so on for the trip successfully. However, it can choose any one among trains, flights, buses and etc.

Now, we will give the extended language by adding dynamic nature. The syntax of this language is as follows:

$$S ::= P \mid S;S \mid S \lhd b \rhd S \mid \underbrace{S \text{ or } S \text{ or } S \ldots \text{ or } S}_{\tau \text{ or } s}$$
$$\mid \underbrace{S \text{ parl } S \text{ parl } S \ldots \text{ parl } S}_{\tau \text{ parls}}$$
$$\mid \underbrace{S \text{ else } S \text{ else } S \ldots \text{ else } S}_{\tau \text{ elses}}$$
$$\mid \underbrace{S \text{ cpens } S \text{ cpens } S \ldots \text{ cpens } S}_{\tau \text{ cpenss}}.$$

More explanations about the meanings of these constructions see [20]–[23].

3 QUANTIFICATION OF DYNAMIC WEB SERVICE TRUSTWORTHINESSES

In paper [20], Chen et al propose the concept of success degree for a Web service. Here, we will generalize this concept. And we think that it requires kinds of factors to make a trustworthy Web service [21]. So, we have the trustworthiness degree of an atomic Web service as follows:

$$Q[P](c,\omega,t) = \beta_P(c,\omega,t) \in [0,1],$$

and the trustworthiness quantification of a Web service as follows:

$$\mathcal{Q}[S] : \mathfrak{A} \times \mathcal{W}_t \times [0,+\infty) \mapsto [0,1].$$

The analysis of these two formulas and service trustworthiness metrics see [20–23].

And we consider that it will need time to execute a single Web service, and the discussions on the time taken for the various Web services to execute see [20–23].

4 REPUTATION DEGREE AND TRUSTWORTHINESS RATIO

In this section, the analysis of trustworthiness expectation will contain two aspects: reputation degree and trustworthiness ratio.

Accordingly, we give the following definitions:

Definition 4.1. The reputation degree μ of ω that provides the Web service S, is defined as follows. For any t with $t-t_{\omega(S)} \ge t_S$,

$$\mu^S_{t-t_\omega(S)}(\omega) = \frac{\int_{t_w(S)}^{t-t_S} \frac{\sum_{k=1}^{M(S,v,\omega)} Q[S](c_k,\omega,v)}{M(S,v,\omega)^{\gamma_{\{0\}^c}(M(S,v,\omega))}} \pi(dv)}{t-t_\omega(S)},$$

here, $M(S,t,\omega)$ stands for the number of services' clients who get S from ω at the time t, and let $\sum_{k=1}^{M(S,t,\omega)} Q[S](c_k,\omega,t) = 0$ when $M(S,t,\omega) = 0$. π is the counting measure.

Definition 4.2 (Reputation Comparability Conditions-RCC). We think that the Web services are reputation comparability at the website ω and the time t, if the numbers of the clients of the same class for different operating Web services (for example S_1, S_1; S_2) at the website ω and the time t are always the same.

Definition 4.3. We think that for a given time interval $[t_1,t_2]$, the service S_1 provided by ω has a better absolute reputation than the service S_2 also provided by ω, denoted by $S_1 \succeq S_2\begin{bmatrix} t_2 - t_1 \\ \omega \end{bmatrix}$, if

$$\mu^{S_1}_{t_2-t_1}(\omega) \ge \mu^{S_2}_{t_2-t_1}(\omega).$$

Theorem 4.1. For the given time interval $[t_1,t_2]$ and a website ω. If $\forall t \in [t_1,t_2]$, the services satisfy RCC at ω. Then, we will have

(1) $S_1 \succeq S_1; S_2\begin{bmatrix} t_2 - t_1 \\ \omega \end{bmatrix}$.

(2) If $\forall v \in [\max\{t_1, s_{\omega(S_1;S_2)}\}, t_2 - (t_{S_1} + t_{S_2})]$, $M(S_2,v,\omega) = M(S_2,v+t_{S_1},\omega)$, then $S_2 \succeq S_1;S_2\begin{bmatrix} t_2 - t_1 \\ \omega \end{bmatrix}$.

(3) If $t_b = 0$ and $\forall v \in [\max\{t_1, s_\omega(S_1;S_2)\}, t_2 - (t_{S_1} + t_{S_2})]$, $M(S_2,v,\omega) = M(S_2,v+t_{S_1},\omega)$, then $S_1 \lhd b \rhd S_2 \succeq S_1;S_2\begin{bmatrix} t_2 - t_1 \\ \omega \end{bmatrix}$.

4. If $\forall n \in N'$, $\forall t \in [\max\{t_1, s_{\omega(S_1)}\}, \max\{t_1, s_{\omega(S_1 \text{ else} \cdots \text{else } S_n)}\}]$, $Q[S_1](c,\omega,t) = 0$, then $S_1 \text{ else } S_2 \cdots \text{else } S_{n-1} \text{ else } S_n \succeq S_1\left[\frac{t_2-t_1}{\omega}\right], \forall n \in N'$.
5. If $\forall n \in N'$, $\forall t \in [\max\{t_1, s_{\omega(S_1)}\}, \max\{t_1, s_{\omega(S_1 \text{ cpens } S_2 \cdots \text{cpens } S_n)}\}]$, $Q[S_1](c,\omega,t) = 0$, then $S_1 \text{ cpens} \cdots \text{cpens } S_n \succeq S_1\left[\frac{t_2-t_1}{\omega}\right], \forall n \in N'$.
6. $S_i \succeq S_1 \cdots \text{parl} S_n\left[\frac{t_2-t_1}{\omega}\right], \forall n \in N', i = 1, \ldots, n$.

And furthermore, if $\forall n \in N', Q[S_k](c,\omega,t) \geq Q[S_j](c,\omega,t) \Rightarrow t_{S_k} \leq t_{S_j}, k, j = 1, 2, \ldots n$, then

7. If $\forall n \in N'$, $\forall t \in [\max\{t_1, s_{\omega(S_i)}\}, \max\{t_1, s_{\omega(S_1 \text{ or } S_2 \cdots \text{or } Sn)}\}]$ $Q[S_i](c,\omega,t) = 0, i = 1, 2 \ldots n$, then $S_1 \text{ or} \ldots \text{or } S_n \succeq S_i\left[\frac{t_2-t_1}{\omega}\right], \forall n \in N', i = 1, \ldots, n$

Proof. 1. Case 1: $0 \leq t_2 - s_{\omega(S_1;S_2)} < t_{S_1} + t_{S_2}$.
For this case, we have $\mu_{t_2-t_1}^{S_1;S_2}(\omega) = 0$.
So, by Definition 4.3 we get $S_1 \succeq S_1;S_2$.
Case 2: $t_2 - S_{\omega(S_1;S_2)} \geq t_{S_1} + t_{S_2} \geq 0$.
For this case, we have

$$\mu_{t_2-t_1}^{S_1;S_2}(\omega) = \left(\int_{\max\{t_1, s_{\omega(S_1;S_2)}\}}^{t_2 - t_{S_1;S_2}} \frac{\sum_{k=1}^{M(S_1;S_2,\upsilon,\omega)} Q[S_1;S_2](c_k,\omega,\upsilon)}{M(S_1;S_2,\upsilon,\omega)^{\gamma\{0\}^{c(M(S_1;S_2,\upsilon,\omega))}}} \times \pi(d\upsilon) \right) / (t_2 - t_1)$$

$$\leq \left(\int_{\max\{t_1, s_{\omega(S_1)}\}}^{t_2 - t_{S_1}} \frac{\sum_{k=1}^{M(S_1,\upsilon,\omega)} Q[S_1](c_k,\omega,\upsilon)}{M(S_1,\upsilon,\omega)^{\gamma\{0\}^{c(M(S_1,\upsilon,\omega))}}} \pi(d\upsilon) \right) \times \frac{1}{t_2 - t_1}$$

$$= \mu_{t_2-t_1}^{S_1}(\omega).$$

2. Case 1: $0 \leq t_2 - S_{\omega(S_1;S_2)} < t_{S_1} + tS_2$.

For this case, we have $\mu_{t_2-t_1}^{S_1;S_2}(\omega) = 0$. So, by Definition 4.3 we get $S_2 \succeq S_1;S_2$.
Case 2: $t_2 - S_{\omega(S_1;S_2)} \geq t_{S_1} + t_{S_2} \geq 0$.
By assumption, we have

$$\mu_{t_2-t_1}^{S_1;S_2}(\omega) = \left(\int_{\max\{t_1, s_{\omega(S_1;S_2)}\}}^{t_2 - (t_{S_1} + t_{S_2})} \frac{\sum_{k=1}^{M(S_1;S_2,\upsilon,\omega)} Q[S_1;S_2](c_k,\omega,\upsilon)}{M(S_1;S_2,\upsilon,\omega)^{\gamma\{0\}^{c(M(S_1;S_2,\upsilon,\omega))}}} \times \pi(d\upsilon) \right) / (t_2 - t_1)$$

$$\leq \left(\int_{\max\{t_1, s_{\omega(S_2)}\}}^{t_2 - t_{S_2}} \frac{\sum_{k=1}^{M(S_2,\upsilon,\omega)} Q[S_2](c_k,\omega,\upsilon)}{M(S_2,\upsilon,\omega)^{\gamma\{0\}^{c(M(S_2,\upsilon,\omega))}}} \pi(d\upsilon) \right) \times \frac{1}{t_2 - t_1}$$

$$= \mu_{t_2-t_1}^{S_2}(\omega).$$

So, by Definition 4.3 we get $S_2 \succeq S_1;S_2$.

3. Case 1: $0 \leq t_2 - S_{\omega(S_1;S_2)} < t_{S_1} + t_{S_2}$.
For this case, we have $\mu_{t_2-t_1}^{S_1;S_2}(\omega) = 0$.
So, by Definition 4.3 we get $S_1 \lhd b \rhd S_2 \succeq S_1;S_2$.
Case 2: $t_2 - S_{\omega(S_1;S_2)} \geq t_{S_1} + t_{S_2} \geq 0$.
For this case, by assumption we have

$$\mu_{t_2-t_1}^{S_1;S_2}(\omega) = \left(\int_{\max\{t_1, s_{\omega(S_1;S_2)}\}}^{t_2 - (t_{S_1} + t_{S_2})} \frac{\sum_{k=1}^{M(S_1;S_2,\upsilon,\omega)} Q[S_1;S_2](c_k,\omega,\upsilon)}{M(S_1;S_2,\upsilon,\omega)^{\gamma\{0\}^{c(M(S_1;S_2,\upsilon,\omega))}}} \times \pi(d\upsilon) \right) / (t_2 - t_1)$$

$$\leq \left(\int_{\max\{t_1, s_{\omega(S_1)}\}}^{t_2 - t_{S_1}} \frac{\sum_{k=1}^{M(S_1,\upsilon,\omega)} [\![b]\!](c_k,\omega,\upsilon) Q[S_1](c_k,\omega,\upsilon)}{M(S_1,\upsilon,\omega)^{\gamma\{0\}^{C(M(S_1,\upsilon,\omega))}}} \pi(d\upsilon) \right) \times \frac{1}{t_2 - t_1}$$

$$+ \left(\int_{\max\{t_1, s_{\omega(S_2)}\}}^{t_2 - t_{S_2}} \frac{\sum_{k=1}^{M(S_2,\upsilon,\omega)} (1 - [\![b]\!](c_k,\omega,\upsilon)) Q[S_2](c_k,\omega,\upsilon)}{M(S_2,\upsilon,\omega)^{\gamma\{0\}^{C(M(S_2,\upsilon,\omega))}}} \times \pi(d\upsilon) \right) / (t_2 - t_1)$$

$$= \mu_{t_2-t_1}^{S_1 \lhd b \rhd S_2}(\omega).$$

So, by Definition 4.3 we get $S_1 \lhd b \rhd S_2 \succeq S_1;S_2$.

4. Case 1: $0 \leq t_2 - S_{\omega(S_1)} < t_{S_1}$.
For this case, we have $\mu_{t_2-t_1}^{S_1}(\omega) = 0$. So, by Definition 4.3 we get S_1 else S_2 else … S_{n-1} else $S_n \succeq S$.
Case 2: $t_2 - S_{\omega(S_1)} \geq t_{S_1} \geq 0$.
We have that, if $\exists t_0 \in [\max\{t_1, s_\omega(S_1 \text{ else } S_2 \ldots \text{else } S_n)\}, t_2 - t_{S_1}]$ and $\exists i_0 \in \{1, 2, \ldots M(S_1 \text{ else } S_2 \text{ else} \cdots \text{else } S_n, t_0, \omega)\}$, such that $Q[S_1](C_{i_0}, \omega, t_0) = 0$
Obviously, we get

$$S_1 \text{ else } S_2 \text{ else} \ldots S_{n-1} \text{ else } S_n \succeq S_1, \forall n \in N'.$$

Therefore, we only need to show that $\forall t \in [\max\{t_1, s_{\omega(S_1 \text{ else } S_2 \cdots \text{else } S_n)}\}, t_2 - t_{S_1}]$ and $\forall_i \in \{1, 2, \ldots M(S_1 \text{ else } S_2 \text{ else} \ldots \text{else } S_n, t, \omega)\}$, $Q[S_1](c_i, \omega, t) \neq 0$, we have S_1 else S_2 else … S_{n-1} else $S_n \succeq S_1$, $\forall n \in N'$.

And then, by the assumption we get

$$\mu_{t_2-t_1}^{S_1 \text{ else } S_2 \text{ else} \cdots S_{n-1} \text{ else } S_n}(\omega) = \left(\int_{\max\{t_1, s_{\omega(S_1)}\}}^{t_2 - t_{S_1}} \frac{\sum_{k=1}^{M(S_1,\upsilon,\omega)} Q[S_1](c_k,\omega,\upsilon)}{M(S_1,\upsilon,\omega)^{\gamma\{0\}c(M(S_1,\upsilon,\omega))}} \pi(d\upsilon) \right) \times \frac{1}{t_2 - t_1}$$

$$= \mu_{t_2-t_1}^{S_1}(\omega).$$

So, the proof is completed.

5. Using a reasoning similar to one in (4), the proof follows.
6. Using a reasoning similar to one in (4), the proof follows.
7. Case 1: $0 \le t_2 - S_{\omega(S_i)} < t_{S_i}, \forall i \in \{1,2,\dots n\}$.
   For this case, we have $\mu_{t_2-t_1}^{S_i}(\omega) = 0, i = 1, \dots n$.
   Then, the proof comes from Definition 4.3.
   Case 2: $t_2 - S_\omega(S_{i_0}) \ge t_{S_{i_0}} \ge 0, \exists i_0 \in \{1,2,\dots n\}$.
   Now we will use the mathematical induction.
   $n = 2$, by assumptions we get

$$\mu_{t_2-t_1}^{S_1 \text{ or } S_2}(\omega) \ge \left(\int_{\max\{t_1, s_{\omega(S_1)}\}}^{t_2 - t_{S_i}} \frac{\sum_{k=1}^{M(S_i,v,\omega)} Q[S_i](c_k,\omega,v)}{M(S_i,v,\omega)^{\gamma_{\{0\}} c(M(S_i,v,\omega))}} \pi(dv) \right) \times \frac{1}{t_2 - t_1} = \mu_{t_2-t_1}^{S_i}(\omega), \quad i = 1,2.$$

Suppose that $n = l$, we have $S_1 \text{ or} \dots \text{or } S_l \succeq S_i, i = 1, 2, \dots, l$. And then, $n = l + 1$, we have

$$\mu_{t_2-t_1}^{S_1 \text{ or } S_2 \dots S_l \text{ or } S_{l+1}}(\omega) \ge \left(\int_{\max\{t_1, s_\omega(S_1 \text{ or} \dots S_l \text{ or } S_{l+1})\}}^{t_2 - \min\{t_{S_l}, t_{S_{l+1}}\}} \frac{\sum_{k=1}^{\Delta_1} V_{j=1}^{l+1} Q[S_j](c_k,\omega,v)}{\Delta_1^{\gamma_{\{0\}} c(\Delta_1)}} \times \pi(dv) \right) / (t_2 - t_1)$$

$$\ge \begin{cases} \left(\int_{\max\{t_1, s_\omega(S_1 \text{ or} \dots S_l)\}}^{t_2 - \left(\sum_{j=1}^{l} t_{S_j}\right)} \frac{\sum_{k=1}^{\Delta_2} V_{j=1}^{l} Q[S_j](c_k,\omega,v)}{\Delta_2^{\gamma_{\{0\}} c(\Delta_2)}} \times \pi(dv) \right) / (t_2 - t_1) \\ \left(\int_{\max\{t_1, s_\omega(S_{l+1})\}}^{t_2 - t_{S_{l+1}}} \frac{\sum_{k=1}^{M(S_{l+1},v,\omega)} Q[S_{l+1}](c_k,\omega,v)}{M(S_{l+1},v,\omega)^{\gamma_{\{0\}} c(M(S_{l+1},v,\omega))}} \times \pi(dv) \right) / (t_2 - t_1) \end{cases}$$

$$= \mu_{t_2-t_1}^{S_i}(\omega), \quad i = 1,2,\dots,l,l+1,$$

here, let $\Delta_1 = M(S_1 \text{ or } S_2 \dots S_l \text{ or } S_{l+1}, v, \omega)$, $\Delta_2 = M(S_1 \text{ or } S_2 \dots S_{l-1} \text{ or } S_l, v, \omega)$.

So, it complete the proof.

Definition 4.4. We will say that for the given time interval $[t_1,t_2]$, the Web service S_1 provided by ω has a better relative reputation than the Web service S_2 also provided by ω, denoted by $S_1 \boxed{\succeq} S_2 \begin{bmatrix} t_2 - t_1 \\ \omega \end{bmatrix}$, if $\mu_{t_2-\max\{t_1, s_\omega(S_1,S_2)\}}^{S_1}(\omega) \ge \mu_{t_2-\max\{t_1, s_\omega(S_1,S_2)\}}^{S_2}(\omega)$. Here, $s_\omega(S_1,S_2)$ stands for the time that ω begins providing the Web services S_1 and S_2.

Theorem 4.2. For the given time interval $[t_1,t_2]$ and a Web website ω. If $\forall t \in [t_1,t_2]$, the Web services satisfy RCC at ω. Then, we get

1. $S_1 \boxed{\succeq} S_1; S_2 \begin{bmatrix} t_2 - t_1 \\ \omega \end{bmatrix}$.
2. If $\forall v \in [\max\{t_1, s_{\omega(S_1;S_2)}\}, t_2 - (t_{S_1} + t_{S_2})]$, $M(S_2,v,\omega) = M(S_2, v + t_{S_1}, \omega)$, then $S_2 \boxed{\succeq} S_1; S_2 \begin{bmatrix} t_2 - t_1 \\ \omega \end{bmatrix}$.
3. If $t_b = 0$ and $\forall v \in [\max\{t_1, s_{\omega(S_1;S_2)}\}, t_2 - (t_{S_1} + t_{S_2})]$, $M(S_2,v,\omega) = M(S_2, v + t_{S_1}, \omega)$, then $S_1 \lhd b \rhd S_2 \boxed{\succeq} S_1; S_2 \begin{bmatrix} t_2 - t_1 \\ \omega \end{bmatrix}$.
4. S_1 else $S_2 \dots S_{n-1}$ else $S_n \boxed{\succeq} S_1 \begin{bmatrix} t_2 - t_1 \\ \omega \end{bmatrix}$, $\forall n \in N'$.
5. S_1 cpens…cpens $S_n \boxed{\succeq} S_1 \begin{bmatrix} t_2 - t_1 \\ \omega \end{bmatrix}$, $\forall n \in N'$.
6. $S_i \boxed{\succeq} S_1 \dots$parl $S_n \begin{bmatrix} t_2 - t_1 \\ \omega \end{bmatrix}$, $\forall n \in N', i = 1, \dots, n$.

And furthermore, if $\forall n \in N'$, $Q[S_k](c,\omega,t) \ge Q[S_j](c,\omega,t) \Rightarrow t_{S_k} \le t_{S_j}, k, j = 1, \dots n$, then

7. S_1 or…$S_n \boxed{\succeq} S_i \begin{bmatrix} t_2 - t_1 \\ \omega \end{bmatrix}$, $\forall n \in N', i = 1, \dots, n$

Proof. By a similar reasoning in theorem 4.1, the proof follows.

Note 4.2. The items (1), (2), (3) and (6) show that the services composition such as S_1; S_2 and S_1 parl S_2 have a lower reputation degree than single services (S_1; S_2). The items (4) and (5) show that the composition of services which have low reputation degree can gain relatively high reputation degree by else composition or cpens composition. The item (7) shows that if we think the longer time services spend will have a lower trustworthiness degree, then or composition helps to improve the services' reputation.

Definition 4.5 (Trustworthiness Comparability Conditions-SCC). We think that the Web services are trustworthiness comparability for the client c at the time t, if

1. The numbers of websites that provide different operating services (for example S_1, S_1; S_2) for the client c at the time t are the same;
2. For a given website ω, c has the same level for the different operating services at the website ω and the time t.

Definition 4.6. For a given Web service S, its trustworthiness ratio is defined as follows:

$$= \sum_{\omega \in \mathcal{W}_{t-t_S}(S)}^{\eta_t(S)} Q[S](c,\omega,t) \times \mu_{t-t_{w(S)}}^{S}(\omega) + \sum_{\omega \in \mathcal{W}_t(S) - \mathcal{W}_{t-t_S}(S)} \frac{1}{4}.$$

Note 4.3. Taking into account the practical problems, we suppose that the client will think S

has a good and bad reputation, and has an even chance of trustworthiness, if the operation time of the service is less than its execution time.

Definition 4.7. We think that the Web service S_1 is better than the Web service S_2, denoted by $S_1 \sqsupseteq S_2\left[\begin{smallmatrix}c\\t\end{smallmatrix}\right]$, if $\eta_t^c(S_1) \geq \eta_t^c(S_2)$.

Theorem 4.3. For a client c and a given time t. If the services satisfy SCC for c at t, and if for any $s \in [0,t]$ and $\omega \in \mathcal{W}_s$, the Web services satisfy RCC at ω. Then, we get

1. $S_1 \sqsupseteq S_1;S_2\begin{bmatrix}c\\t\end{bmatrix}$.
2. If for each $\omega \in \mathcal{W}_t$, $\forall \upsilon \in [s_{\omega(S_1;S_2)}, t-(t_{S_1}+t_{S_2})]$, $M(S_2,\upsilon,\omega) = M(S_2,\upsilon+t_{S_1},\omega)$ and $Q[S_2](c,\omega,t) = Q[S_2](c,\omega,t+t_{S_1})$, then $S_2 \sqsupseteq S_1;S_2\left[\begin{smallmatrix}c\\t\end{smallmatrix}\right]$.
3. If $t_b = 0$ and for each $\omega \in \mathcal{W}_t$, $\forall \upsilon \in [s_{\omega(S_1;S_2)}, t-(t_{S_1}+t_{S_2})]$, $M(S_2,\upsilon,\omega) = M(S_2,\upsilon+t_{S_1},\omega)$, and $Q[S_2](c,\omega,t) = Q[S_2](c,\omega,t+t_{S_1})$, then $S_1 \lhd b \rhd S_2 \sqsupseteq S_1;S_2\left[\begin{smallmatrix}c\\t\end{smallmatrix}\right]$.
4. If $\forall n \in N'$, for each $\omega \in \mathcal{W}_t$, $\forall \upsilon \in [s_{\omega(S_1)}, s_{\omega(S_1 \text{ else } S_2 \cdots \text{else } S_n)}]$, $Q[S_1](c,\omega,\upsilon) = 0$, then S_1 else ... else $S_n \sqsupseteq S_1\left[\begin{smallmatrix}c\\t\end{smallmatrix}\right]$, $\forall n \in N'$.
5. If $\forall n \in N'$, for each $\omega \in \mathcal{W}_t, \forall \upsilon \in [s_{\omega(S_1)}, s_{\omega(S_1 \cdots \text{cpens } S_n)}]$, $Q[S_1](c,\omega,\upsilon) = 0$, then S_1 cpens$\cdots$cpens $S_n \sqsupseteq S_1\left[\begin{smallmatrix}c\\t\end{smallmatrix}\right]$, $\forall n \in N'$.
6. $S_i \sqsupseteq S_1 \cdots \text{parl } S_n\begin{bmatrix}c\\t\end{bmatrix}, \forall n \in N', i = 1, 2, \ldots, n$.

   And furthermore, if for any $s \in [0,t]$ and $\omega \in \mathcal{W}_s$ $\forall \upsilon \geq 0, Q[S_k](c,\omega,\upsilon) \geq Q[S_j](c,\omega,\upsilon) \Rightarrow t_{S_k} \leq t_{S_j}$, $k,j = 1,2, \ldots n$, then
7. If $\forall n \in N'$, $\forall \upsilon \in [s_{\omega(S_i)}, s_{\omega(S_1 \text{ or } S_2 \cdots \text{or } S_n)}]$, $Q[S_i](c,\omega,\upsilon) = 0, i = 1,2 \ldots n$, then $S_1 \ldots S_{n-1}$ or $S_n \sqsupseteq S_i\left[\begin{smallmatrix}c\\t\end{smallmatrix}\right]$, $\forall n \in N', i = 1, \ldots, n$.

Proof. By a similar reasoning in theorem 4.1, the proof follows.

However, $\eta_t^c(S)$ is a stochastic process because $\mathcal{W}_t(S)$ is a Poisson process. Thus, we will give the expected trustworthiness ratio in the following.

Definition 4.8. For a given service S, its expected trustworthiness ratio (E-trustworthiness ratio) is defined as follows:

$$E\left[\eta_t^c(S)\right] = E\left[\sum_{\omega \in \mathcal{W}_{t-t_S}(S)} Q[S](c,\omega,t) \cdot \mu_{t-t_{\omega(S)}}^S(\omega) + \sum_{w \in \mathcal{W}_t(S) - \mathcal{W}_{t-t_S}(S)} \frac{1}{4}\right].$$

Definition 4.9. We say that Web service S_1 is expectantly better than Web service S_2, given by $S_1 \sqsupseteq_E S_2\left[\begin{smallmatrix}c\\t\end{smallmatrix}\right]$, if $E[\eta_t^c(S_1)] \geq E[\eta_t^c(S_2)]$.

Proposition 4.1. $\mathcal{W}_t(S_1)$ and $\mathcal{W}_t(S_1;S_2)$ are both Poisson processes, let $\mathcal{W}_t(S_1(S_1;S_2)) = \mathcal{W}_t(S_1) + \mathcal{W}_t(S_1;S_2)$. Then, we have that $\mathcal{W}_t(S_1(S_1;S_2))$ is a Poisson process.

Proof. The proof immediately follows from the definition of Poisson process in [24].

Theorem 4.4. For a given time t and a client c. If the services satisfy SCC for c at t, and if for any $s \in [0,t]$ and $\omega \in \mathcal{W}_s$, the services satisfy RCC at ω. Then, we have

1. $S_1 \sqsupseteq_E S_1;S_2\begin{bmatrix}c\\t\end{bmatrix}$.
2. If for each $\omega \in \mathcal{W}_t$, $\forall \upsilon \in [s_{\omega(S_1;S_2)}, t-(t_{S_1}+t_{S_2})]$, $M(S_2,\upsilon,\omega) = M(S_2,\upsilon+t_{S_1},\omega)$ and $Q[S_2](c,\omega,t) = Q[S_2](c,\omega,t+t_{S_1})$, then $S_2 \sqsupseteq_E S_1;S_2\left[\begin{smallmatrix}c\\t\end{smallmatrix}\right]$.
3. If $t_b = 0$ and for each $\omega \in \mathcal{W}_t$, $t-(t_{S_1}+t_{S_2})]$, $M(S_2,\upsilon,\omega) = M(S_2,\upsilon+t_{S_1},\omega)$, and $Q[S_2](c,\omega,t) = Q[S_2](c,\omega,t+t_{S_1})$, then $S_1 \lhd b \rhd S_2 \sqsupseteq_E S_1;S_2\left[\begin{smallmatrix}c\\t\end{smallmatrix}\right]$.
4. If $\forall n \in N'$, for each $\omega \in \mathcal{W}_t$, $\forall \upsilon \in [s_{\omega(S_1)}, s_{\omega(S_1 \text{ else } S_2 \cdots \text{else } S_n)}]$, $Q[S_1](c,\omega,\upsilon) = 0$, then S_1 else$\cdots$else $S_n \sqsupseteq_E S_1\left[\begin{smallmatrix}c\\t\end{smallmatrix}\right]$, $\forall n \in N'$.
5. If $\forall n \in N'$, for each $\omega \in \mathcal{W}_t$, $\forall \upsilon \in [s_{\omega(S_1)}, s_{\omega(S_1 \cdots \text{cpens } S_n)}]$, $Q[S_1](c,\omega,\upsilon) = 0$, then S_1 cpens $S_2 \cdots$cpens $S_n \sqsupseteq_E S_1\left[\begin{smallmatrix}c\\t\end{smallmatrix}\right]$, $\forall n \in N$
6. $S_i \sqsupseteq_E S_1 \ldots \text{parl } S_n\begin{bmatrix}c\\t\end{bmatrix}, \forall n \in N', i = 1, \ldots, n$.

   And furthermore, if for any $s \in [0,t]$ and $\omega \in \mathcal{W}_s$, $\forall \upsilon \geq 0$, $Q[S_k](c,\omega,\upsilon) \geq Q[S_j](c,\omega,\upsilon) \Rightarrow t_{S_k} \leq t_{S_j}$, k, $j = 1,2,\ldots n$, then
7. If $\forall n \in N'$, $\forall \upsilon \in [s_{\omega(S_i)}, s_{\omega(S_1 \text{ or } S_2 \cdots \text{or } S_n)}]$, $Q[S_i](c,\omega,\upsilon) = 0$, $i = 1,2, \ldots n$, then S_1 or ... or $S_n \sqsupseteq_E S_i\begin{bmatrix}c\\t\end{bmatrix}$, $\forall n \in N', i = 1, \ldots,$.

Proof. The proof immediately follows from theorem 4.3.

Note 4.4. Just as the compositions of services have the effect on reputation degree, the compositions of services have the same effect on trustworthiness ratio and E-trustworthiness ratio.

Note 4.5. Here, the first condition in Definition 4.5 can be weakened to "The websites' numbers that provide different operating services (for example S_1, S_1; S_2.) for the client c at the time t are expectantly the same".

However, the E-trustworthiness ratio is formally discussed above. In practice, we usually need to calculate the E-trustworthiness ratio. The following proposition 4.2 tells us that the E-trustworthiness ratio can be calculated for some special cases. For any given service discussed above, we should accept that there are good websites and bad websites as well, considering the quality of service, etc.. Here, for the introduction of proposition 4.2 we presuppose:

1. We can sort the websites $\omega \in \mathcal{W}_t$ in descending order of their relative service quality.
2. For simplicity, let $\mu^{S}_{t-t_{\omega(S)}}(\omega) \equiv 1/2$.

Proposition 4.2. Let $Q[S]$ $(c, \omega_k, t) = \rho^{k/d} \cdot Q[S]$ (c,t), $0 < \rho \leq 1$, $k = 1,2,\ldots,\left|\mathcal{W}_{t-t_S}(S)\right|$, $d \geq 1$.

1. $\rho = 1$, we have

$$E\left[\eta^c_t(S)\right] = \alpha_S\left(t - t_S - t_{\omega(S)}\right)\cdot\frac{Q[S](c,t)}{2} + \alpha_S t_S \cdot \frac{1}{4}.$$

2. $\rho \neq 1$, we have

$$E\left[\eta^c_t(S)\right] = \frac{\rho^{1/d}}{1-\rho^{1/d}}\cdot\frac{Q[S](c,t)}{2} \cdot\left(1 - e^{\left(\rho^{\frac{1}{d}}-1\right)\alpha_S\left(t-t_S-t_{\omega(S)}\right)}\right) + \frac{\alpha_S t_S}{4}.$$

Here, $\omega_1, \omega_2, \ldots, \omega_k, \ldots$ denotes the sorting of the websites $\omega \in \mathcal{W}_t$ in descending order of their relative service quality at the time t (for the client c), ρ denotes the element factor of the service quality, d denotes the distinguishability of the service quality.

Proof. Condition the expectation on the variable $|\mathcal{W}_t|$:

(4.1)

$$E\left[\eta^c_t(S)\right] = \sum_{n=0}^{\infty}\left\{\sum_{k=1}^{n} E\left[\rho^{k/d}\cdot\frac{Q[S](c,t)}{2}\right]\right\}\cdot P(|\mathcal{W}_{t-t_S}(S)| = n) + \frac{\alpha_S t_S}{4}$$

1. If $\rho = 1$, then

$$(4.1) = \alpha_S\left(t - t_S - t_{\omega(S)}\right)\cdot\frac{Q[S](c,t)}{2} + \alpha_S t_S \cdot \frac{1}{4}.$$

2. If $\rho \neq 1$, then

(4.1)

$$= \sum_{n=0}^{\infty}\left\{\frac{\rho^{1/d}\cdot(1-\rho^{n/d})}{1-\rho^{1/d}}\frac{Q[S](c,t)}{2}\right\} \cdot P(|\mathcal{W}_{t-t_S}(S)| = n) + \frac{\alpha_S t_S}{4}$$

$$= \frac{\rho^{1/d}}{1-\rho^{1/d}}\cdot\frac{Q[S](c,t)}{2}\cdot\left(1 - e^{(\rho^{\frac{1}{d}}-1)\alpha_S\left(t-t_S-t_{\omega(S)}\right)}\right)\frac{\alpha_S t_S}{4}.$$

So, the proof is completed.

Example 4.1. S_1 is provided since $t = 0$ and $S_1; S_2$ is provided since $t = 2$. The execution time of S_1 and $S_1; S_2$ are $t_{S_1} = 1$ and $t_{S_1;S_2} = 4$, respectively. And $d = 1000$, $\rho = 0.3$, $\alpha_{S_1} = 0.5$, $\alpha_{S_1;S_2} = 0.55$, $Q[S_1](c,t) = 0.9$, $Q[S_2](c,t) = 0.87$. The current time is $t = 22$.

Furthermore, we assume that we have got the service quality of each website $\omega \in \mathcal{W}_t(S_1)$ and $\omega \in \mathcal{W}_t(S_1;S_2)$ at the time t, and these websites (providing S_1 and $S_1;S_2$) can be respectively sorted according to their service quality as follows:

$$\omega_1\left(0.3^{\frac{1}{1000}}\right), \omega_2\left(0.3^{\frac{2}{1000}}\right), \omega_3\left(0.3^{\frac{3}{1000}}\right)\ldots,$$

and

$$\omega_1\left(0.3^{\frac{2}{1000}}\right), \omega_2\left(0.3^{\frac{4}{1000}}\right), \omega_3\left(0.3^{\frac{6}{1000}}\right)\ldots.$$

Now, we can compute the E-trustworthiness ratios of S_1 and $S_1;S_2$.

By proposition 4.2, we have

$$E\left[\eta^c_t(S_1)\right] = \frac{\rho^{1/d}}{1-\rho^{1/d}}\times\frac{Q[S_1](c,t)}{2}\times\left(1 - e^{\left(\rho^{\frac{1}{d}}-1\right)\alpha_{S_1}\left(t-t_{S_1}-t_{\omega(S_1)}\right)}\right) + \frac{\alpha_{S_1} t_{S_1}}{4} \approx 4.8,$$

$$E\left[\eta^c_t(S_1;S_2)\right] = \frac{\rho^{2/d}}{1-\rho^{2/d}}\times\frac{Q[S_1;S_2](c,t)}{2} \times\left(1 - e^{\left(\rho^{\frac{2}{d}}-1\right)\alpha_{S_1;S_2}\left(t-\left(t_{S_1}+t_{S_2}\right)-t_{\omega(S_1;S_2)}\right)}\right) + \frac{\alpha_{S_1;S_2}\left(t_{S_1}+t_{S_2}\right)}{4} \approx 4.0.$$

And then, we get $E[\eta^c_t(S_1)] \sqsupseteq_E E[\eta^c_t(S_1;S_2)]$.

5 TMRWS

In this section, a tool based on our theory is given, which is developed by Java language. In Figure 1, it shows that the developing framework of TMRWS has three main parts.

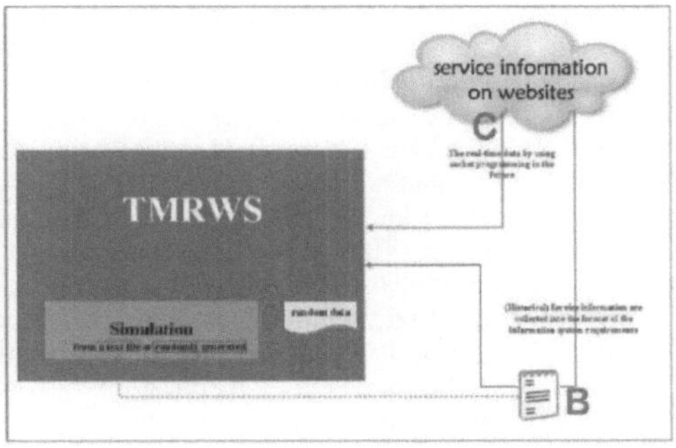

Figure 1. The developing framework of TMRWS.

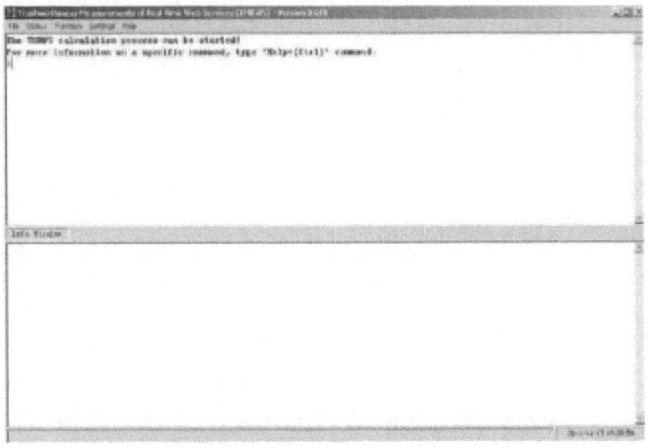

Figure 2. The UI of TMRWS.

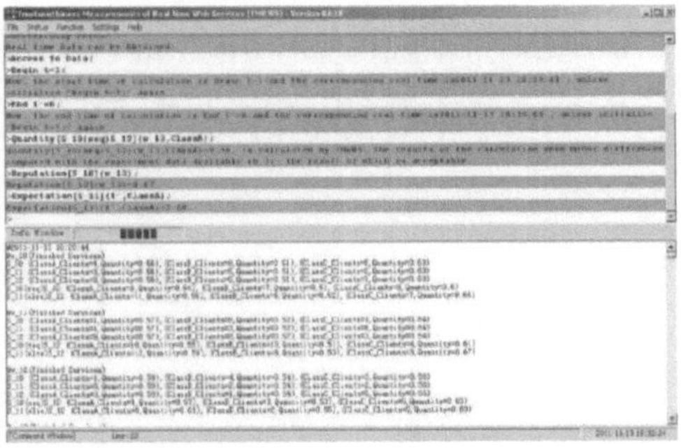

Figure 3. The operating state of TMRWS.

Part A. Simulation. It simulates the real-time service of websites on Internet and randomly generates service information, then TMRWS can analyse and calculate these information interactively.

Part B. Convert services information into the document's format which TMRWS can identify. The services information mainly include the historical service information on the websites.

Part C. Get services information of websites on the Internet. The real-time service information (such as, customer evaluation information) will be obtained through Socket programming. But the prerequisite is that we will have reached an agreement with websites. Now, it is underway.

We use TMRWS to randomly generates service information and show that how this tool is used to analyse and calculate these generated information (see Figs. 2 and 3).

Note 5.1. 1. Website operators can easily know their own services' trustworthiness and services' trustworthiness on other websites by TMRWS.

2. TMRWS provides some useful experience for clients to decide to choose services.
3. It can use TMRWS to test and verify the definition of the trustworthiness degree, see Figure 3.

6 CONCLUSIONS

In this paper, we think that clients, websites and time directly affect the trustworthiness of a Web service S. So, based on the previous work [20]–[23], we give a real-time Web service-based formal language and discuss the trustworthiness evaluation of real-time Web services. We use reputation degree to evaluate the trustworthiness expectation of a Web service offered by some website during a given time interval, and use trustworthiness ratio to evaluate the trustworthiness expectation of a Web service offered by all websites for some client at some time. Furthermore, a tool TMRWS based on our theory is developed.

However, there exist some problems, for instance, how to get an explicit formulation of $\beta(c,\omega,t)$. What's more, if the numbers of clients who get these different services (such as $S_1,S_1;S_2$) provided by ω at the time t are not the same, then how to compare the Web services' reputation degrees of these different Web services provided by ω. We will consider these problems and try to better apply our measurement model in this paper to solve practical problems in our future work.

REFERENCES

[1] Leymann F. Web Services Flow Language (WSFL 1.0). http://www-306.ibm.com/software/solutions/webservices/pdf/WSFL.pdf, (2001).

[2] Butler M, Ferreira C. An operational semantics for StAC, a language for modelling long-running business transactions. Coordination models and languages: 6th international conference, COORDINATION 2004, Pisa, Italy, Feb. 24–27, 2004: proceedings, Springer-Verlag New York Inc: 87–104 (2004).

[3] Thatte S. XLANG Web Services for Business Process Design. http://www.gotdotnet.com/team/xml wsspecs/xlang-c/default.htm, (2001).

[4] Andrews T, Curbera F, Dholakia H, Goland Y, Klein J, Leymann F, Liu K, Roller D, Smith D, Thatte S, Trickovic I, Weerawarana S. Business Process Execution Language for Web Services (BPEL4 WS). http://www–106.ibm.com/developerworks/library/ws-bpel/, (2004).

[5] Qiu ZY, Wang SL, Pu GG, Zhao XP. Semantics of BPEL4 WS-like Fault and Compensation Handling. In Proc. of FM 2005, New Castle, UK, 2005.7. LNCS 3582: 350–365 (2005).

[6] Xiao QJ, Rao R., You JY. A Language for Reliable Service Composition. SOFSEM 2007, LNCS 4362: 554–565 (2007).

[7] Zhu HB, Pu GG, He JF. A Denotational Approach to Scope-based Compensable Flow Language for Web Services. LNCS Asian 2006, LNCS 4435: 28–36 (2007).

[8] He JF. Modelling Coordination and Compensation. ISOLA 2008, CCIS 17: 15–36 (2008).

[9] Dijkstra EW. A Discipline of Programming. Prentice Hall, Englewood Cliffs (1976).

[10] Hoare C, He J. Unifying theories of programming. Prentice Hall (1998).

[11] He JF. UTP Semantics for Web Services. IFM 2007, LNCS 4591: 353–372 (2007).

[12] Fenton N, Neil M. Software Metrics: Successes, Failures, and New Directions. In Journal of Systems and Software, 47: 149–157 (1999).

[13] Fenton N, Pfleeger S. Software Metrics: A Rigorous and Practical Approach (Second Edition). PWS (1998).

[14] Gates B. Trustworthy Computing. Wired News, http://www.wired.com/techbiz/media/news/2002/01/49826(Jan. 17, 2002).

[15] Yang SJ, Hsieh JS, Lan BC, Chung J. Composition and evaluation of trustworthy Web services. In Proceedings of the IEEE Eee05 international Workshop on Business Services Networks(Hong Kong, March 29–29,2005). ACM International Conference Proceeding Series, vol. 87. IEEE Press, Piscataway, NJ: 5–5 (2005).

[16] Bianculli D, Binder W, Drago L and Ghezzi C. Transparent reputation management for composite Wen services. In Proceedings of the 2008 IEEE international Conference on Web Services (September 23–26, 2008). ICWS. IEEE Computer Society, Washington, DC: 621–628 (2008).

[17] Rajbhandari S, Rana OF, Wootten I. A fuzzy model for calculating workflow trust using provenance data. IN Proceedings of the 15th ACM Mardi Gras Conference: From Lightweight Mash-Ups To Lambda Grids: Understanding the Spectrum of Distributed Computing Requirements, Applications, Tools, infrastructures, interoperability, and then incremental Adoption of Key Capabilities (Baton Rouge, Louisiana, January 29-February 03, 3008). MG'08, vol. 320. ACM, New York, NY:1–8 (2008).

[18] Cardoso J, Miller J, Sheth A and Arnold J. Modeling Quality of Service for Workflows and Web Service Processes. Web Semantics Journal: Science, Services and Agents on the World Wide Web, 1(3), Elsevier Inc, MA, USA: 281–308 (2004).

[19] Golbeck J. Weaving a Web of trust. Science, vol. 321, no. 5896: 1640–1641 (2008).

[20] Chen YX, Wu XX. Success measurement of Web services with BPEL. In 5th IEEE International Workshop on Service-Oriented System Engineering (SOSE2010), Jun. 4–5, 2010, Nanjing: 86–90 (2010).

[21] Wu XX, Chen YX. Trustworthiness Expectation of Real-time Web Services. The 7th International Conference on Autonomic and Trusted Computing (ATC 2010), 2010 Symposia and Workshops on Ubiquitous, Autonomic and Trusted Computing, Oct. 26–29, 2010, Xi'an: 292–298 (2010).

[22] Wu XX, Chen YX. Trustworthiness Measurements of Real-time Web Services. Manuscript (2010).

[23] Wu XX, Chen YX. Trustworthiness Measurements of Web Services. Fuzzy Systems and Mathematics, Vol. 24(Supplement):128–137 (2010) (In Chinese).

[24] Wentzell AD, Translated by Chomet S. A Course in the Theory of Stochastic Processes. McGraw-Hill Inc (1981).

E-Commerce, E-Business and E-Service – Lee (Ed)
© 2014 Taylor & Francis Group, London, ISBN 978-1-138-02646-9

Numerical back analysis method for in-situ stress field distribution in deep river valley region

H.G. Xing
College of Architecture and Environment, Sichuan University, China

F.G. Xu & J.W. Zhou
State Key Laboratory of Hydraulics and Mountain River Engineering, Sichuan University, China

ABSTRACT: This paper presents a three-dimensional nonlinear finite element-based back analysis method for the in-situ stress field distribution in deep river valley region. The proposed approach incorporates a three-dimensional finite element model, monitoring results for the initial in-situ stress field, and multiple regression analysis. The in-situ stress field is assumed to comprise gravitational (lithostatic) stresses and tectonic stresses, and river erosion effect is considered during the numerical back analysis process of in-situ stress field. The tectonic stresses are separated into five components. The paper further applies the three-dimensional nonlinear finite element back analysis method to ascertain the in-situ stress field distribution in the dam site region of Jinping I Hydropower Station. The computed results show that the back analysis method can effectively determine the initial stresses field, and that the computed in-situ stress field is close to that of the test data.

1 INTRODUCTION

Under increasing worldwide socioeconomic development, especially in China, a large number of Hydropower Stations are being constructed or planned. Most of the Hydropower Stations are constructed at the mountains and canyons region. A lot of geotechnical problems are existed for the designers and engineers, but numerical analysis can provide a realistic representation of the field conditions, the most widely used technique in numerical modeling of geotechnical problems is the finite element method (Zhou et al., 2007). However, the in-situ stress field is difficult to determine, and consequently the methods or techniques for determining this important computational parameters have been the subject of considerable research (Chen et al., 2001; Fakhimi et al., 2004).

In the numerical simulation of dams, slopes or underground caverns, the computational results are clearly influenced by the initial stresses field and mechanical parameters (Arai, 1984; Feng et al., 2000). This paper will study the back analysis method for ascertaining the in-situ stress field in deep river valley region. Back analysis is an indirect technique that has been used to determine the in-situ stress field by using field measurements of displacements, strains or stresses (Li et al., 2009). In recent years, a number of back analysis methods have been proposed and applied in the design and construction of underground engineering, slope engineering, water conservation and hydroelectric power engineering projects (Ferrero et al., 2013).

For the in-situ stress field, the back analysis methods can be divided into two categories. In the first approach, known as the multiple regression analysis of the in-situ stress field, the back analysis requires a small number of measured values of the in-situ stress field. Results from the multiple regression analysis of these measurements are then fed into a three-dimensional finite element model to determine the in-situ stress field (Kavanagh and Clough, 1972; Brady and Brown, 1985). In the second approach, known as the displacement back analysis method, the back analysis of the initial stresses field is based on the monitored displacement field. For general underground engineering excavation in rock masses, a measuring system for displacements is usually established to ensure the safety of the construction, and the displacements in the surrounding rock can be used in the back analysis of rock mass mechanical parameters (Kaiser et al., 1990; Swoboda et al., 1999; Yang et al., 2001). For this reason, it is not always necessary to establish an extra measuring system for the back analysis.

The main objective of this paper is to present a three-dimensional nonlinear finite element back analysis method to ascertain the in-situ stress field in deep river valley region. Specifically, the multiple regression analysis method and associated design techniques are developed for the back analysis of in-situ stress field in the numerical simulation process.

2 IN-SITU STRESS FIELD

In the numerical simulation of the tunnel excavation, the decision whether to simulate the in-situ stress field directly affects the quality of numerical results. We first note that "in-situ stress field" relies on the concept of a "field", and on the continuity and homogeneity of the rock mass. Secondly, confined by the actual rock mass conditions, the current stress test results can only guarantee a certain reliability in the calculations for a given point, and can not guarantee the relevance of the measured values of any particular sampling point in the stress field.

Hydropower stations are generally constructed in deep river valley regions, which have generally been subjected to strong surface denudation and strong erosion during their geological history, thereby forming terrain with steep topography. Therefore, when compared with the terrain of a flat region, the geological history and steep topography play a prominent role in the transformation of the in-situ stress field during the region's modern geological history. These factors can even completely change the original stress state within a certain range in the valley, forming localized patterns in the stress field. The characteristics of the in-situ stress field of the valley region are shown in Figure 1, and are mainly influenced by three main factors: the original stress state before the valley formation, recent tectonic movement, and the morphological characteristics of the valley (reflected in the numerical simulation process as the geometry of the computational model).

The in-situ stress field before the valley formation is generally considered as a self-weight stress field, but during the valley evolution process, the stress field changes and becomes much more complicated during its long evolutionary history, making it difficult to provide a reasonable representation of the in-situ stress field. In order to reflect the true state of the in-situ stress field in the numerical simulation process, the role of tectonic

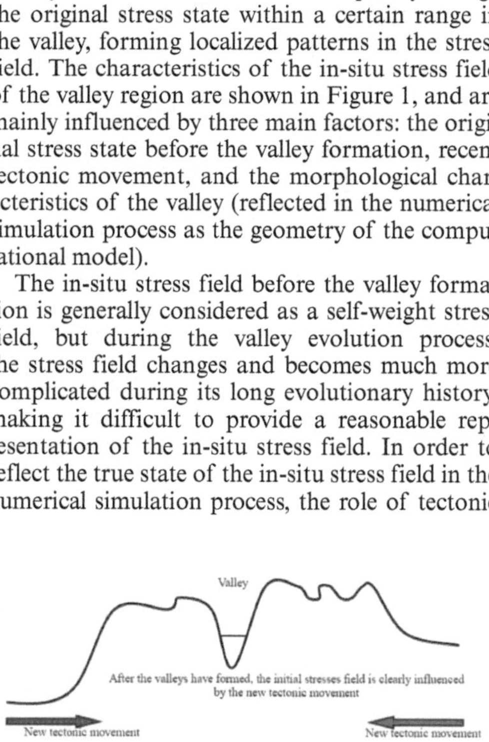

Figure 1. In-situ stress field influenced by new tectonic movement.

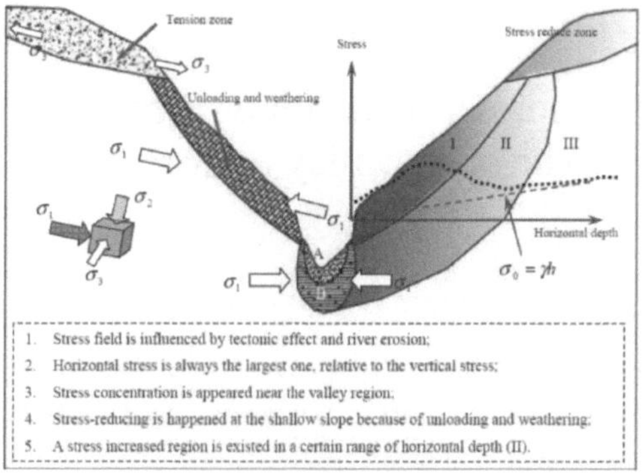

Figure 2. In-situ stress field distribution characteristics in deep river valley region.

stresses is considered, as well as the impact of recent tectonic movement. The valley stress field is therefore composed of two parts: the self-weight stress field and the tectonic stress field, with the latter including the tectonic stresses field generated during valley evolution and that caused by recent tectonic movement. The valley stress field is also influenced by the river erosion effect; unloading and weathering are happened in the shallow slope, stress-reducing is appeared in the shallow rock masses. Figure 2 shows the in-situ stress field distribution characteristics in deep river valley region.

As shown in Figure 2, at the bottom of the river valley, stress concentration is happened and causes the cake crack in rock core. The in-situ stress in river valley is always very large, especially in the horizontal direction. But in the topper mountain, a tension zone is appeared because of stress reducing, and the in-situ stress is always very little in this region. At the shallow slope, unloading and weathering effect is impact on the in-situ stress field, combined with the river erosion effect, the stress distribution characteristics is very complicated (I region in Fig. 2). With the increasing horizontal depth, a stress increased region is existed in a certain range of horizontal depth (II region in Fig. 2). When the horizontal depth is larger than a certain value, the in-situ stress is basically closed to the gravity stress (III region in Fig. 2). As can be seen from the above analysis, two key issues should be considered in the numerical analysis of in-situ stress field in deep river valley region.

3 NUMERICAL BACK ANALYSIS METHOD

In this section, firstly, the importance and composition of the in-situ stress field are described.

Then, the three-dimensional numerical back analysis method based on Finite Element Method (FEM) for the in-situ stress field is presented.

3.1 *Importance of in-situ stress field*

A large number of engineering studies have shown that the in-situ stress field is mainly controlled by self-weight stresses, geological tectonic movements and erosion of the earth's surface. The impact of closure stresses is relatively small, and difficult to quantify in the in-situ stress field analysis. Thus, during the three-dimensional back analysis of the in-situ stress field, the following two factors are considered. First is the self-weight stress. Second is the tectonic stresses, which can be divided into five parts: the compressive tectonic stress σ_{xx} in the horizontal direction x, the compressive tectonic stress σ_{yy} in the horizontal direction y, the uniform shear tectonic stress τ_{xy} within the horizontal plane, the uniform shear tectonic stress τ_{yz} within the vertical plane perpendicular to the direction x and the uniform shear tectonic stress τ_{xz} within the vertical plane perpendicular to the direction y. Tectonic stresses is very important to numerical analysis of the in-situ stress field distribution in deep river valley, and should be superimposed the gravity stress. Furthermore, river erosion should be considered in the numerical analysis process.

The back analysis method for in-situ stress field presented in this paper is mainly based on the test data. Next, a three-dimensional finite element model is established, considering the role of self-weight stresses and tectonic stresses, and using multi-regression analysis method for the back analysis of the in-situ stress field. The back analysis process is as follows: (1) according to the geological engineering conditions in the computation region, establish a three-dimensional finite element model; (2) assume values for undetermined parameters that may influence the initial stresses field (for example gravity and tectonic movement). For each undetermined parameter, the stress value of all test points can obtained by the numerical simulation method, and the multiple regression equation between the computed result obtained by each undetermined parameter and the test values of in-situ stress field can be established; (3) use the least-squares method, invoking the principle of minimum residual sum of squares between the computed values and test values, to obtain the optimal solution for the respective variable coefficients of the multiple regression equation; and finally (4) simulate the in-situ stress field in the three-dimensional numerical model.

3.2 *Numerical back analysis process*

In the three-dimensional finite element method, the in-situ stress field of the numerical model is a function of the following variables:

$$[\sigma] = f\left(x, y, z, E, \mu, \gamma, \sigma_g, \sigma_{xx}, \sigma_{yy}, \tau_{xy}, \tau_{xz}, \tau_{yz}\right) \tag{1}$$

where $f(\cdot)$ is a nonlinear function; $[\sigma]$ is the initial stresses field, containing the six stress components in the three-dimensional problem; x, y and z are the spatial location co-ordinates; E, μ and γ are the elastic modulus, Poisson's ratio and bulk density of the rock mass, respectively; σ_g is the self-weight stress; and σ_{xx}, σ_{yy}, τ_{xy}, τ_{xz} and τ_{yz} are the five tectonic stress components.

An elastic constitutive model is applied in the numerical simulation of the in-situ stress field by the three-dimensional finite element method. Generally, the self-weight stress σ_g and tectonic stress components (σ_{xx}, σ_{yy}, τ_{xy}, τ_{xz} and τ_{yz}) are influenced by the elastic modulus (E), Poisson's ratio (μ) and bulk density (γ) of the rock mass, so that:

$$\left[\sigma_g, \sigma_{xx}, \sigma_{yy}, \tau_{xy}, \tau_{xz}, \tau_{yz}\right] = f(E, \mu, \gamma) \tag{2}$$

where $f(\cdot)$ is a nonlinear function.

From the three-dimensional finite element method, the in-situ stress field at each point can be obtained:

$$\begin{aligned}[\sigma] = {} & m_1\sigma_g + m_2\sigma_{xx} + m_3\sigma_{yy} \\ & + m_4\tau_{xy} + m_5\tau_{xz} + m_6\tau_{yz} + \cdots + \varepsilon_k\end{aligned} \tag{3}$$

where m_i $(i = 1, 2, \ldots, n)$ are the regression coefficients and ε_k is the observation error; ε_k is a random variable.

3.3 *Multiple regression analysis*

According to the principles of multiple regression analysis, the value of $\tilde{\sigma}_{jk}$ determined in the regression is a dependent variable, and the computed value of σ^i_{jk} is an independent variable corresponding to the self-weight stress and tectonic stresses under the control of various sub-factors obtained by the finite element method. The multiple regression equation is therefore written as:

$$\tilde{\sigma}_{jk} = \sum_{i=1}^{n} m_i \sigma^i_{jk} \tag{4}$$

where k is the monitoring points index; $j = 1, 2, \ldots, n$ are the numbers of the stress components, corresponding to the stress components of n, respectively; $\tilde{\sigma}_{jk}$ is the multiple regression calculation value of the stress component j at monitoring point k; m_i is the multiple regression coefficient corresponding to the independent variable; σ^i_{jk} is the numerical simulation result of the stress component j at the monitoring point k under the load pattern of i; and n is the load pattern number of the self-weight stress and tectonic stresses.

Because the tectonic stresses in each direction are compatible, the introduction of one factor will lead to a weaker relationship with the remaining factors, so the multiple correlation coefficients and partial correlation coefficients of the tectonic stresses in each direction need to be verified. Next, the multiple regression correlation coefficients are significance tested, and the insignificant factors are then removed. The multiple regression is then repeated without these insignificant factors, and the distribution of the in-situ stress field for the three-dimensional numerical model can be obtained.

4 CASE STUDY

In this section, the test results of the in-situ stress field at the Jinping I Hydropower Station are described. The test results and monitoring data will be used in the following section for the back analysis of the in-situ stress field. Finally, comparison analysis is carried out for the back analysis results and field test results.

4.1 *Background and geological conditions*

Jinping I Hydropower Station is located at the Liangshan autonomous prefecture, Sichuan province, southwest China. It's an important Hydropower Station located at the Yalong River. The installed capacity of power station is 3600 MW, with a height of 305 m arch dam. The dam site region is located at the Qinghai-Tibet Plateau to the Sichuan Basin, the terrain performance of mountains and canyons, a typical deep "V" shaped valley, and the relative elevation is about 1000 m. Figure 3 shows the topographic and geomorphic conditions of the dam site region at the Jinping I Hydropower Station.

As shown in Figure 3, the slope inclination is mostly larger than 40°, the left bank is a reverse slope and the right bank is a dip slope. The main rock masses are thick-bedded sandstone and marble, Table 1 shows the mechanical parameters of the main rock masses at the dam site region.

These above parameters are used for the numerical back analysis of the in-situ stress field of the dam site region at Jinping I Hydropower Station. In this region, because of the strong tectonic effect the river erosion effect, the stress distribution characteristics is very complicated and stress concentration is happened at the bottom of river bed and middle of slope.

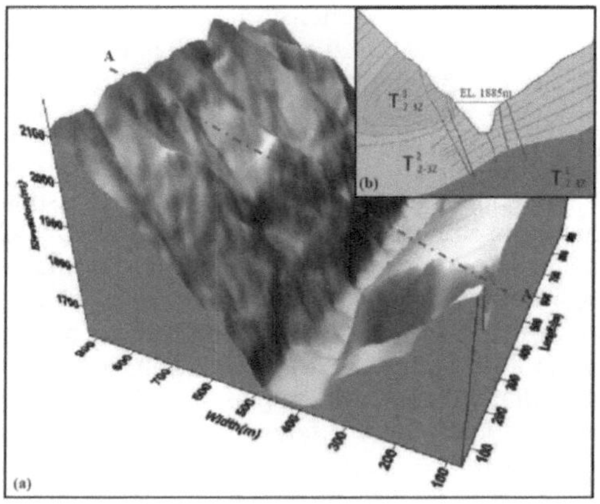

Figure 3. Topographic and geomorphic conditions of the dam site region at the Jinping I Hydropower Station: (a) three-dimensional visualization of the dam site region and (b) main cross section located at the dam site.

Table 1. Mechanical parameters of the main rock masses at the dam site region.

Rock mass	E (GPa)	v	γ (kg/m^3)	c (MPa)	φ (°)
Weak weathered	11.00	0.250	2700	1.50	46.94
Strong weathered	3.00	0.300	2700	0.80	38.66
Faults	1.00	0.300	2400	0.40	30.96

4.2 *Test results of the in-situ stress field*

As an in-situ stress field test in the project region, the in-situ stress field relief method was carried out at the two banks of slope. There are 17 points in right bank and 7 points in left bank of slope. Test results show that the distribution of maximum principal stresses is consistent with the regional geological conditions. At large depths, the natural stress field is mainly composed of the horizontal tectonic stresses, and the value of the in-situ stress field increased with increasing depth, but did not follow a simple linear relationship. Figure 4 shows the field test results of the in-situ stress at the dam site region.

As shown in Figure 4(a), the test points are mainly located at the elevation of 1650–1830 m, the horizontal depths are located at the range of 40–500 m. The principal stresses are increased with the increasing of horizontal depth (less then 350 m), when the horizontal depth is larger than 350 m, the in-situ stress is basically equal to the gravity stress. The maximum principal stress can reach to 35.7 MPa: when the horizontal depth is

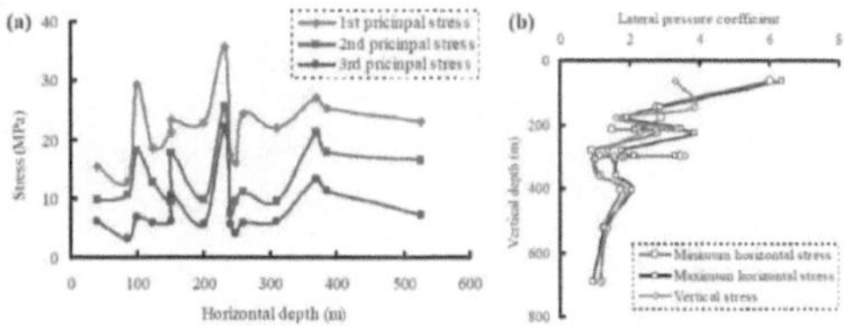

Figure 4. Field test results of the in-situ stress at the dam site region: (a) principal stresses varied with the horizontal depth and (b) lateral pressure coefficient varied with the vertical depth.

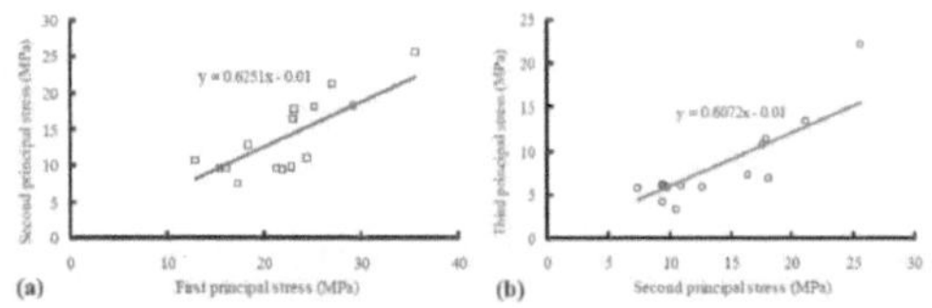

Figure 5. Relationship between the different principal stresses: (a) first principal stress vs. second principal stress and (b) second principal stress vs. third principal stress.

less than 100 m, the first principal stress is about 12.96–15.42 MPa; when the horizontal depth is about 100–350 m, the first principal stress is about 16.13–35.70 MPa; when the horizontal depth is larger than 350 m, the first principal stress is about 23.02–27.11 MPa. As shown in Figure 4(b), the lateral pressure coefficient is decreased with the increasing vertical depth, that's mean the in-situ stress is gradually equal to the gravity stress in the deep region. The in-situ stress field in the shallow depth is influenced by the river erosion and tectonic effect obviously. Figure 5 shows the relationship between the different principal stresses.

As shown in Figure 5, there is an approximate fitting relationship between different principal stresses, can be described as follow:

$$\frac{\sigma_2}{\sigma_1} \cong \frac{\sigma_3}{\sigma_2} \cong 0.61 \sim 0.63 \tag{5}$$

The second principal stress is about 0.61–0.63 times of the first principal stress, and also the third principal stress is about 0.61–0.63 times of the second principal stress.

4.3 *Numerical simulation models*

The numerical model is considered the river erosion process during the back analysis of in-situ stress field in the deep river valley region. Figure 6 shows the numerical simulation of the river erosion process.

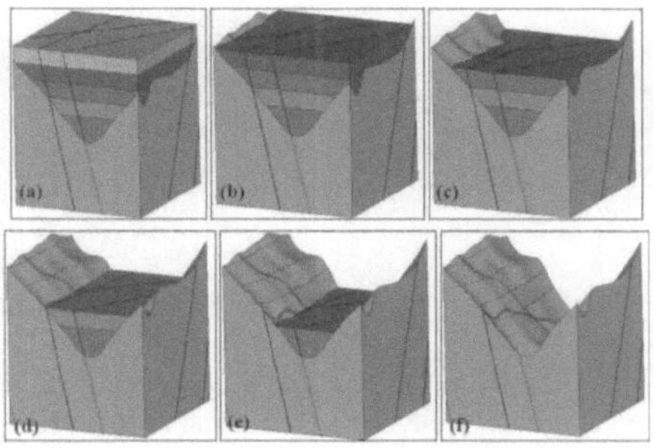

Figure 6. Numerical simulation of the river erosion process: (a) step 1, the ancient terrain; (b)–(e) different river erosion stage and (c) today's valley topography.

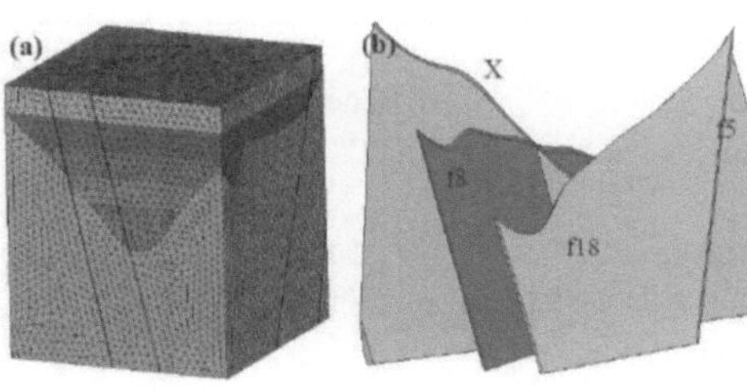

Figure 7. Numerical model for the back analysis of in-situ stress field: (a) mesh of the numerical model and (b) main faults considered in the numerical model.

As shown in Figure 6, firstly, a horizontal plane is existed at the ancient stage, after many years of river erosion, formed the deep river valley at the dam site region. The river erosion process is considered five cutting processes.

Figure 7 shows the numerical model for the back analysis of in-situ stress filed at the dam site region of Jinping I Hydropower Station. There are four faults are considered in the numerical model: X fault, f5 fault, f8 fault and f18 fault. The in-situ stress filed is influenced by these four faults obviously.

4.4 *Back analysis results and comparison analysis*

According to the back analysis method for the in-situ stress field described in Section 3, which is combined with the in-situ stress field test results in Section 4.2, the multiple regression analysis method was used in the back analysis of the in-situ stress field. In the multiple regression analysis, several measuring points were selected and divided into different combinations, and the best combination was identified by statistical methods. Figure 8(a) shows the numerical simulation results of the first principal stress, and Figure 8(b) shows the numerical simulation results of the third principal stress.

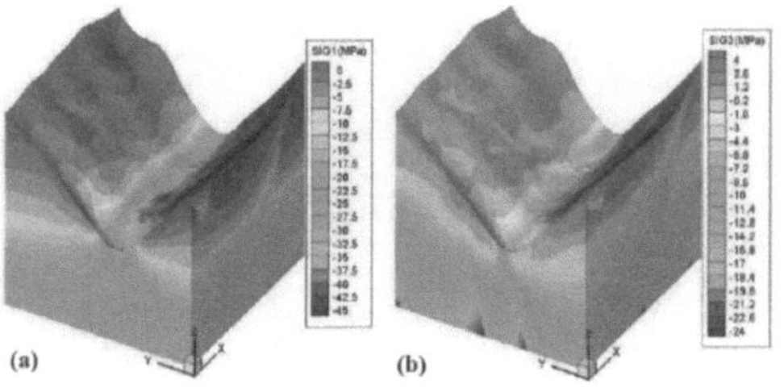

Figure 8. Back analysis results of the in-situ stress field: (a) first principal stress and (b) third principal stress.

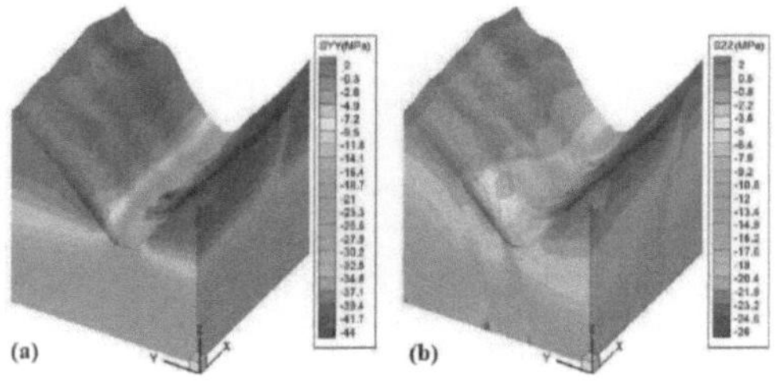

Figure 9. Back analysis results of the in-situ stress field: (a) maximum horizontal stress and (b) vertical stress.

Table 2. Comparison of the simulated results and field test data.

Monitoring points	σ_{xx} (MPa)	σ_{yy} (MPa)	σ_{zz} (MPa)	τ_{xy} (MPa)	τ_{yz} (MPa)	τ_{zx} (MPa)
$\sigma_{14\text{-}2}$						
Test data	15.109	11.404	23.407	–4.528	4.214	–4.201
Simulated result	13.211	14.368	22.578	–3.989	4.132	–4.421
$\sigma_{54\text{-}1}$						
Test data	6.579	18.436	12.066	–0.158	5.327	–1.136
Simulated result	6.211	16.231	11.562	–0.312	4.987	–1.212
$\sigma_{19\text{-}1}$						
Test data	9.099	12.328	17.023	–4.292	–6.319	2.466
Simulated result	8.564	13.214	15.369	–5.012	–5.689	2.721
$\sigma_{45\text{-}1}$						
Test data	14.758	14.106	12.526	–8.503	–3.180	2.913
Simulated result	12.967	15.687	12.867	–8.312	–3.231	3.021

As shown in Figure 8(a), the first principal stress is increased with the vertical depth, and stress reducing is happened at the shallow slope. The maximum value of first principal stress is about 45 MPa, appeared at the bottom of river bed. The third principal stress is influenced by the faults, and stress concentration is happened at the nearby of faults. Figure 9(a) shows the numerical simulation results of the maximum horizontal stress, and Figure 9(b) shows the numerical simulation results of the vertical stress.

As shown in Figure 9(a), the maximum value of the maximum horizontal stress is about 44 MPa, appeared at the bottom of the river bed. But the maximum value of vertical stress is only 26 MPa and appeared at the nearby of fault. The tectonic effect and river erosion are impact on the in-situ stress filed very obviously.

Table 2 shows a comparative analysis between the computational results and measured values of the in-situ stress field, revealing that the error between the in-situ stress field computed by back analysis and the test results is very small. The computed results show that the three-dimensional nonlinear finite element back analysis method proposed in this paper can be applied in real engineering applications.

5 CONCLUSIONS

This paper has presented a three-dimensional nonlinear finite element-based back analysis method for the in-situ stress field in a deep river valley region. The back analysis method for the in-situ stress field is a combination of multiple regression analysis techniques, the nonlinear finite element method and monitoring data for the in-situ stress field. Finally, the presented back analysis method is used to ascertain the in-situ stress field at the dam site region of the Jinping I Hydropower Station. The computed results of the in-situ stress field presented in this paper have demonstrated that the back analysis method can effectively determine the initial stresses field, and that the computed in-situ stress field is close to that of the test data. However, the computational efficiency of the back analysis method is poor, so that finding a faster back analysis method will be of high value.

ACKNOWLEDGEMENTS

This work is supported by the National Natural Science Foundation of China (51209156).

REFERENCES

Arai, R. 1984. An inverse problem approach to the predication of multi-dimensional consolidation behavior. *Soil and Foundation* 24: 95–108.

Brady, B.H.G. and Brown, E.T. 1985. *Rock mechanics for underground mining*. George Allen & Unwin Ltd, London.

Chen, S.H., Chen, S.F., Shahrour, I. and Egger, P. 2001. The feedback analysis of excavated rock slope. *Rock Mechanics and Rock Engineering* 34: 39–56.

Fakhimi, A., Salehi, D. and Mojtabai, N. 2004. Numerical back analysis for estimation of soil parameters in the Resalat Tunnel engineering. *Tunnelling and Underground Space Technology* 19: 57–67.

Feng, X.T., Zhang, Z.Q. and Sheng, Q. 2000. Estimating mechanical rock masses parameters relating to the Three Gorges Engineering permanent shiplock using an intelligent displacement back analysis method. *International Journal of Rock Mechanics & Mining Sciences* 37: 1039–1054.

Ferrero, A.M., Migliazza, M., Segalini, A. and Gullì, D. 2013. In situ stress measurements interpretations in large underground marble quarry by 3D modeling. *International Journal of Rock Mechanics & Mining Sciences* 60: 103–113.

Kaiser, P.K., Zou, D. and Lang P.A. 1990. Stress determination by back analysis of excavation-induced stress changes—A case study. *Rock Mechanics and Rock Engineering* 3: 185–200.

Kavanagh, K.T. and Clough, R.W. 1972. Finite element application in the characterization of elastic solids. *International Journal of Solids and Structures* 7: 11–23.

Li, G., Mizuta, Y., Ishida, T., Li, H., Nakama, S. and Sato, T. 2009. Stress field determination from local stress measurements by numerical modeling. *International Journal of Rock Mechanics & Mining Sciences* 46: 138–147.

Swoboda, G., Ichikawa, Y. and Dong, Q. 1999. Back analysis of large geotechnical models. *International Journal for Numerical and Analysis Methods in Geomechanics* 23: 1455–1472.

Yang, Z.F., Wang, Z.Y., Zhang, L.Q., Zhou, R.G. and Xing, N.X. 2001. Back analysis of viscoelastic displacement in a soft rock road tunnel. *International Journal of Rock Mechanics & Mining Sciences* 38: 331–341.

Zhou, J.W., Xu, W.Y., Tong, F.G., Chu, W.J. and Liu, X.N. 2007. Back analysis for the No. 2 diversion tunnel of Nuozhadu Hydropower Station by use of 3D nonlinear finite element method. *Chinese Journal of Geotechnical Engineering* 29: 1527–1535. (in Chinese).

E-Commerce, E-Business and E-Service – Lee (Ed)
© 2014 Taylor & Francis Group, London, ISBN 978-1-138-02646-9

Questionnaire generator: Web application using cloud computing

M. Hoshi & V. Klyuev
University of Aizu, Japan

ABSTRACT: There are many ways to develop Web applications. This paper focuses on Google App Engine with Java. We present a questionnaire application using Google App Engine. We discuss the key technologies used and details of algorithms implemented. We demonstrate the preliminary results of the evaluation of the application created.

1 INTRODUCTION

Nowadays, there are many programming languages to develop a web application; for example PHP, Ruby, Python, Java, and Go. Each language has different requirements for its environment, database, server, infrastructure, etc. A good development tool is a key to create a reliable solid application.

One example of developing web applications is using cloud computing. Cloud computing is a utility form using computer resources based on the Internet. Cavalcant (2012) noticed that this topic is becoming very famous in information technology. If we have a machine that connects to the Internet we can use many functions. According to Tsai (2010), there are three types of services:

- Software as a Service,
- Platform as a Service, and
- Infrastructure as a Service.

Software as a Service (SaaS) allows us to use application software through the Internet. Examples of SaaS are DropBox, Google Apps, etc.

Platform as a Service (PaaS) allows us to prepare applications using virtualized databases, servers, etc and we can run applications on the Internet. Example of PaaS are Google App Engine, Amazon S3, etc.

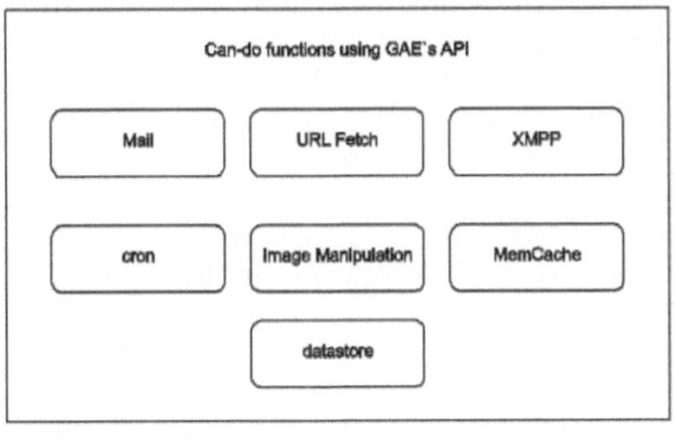

Figure 1. Can-do functions using GAE's API.

Infrastructure as a Service (IaaS) gives us the infrastructure that is based on a computer system to manage and set up through the Internet. Example of IaaS are Amazon EC2, etc.

Cloud computing has many advantages including low cost, virtualization, high scalability etc. Those features are very convenient for users and developers. For example, if the application grows, then the cloud computing will accommodate it.

This paper focuses on Google App Engine for Java. There are two reasons. Java is one of the most famous programming languages and is object oriented. Google App Engine (GAE) is a PaaS provided by Google in April 2008 and it can use technologies including data store, server etc. Furthermore, GAE can use JavaServer Programs (JSP) and Servlet technologies.

The GAE could use only Python at first in April 2008; in 2009, it was enhanced for use with Java; in 2011, it was upgraded for use with Go; and in 2013, it was upgraded for use with PHP. And GAE is a basic development web application using cloud computing. The GAE is free, but there are some limits. So, if we are managing big web applications we have to pay money. Figure 1 illustrates can-do functions using GAE's API. If the developer uses GAE's API, s/he can use many functions.

The purpose of this paper is developing a questionnaire application with a Google App Engine for Java.

2 RELATED WORK

There are many tools to use cloud computing. Some of them introduced by Yin (2010).

There are many ways to develop questionnaire applications. One example is using a relational database management system presented in DBPowder-web (2006). The data of that application were stored in a relational database.

Generally, web applications use Relational Databases (RDB) to store data: for example Oracle Database, MySQL, PostgreSQL etc. RDB's are databases based on the relational model.

A questionnaire application using the GAE uses another type of database that is based on NoSQL. This database can use data as objects. This questionnaire application uses cloud computing. So, we don't have to think about database storage limits.

3 ARCHITECTURE OF THE APPLICATION

Figure 2 illustrates the architecture of our questionnaire application.

At first, a user must log in to this application via a Google account. After that, a user can select three functions: creating a questionnaire, answering the questionnaire and seeing results.

These functions receive data and register data for GAE's data store. Creating a questionnaire phase registers the questionnaire data. Answering the questionnaire phase receives the questionnaire titles and registers the answer data. The seeing results phase receives the questionnaire titles and the answered data.

4 USAGE OF THE APPLICATION

4.1 *Login*

At first, we must logs in to this application. This application uses only Google account. Once a user is logged in that can select creating and answering a questionnaire and seeing result of questionnaire.

4.2 *Creating a questionnaire*

The administrator of the questionnaire must decide from among drop-down lists of two question types; checkbox type question and text type question. Within each question type, there are five questions to choose from. After deciding, the administrator is taken to the questions page. Here,

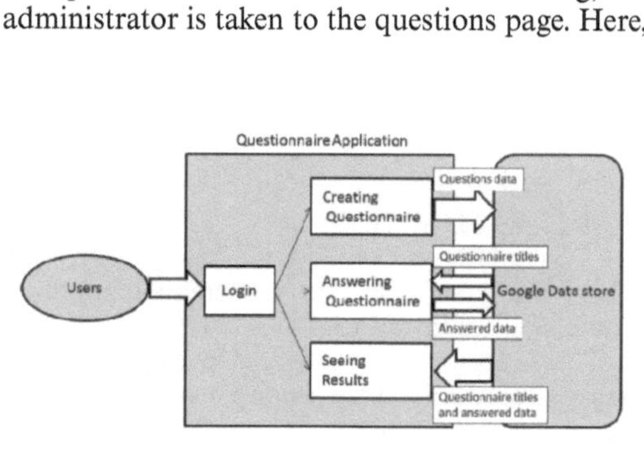

Figure 2. Architecture of the application.

four items are available: the questionnaire title, a textbox with the question type, and for the checkbox type question, four answers are displayed. Once the sdministrator is finished inputting, they submit the questionnaire. The data are then registered in GAE's database.

4.3 *Answering a questionnaire*

There is a list of questionnaire titles that comes from the GAE's data store. A user chooses from among this list of questionnaire titles. The user must create a key that the database recognizes as unique for that response. Once finishing, the user submits the response.

4.4 *Seeing results*

This phase is like the previous phase. We select the title that we want. The answers of textbox type question and the number of each checkbox type answer are displayed.

5 TECHNOLOGY

5.1 *OpenID*

One of the features of this web application is its login system which uses OpenID. OpenID presented by Roche (2009) is a user certification system that can use a Google account, a Yahoo! JAPAN account and a mixed account, etc. When opening a screen that needs certifying, one is redirected to the OpenID certification site. Once login is complete, the user is returned to the home page. GAE prepares a UserService Interface to use this OpenID system. This application can use only a Google account.

5.2 *JDO*

This application consists of two types of data class; the Question class and the Answer class.

The Question data class is to store questionnaire data for data store. There are three ways to read and write data to data store according to Nakagaki (中垣 健志) (2011): Java Data Objects (JDO), Java Persistence API (JPA), and low-level API. This questionnaire application uses JDO.

It requires care to choose which data class to use. GAE's data class cannot use two dimensional arrays. However, this application uses two dimensional arrays. If two dimensional data is stored, that data changes to a one-dimensional array.

A code presented below is an abbreviated sample of the Question data class code. @PersistenceCapable means a class is given PersistenceCapable annotation and is defined as data class. @PrimaryKey means this field's value is not duplicate in entity. So, we know this field's value data is determinable. This Question class's primary key is a title because if we know title, we can answer the questionnaire. @Persistent means this field should hold in data store.

```
@PersistenceCapable(
identityType=IdentityType.
APPLICATION)
public class Question {
 @PrimaryKey
 private String title;
 @Persistent
 private ArrayList<String>
 questionsOfTextbox;
 @Persistent
 private ArrayList<String>
 questionsOfCheckbox;
 @Persistent
 private ArrayList<String>
 answerOfCheckbox;
 }
```

The next code is the Answer data class. The primary key of this class is name. This is a special name. This code is similar to a Question class code.

```
public class Answer {
 @PrimaryKey
 private String name;
 @Persistent
 private String title;
 @Persistent
 private ArrayList<String>
 answersOfTextbox;
 @Persistent
 private ArrayList<String>
 answerOfCheckbox;
}
```

The GAE has PersistenceManager interface. This interface is what it takes to control a datastore. And this has some basic method to CRUD control; create, retrieve, update and delete. The GAE has PersistenceManagerFactory (PMF) class to use this interface. This class can reuse code from *other project.*

```
import javax.jdo.JDOHelper;
import
 javax.jdo.PersistenceManager
 Factory;
public final class PMF {
  private static final
  PersistenceManagerFactory
  pmfInstance=
   JDOHelper
   .getPersistenceManagerFactory
    ("transactions-optional");
  private PMF() {}
   public static
    PersistenceManagerFactory get() {
        return pmfInstance;
    }
}
```

This code is to register questions for GAE datastore. In this code, a new Question object is made in order to store that data

```
PersistenceManager pm =
PMF.get().getPersistenceManager();
 try {
  Question question =
  new Question(title,
  questionsOfTextboxArray,
  questionsOfCheckboxArray,
  answersOfCheckboxArray );
  pm.makePersistent(question);
  req.setAttribute("question",
question);
 }finally{
 pm.close();
}
```

6 ALGORITHM

6.1 *Top screen*

This code is written in XHTML and Java. <c:if test = "${not empty user}"> means if a user logs in to an application, it can see

```
<c:if test="${not empty user}">
 <a href="/createquestionnaire">
  Create Questionnaire</a><br />
 <a href="/answertop">
  Answer Questionnaire</a><br />
 <a href="/result"
  Seeing Questionnaire Result</
  a><br />
</c:if>
```

6.2 *Creating the questionnaire*

At first, it selects the number of each type of question via a drop-down list. And those data are stored in each variable.

```
<select name="numberOfCheckbox">
<c:forEach begin="0" end="5"
var="i">
<option
 <c:iftest="${i==numberOfCheckbox}">
  selected</c:if>>
<c:out value="${i }" /></option>
```

```
</c:forEach>
</select>
```

In the following code we set the title, each type of question and answers of checkbox type questions.

```
<input type="text" name="title"
 value="<c:out
 value="${title}"/>"/>
 <% for(int
 i=0;i<numberOfCheckbox;i++){%>
```

Input checkbox questions

```
<%= i+1%><input type="text"
 name="questionsOfCheckbox"
 value="<c:out
 value="${questionsOfCheckbox[i]}"
 />"/>
```

Input answers of question

```
 <% for(int j=0;j<4;j++){%>
  <%= j+1%><input type="text"
  name="answersOfCheckbox"
  value="<c:out
  value="${answersOfCheckbox[i]
  [j]}"/>"/>
 <% }%>
<% }%>
```

Input textbox questions

```
<% for(int
i=0;i<numberOfTextbox;i++){%>
<%= i+1%>
 <input type="text"
  name="questionsOfTextbox"
  value="<c:out
    value="${questionsOfTextbox[i]
    }"/>"/>
<%}%>
```

6.3 *Answering the questionnaire*

At first, the questionnaire titles from data store are automatically displayed. We select questionnaire title and input that title.

```
<%
  PersistenceManager pm =
     PMF.get().
     getPersistenceManager();
  String query =
   "select from "
   + Question.class.getName();
  List<Question> questionnaires=
   (List<Question>) pm.
   newQuery(query).execute();

  for(Question q: questionnaires){
   out.println(q.getTitle());
%>
<br />
<% }%>
```

Questionnaire data are displayed via a textbox or checkbox. This application uses radio buttons in checkbox type questions

6.4 *Seeing results*

At first, the program works to display questionnaire title and selects title that you want to see. All answer data that accord chose title gets from GAE's datastore. Then, it counts checkbox type questions answer.

Finally, it displays all data.

7 EVALUATION

The functionality of this application were compared with the functionality of SurveyMonkey (2014). Results are presented below.

The five users answered a questionnaire created using this application. That questionnaire consists of one textbox type of question and two checkbox type questions. Evaluators are university students majoring in computer science. They have experience to fill out on-line questionnaires.

They suggested several improvements. The following are among them.

- This questionnaire application generates questionnaires of standard type. There are no special features available. These features should be incorporated.
- A user inputs a primary key when the user answers the questionnaire. That key should be made automatically using the user name and the questionnaire title.
- Graphical visualization of the results should be added.

Table 1. Comparison with SurveyMonkey.

Functions	SurveyMonkey	This application
Creating questions	Dynamically	Number of questions should be fixed in advance
Graphical representation of results	Yes	No
Access to the questionnaire results	Administrator	Any respondent

8 CONCLUSION

The Google App Engine is very good to develop web applications using cloud computing. There are many function available to create advanced applications.

The GAE has many advantages for developers. For example, it is not necessary to set up a database, server, there is no need to worry about versions system updates, etc.

A questionnaire application is one example of using functions and advantages of GAE. Cloud computing has a strong influence on information technology.

Basic functions of this questionnaire application work well. Several improvements should be done. The following are among them: adding graph, usability, etc.

ACKNOWLEDGEMENT

This work was supported by JSPS KAKENHI Grant Number 25330410.

REFERENCES

Cavalcant E., Almeida A., Batista C., Cacho N., Lopes F., Delicato F.C., Sena T., and Pires P.F. 2012. Exploiting Software Product Lines to Develop Cloud Computing Applications, SPLC'12, pp. 179–187.

Murakami T. 2006. DBPowder-web: Web Application Development Framework with RDBMS, DEWS2006, pp. 4A–o4, (in Japanese).

Nakagaki K. 2011. Google App Engine for Java Programmers, Shoeisha Ltd, (in Japanese).

Roche K. & Douglas J. 2009. Beginning JavaTM Google App Engine, Apress.

SurveyMonkey. 2014. On-line document. Available at www.surveymonkey.com.

Tsai, Sun X., Balasooiya J. 2010. Service-Oriented Cloud Computing Architecture, 2010 Seventh International Conference on INTG, IEEE, *doi:10.1109/INTG.2010.214*.

Yin H., Han H., Liu J., Hongyun X. 2010. Development and Research of Multimedia Courseware Sharing Platform Based on GAE, 2010 International Conference on EDT, IEEE, doi:*10.1109/EDT.2010.5496366*.

E-Commerce, E-Business and E-Service – Lee (Ed)
© 2014 Taylor & Francis Group, London, ISBN 978-1-138-02646-9

Safety analysis and risk aversion system structure for internet of things

Qiang Wu
Feixian College, Linyi University, Feixian, Shandong, China

ABSTRACT: Internet of things stirs up another new wave in the information industry following computer, internet and mobile communication, which blazes an extensive market for IT and communication. However, it is now faced with serious security issues in terms of information and privacy, so it is necessary to make breakthroughs in core technology and build a comprehensive risk aversion structure from the perspective of uniform protocol and law.

Keywords: internet of things; security model; kill RFID

1 INTERNET OF THINGS AND ANALYSIS OF ITS PRINCIPLE

1.1 *Definition of internet of things*

The concept of internet of things was first put forth by MIT Auto-ID center in 1999, with the basic idea to connect RFID and other sensors to form distributed network of RFID structure. The core and basis of the internet of things is internet, whose client end extends and expands to information exchange and communication between any things. As defined by the European technology platform on smart systems integration (EPoSS), internet of things refers to the internet which connects any article with internet through Radio Frequency Identification (RFID), infrared sensor, global positioning system, laser scanner and other information sensing devices, to realize intelligent identification, positioning, tracking, monitoring and management.

1.2 *Technology platform and principle*

Key technologies of internet of things include Radio Frequency Identification (RFID), infrared sensor, global positioning system, laser scanner and other information sensing devices, EPC technology, internet technology, Physical Markup Language (PML) development technology, sensor network technology, embedded technology, and information security technology.

As for ordinary people, these professional technologies make it hard to understand the internet of things. Actually, it is no magic. It mainly relies on the technology called RFID to operate. Card swiping system for city bus and non-stop toll collection system at the highway toll gates we often see are using RFID system. With the help of RFID, the internet of things connects all the things in the world, and these things can sense each other. The internet of things also processes smartly huge amount of information, integrates the current internet and realizes in the real sense the communication between things.

European committee holds that the internet of things is an integral part of future internet and it can be defined as a dynamic global network basis. Based on the standard and inter-operated communication protocol, the "thing", whether physical or virtual, has identity, physical property and virtual characteristics, possesses the self-configuration ability and uses smart interfaces, thus making it possible to be integrated seamlessly into the information network. The cost to monitor and control one thing practically and in real time is extremely high. However, all the things can be found in the virtual world which can be monitored and controlled at lower cost. As for a real thing, its required applications and operations can be realized only through application and operation of the corresponding virtual things in the virtual world.

2 THE SECURITY ISSUES OF THE INTERNET OF THINGS

2.1 *Information security technology*

"There is no application without security and no development without application", and this is the destiny of the information system and is particularly true for wireless communication. The internet of things is the application of communication technology and normalization of things, as well as the internet linking informative things. Once the internet of things system suffers the information security attack, not only the obtained data and information become meaningless, but also they become harmful and even lead to the collapse or paralysis of system and plunge human society into chaos. The information security technology of the internet of things mainly involves such aspects as intrusion detection, information encryption, malicious nodes identification and eradication.

2.2 *Personal privacy leakage*

A complete internet of things is an inclusive internet, including economic management, production operation, social management, and even personal lives with different political beliefs, nations, cultures and interests. In developed internet of things, the things owned by any person and individual can be connected to the internet and sensed at any time and any place. In such environment, how to ensure the safety and privacy of information and avoid the loss or misappropriation of personal information, business information and property is the issue that every citizen in the internet of the things shall pay attention to. Besides, even through the information is not disclosed, people with higher management authority of internet of things still can know all about you: where you are, who you are with, what you are doing, etc. You are placed in an environment where you are monitored 24 hours a day and have no place to hide or any privacy.

2.3 *Local security issue of machine/sensing nodes*

Since the application of internet of things can replace human beings to finish some complicated, dangerous and mechanical work, the machine/sensing nodes of the internet of things are placed mostly in the environment without human monitoring. Therefore, the attacker can easily contact these devices and cause damage to them, and even replace the hardware/software of the machine through local operation.

2.4 *Security issue of application and management*

Since the internet of things is deployed before it is connected to internet and the nodes of internet of things are not guarded by people, how to conduct remote subscription information and application information allocation for equipment of internet of things is a big issue. Meanwhile, the huge complicated platform of internet of things definitely needs a powerful and uniform safety management platform, so how to manage the security information like logs of machine of internet of things becomes a new issue, which can sever the trust relationship between network and application platform and arouse a new round of security issue.

3 RISK AVERSION SYSTEM STRUCTURE OF INTERNET OF THINGS

3.1 *Core technology aspect*

Security model is established by referencing to mature security model of internet, including adopting Kill safety electronic tag, Faraday net, shield interference and other methods; solution of effective resistance to physical capture and security key encryption distribution technology are designed to prevent unauthorized identification and attack and isolate information interception and malicious tracking and to ensure the security of private information and data of the internet of the things.

The electronic tag consists of coupling element and chip, which is the "identification card" in the internet of the things: A large amount of important information is contained. Due to the lack of point-to-point encryption support and PKI security key exchange function, in the application of the internet of things, the attacker can obtain and make use of the content on the RFID tag. Kill tag is able to perfectly prevent the scanning and tracking. Faraday net is the container formed by metal net or metal foil which cannot be penetrated by radio signal, thus it can be used to protect human privacy.

A feasible solution which combines distributed capture detection and security key distribution shall be formulated: This solution will implant network intrusion detection technology to check closely each node, improve the functions of data security, access control, and privacy protection at the client end and resist the complicated attack. Meanwhile, during the transmission, information server for remote things performs nested encryption for thing information with the public key from the back to the front following the order of nodes in the set route; the encrypted data will be decoded once when passing each route node until the data finally gets to the local information server where the thing information is restored to plain text. Each route node in the transmission can verify the completeness of the received data and the authenticity of the forward path, thus ensuring that no key information or data is leaked to the network after the physical facility in the internet of the things is captured and replaced in the network, which can greatly reduce the possibility of nodes pretending communication after capture.

3.2 *Other technical measures*

Analyzing malicious commands to the transport layer and establishing a disaster recovery mechanism timely. Taking backup measures on secret documents and mobile device files. Cipher text queries, data mining and security-related, secure multi-party computation, secure cloud computing technology etc.

3.3 *Establish a set of uniform technology protocol and standard*

The internet of things is to use the sensor and embedded chip on the thing to transmit and receive the information of the thing, realize the local processing through sensor network and connect to the internet of things. The reliable transmission between tag and reader is realized mainly thorough relevant protocols. Since information reading among different sensor networks is involved, a set of uniform technology protocol and standard concentrated on internet and sensors is necessary. For example, it is wise to establish security addressing protocol which realizes the longitudinal transmission of identity authentication results from the application layer to network layer and identify the access validity of different stages in addressing process.

3.4 *Legal system guarantee and reliable security management*

It's very necessary to formulate relevant laws and regulations of the internet of things, because these laws and regulations will provide a clear judicial interpretation and detailed rules on the various safety violations and disciplinary measures. Strict management system should be drawn up, which can make responsibilities and rights of Internet of things clear in safety management. In addition, normative behavior will avoid mistakes and omissions. To improve safety awareness of users, providing safety training to users and let them learn how to prevent various security issues and solutions. All the "things" in the internet of things are connected to the global network and communicate with each other, which also bring new safety and privacy issues, e.g. credibility, authentication and fusion of data sensed or exchanged by things. The privacy of people and things should be protected effectively. Therefore, it's very important to establish security hazards and risk evaluation mechanism, and system improvement shall be improved.

4 THE PROSPECT OF THE INTERNET OF THINGS

The fundamental impetus and final target of the development of the internet of things are its application in practice to serve people. In that case, our focus should be on its development and application when we are now in the prospering research trend of the internet of things. Otherwise, it would make no sense. But we need to be aware that it is its safety that is a key for the internet of things. Once the system of the internet of things applied is attacked, the users will suffer damage on their interest through the attack on their information caused by the breakdown or collapse of the system.

The internet of things may embed sensors into various systems of things, such as systems of banks, grids, transportations, constructions, oil and gas pipelines, water supply. Thus, the internet of things and information exchanges is formed, and additionally with the integrity with the internet, all things could be sensed and controlled remotely, exchange information with the net it belonged to directly an indirectly, which actually build up a smarter system for production and living. As a consequence, the production efficiency is promoted, and life is more convenient. However, the above is all based on the security on the information and technology of the internet of things. Therefore, regardless of the application background of internet of things, the priority lies in the design on its information and technology safety, especially that of the detection of intrusion, cording of information, recognition and cast of malicious nodes while constructing its application system. For instance, in banking system, online users should be highly vigilant and always mind the security vulnerabilities when paying for certain services. Otherwise, the debugging information of the process of payment could be leaked to hackers when we initiate the debugging of the server for payment service. And because of the loophole, the banking log of secure payment could be downloaded traversed and downloaded, which makes much information of bank cards leaked, like cardholder's name, ID card number, CVV codes, Bin of cards (the first six numbers of a bank card). Traversal means the process of visiting each node, exactly once, in turn in a certain route. In face of this problem, banks need to forecast the potential occurrence of the phenomena alike, do well in each step of the prevention, improve the payment service system when initiating the payment service. For one thing, all records of user's payment would be saved in a text whenever the server needs debugging; for another, the server does not strictly undergo security collocation of baseline, and there exist bugs of catalogue traversal. Therefore, all debugging information during the process of payment is subject to hackers, which will cause the damage of user's property. Admittedly, bugs cannot be prevented completely. What we must aware of is that, however, we should fix them immediately after they emerged, so that the damage is minimized. And if the bug caused property damage on the user, the bank and related service supplier(s) ought to compensate.

In short, although the prospect of internet of things is very wonderful, its development is faced with many difficulties, including backward institution, imperfect technology, huge cost, lack of standards, obscure business model, paralyzed information security, barren address resources and so on. To promote its development in application, therefore, it calls for the involvement of all fields, and particularly the government's leading role as well as the helpful policies, regulations and laws. We are looking forward to the internet of things and the greater wisdom and value it will bring along with.

5 CONCLUSIONS

The huge and complex internet of things requires for a comprehensive and unified management platform. In terms of technology and management, internet of things has many potential security problems. For example, the wireless transmission signals of internet of things can be easily stolen and interfered. Information security and privacy is easily damaged and revealed. Therefore, we should take the following measures: We should put more efforts in the development of internet and establish complete and highly-effective information expressway to ensure the vastness of the internet of things and the smoothness of the path. We should delve into and perfect the key technology in internet of things, improve the internet's own ability to recognize and block the inappropriate information and enhance its ability to defend hackers' invasion so as to ensure information security and protect personal information as well as property privacy. In addition, we should formulate standard agreement and policies for the management of internet of things so that privacy protection and synchronization of information security can be facilitated. Furthermore, we should explore to discover reasonable and highly effective modes applicable to business. We should adhere to human orientation and construct an intelligent globe, thus providing service for the whole human kind with technology.

REFERENCES

CASAGRAS Final Report: RFID and the inclusive model for the Internet of things. 2010.

Commission, I.D.E., Internet of things Strategic Research Roadmap. 2009.

Feng Songlin & Ye Tianchun, Initial Study on Internet of Things/Sensor Network [J]. *Post-IP Age and Internet of Things,* 2010(1):10–13.

ITU Internet Reports 2005: The Internet of Things. 2005, ITU.

Li Wenqing & Guo Zongliang, Overview of Growth and Development of Internet of things [J]. *Network and Application,* 2010(2):27.

Meng Xiangru & Zhang Jingang, Countermeasure Analysis of Application of EPC and Internet of Things in China [J]. *Modern Logistics,* 2009(1):41–42.

Zhu Zhongying, Intensity and Trend of Sensor Network and Internet of things [J]. *Microcomputer Application,* 2010(1):1–3.

E-Commerce, E-Business and E-Service – Lee (Ed)
© 2014 Taylor & Francis Group, London, ISBN 978-1-138-02646-9

A design of conceptual framework for persuasive multimedia learning application in enhancing children's awareness of child sexual abuse

Azliza Othman & Wan Ahmad Jaafar Wan Yahaya
Centre for Instructional Technology and Multimedia, Universiti Sains Malaysia(USM), Pulau Pinang, Malaysia

ABSTRACT: This paper presents a design of conceptual framework for persuasive multimedia learning application in educating children about child sexual abuse. Currently, children are not fully aware that it is wrong for an adult to abuse and neglect his or her body. They by nature are particularly vulnerable to all types of abuse because of their temporary helplessness and dependence on adults and their lack of knowledge about danger and self-protection. In such situation, having an appropriate knowledge and skill is seen as necessary to help children to avoid from sexual abuse. Basically, the conceptual framework design in this study integrated persuasive technology and multimedia learning in providing knowledge about child sexual abuse and afford guidance to children as a skill to protect them from sexual abuse. The implementation of this conceptual framework is important as it will map and integrate all the relevant theories, persuasive principles and multimedia principles together to form a concrete pathway for the development of persuasive multimedia learning application.

Keywords: child sexual abuse; persuasive principle; multimedia learning theory; awareness

1 INTRODUCTION

Child abuse has raised worldwide attention in the recent years. Among all types of child abuse, sexual abuse gives a great negative impact to the children during their childhood in physical, emotional and psychological context. Sexual abuse victim faces the greatest trauma in life than other type of child abuse Darshini and Teoh (2009) and Nor Maizura and Salwana (2010). This serious social problem requires effective prevention strategies. Even though countless work has been done by researchers around the world to promote the awareness, prevention and treatment of sexual abuse, however work in the area of prevention remains inadequate (Finkelhor, 2009).

In response to the growing public and professional concerns on child abuse and other crimes against children, schools and other community organizations began to develop and implement programs aimed at helping children to avoid and to report sexual abuse and other victimizations (Finkelhor and Leatherman, 1995). Presently, the prevention attempts have focused primarily on educating children about sexual abuse Renk et al., 2002). According to Rebecca (2003) for prevention program at reducing CSA to be effective, children must be able to provide skills and education. Therefore, skills presented to children must be broad enough in covering the spectrum of potential abuse situations, yet must be specific enough so that children feel competent in using the skills in the potential abuse situations. The evaluations of sexual abuse prevention programs have proven some increase in children knowledge. Unfortunately, the extent to which the knowledge will influence children's behavior and how long this knowledge will be retained remains unclear (Rebecca, 2003).

Current practice in Malaysia, program in schools that educate children in a child-friendly way about bad touches and preventing themselves from perpetrator is far lacking. The child is not fully aware that it is wrong for an adult to abuse and neglect his or her body. As a result the child suffers in silence (UNICEF Malaysia, 2010). According to Hitrec (2011), learning about sexual abuse prevention is necessary, but it is not enough to ensure children's safety. Hitrec also added that children by nature are particularly vulnerable to all types of abuse because of their temporary helplessness and dependence on adults and their lack of knowledge about danger and self-protection. Prevention of this type of abuse and teaching about it are the responsibility of adults. Rebecca (2003) also supports that child abuse prevention programs are, for the most part successful at teaching children concepts about sexual abuse and skills in self protection.

With regard to this, children should be exposed to appropriate educational programs so that they are aware of any situations that might negatively affect their personal safety and know the appropriate actions require when these situations arise. So, the lack of knowledge and awareness among children is the main stumbling block that needs to be tackled by the researcher. Applying Persuasive technology integrated with multimedia learning could be help in increasing children's awareness of any situations that might negatively affect their personal safety and act appropriately when these situations arise.

The rest of the paper is structured as follows. First, a brief literature analysis is provided. Second, theories and principle involved are described. Then, a theoretical framework of the research is explained. Finally, some concluding remarks are made and future directions of the research are suggested.

2 LITERATURE REVIEW

Literature analysis focus on three specific aspects which are important to address the research concern: child sexual abuse, the use of persuasive technology and multimedia learning.

2.1 *Child sexual abuse*

Child Sexual Abuse (CSA) is defines as inappropriately exposing or subjecting a child to sexual contact, activity, or behavior (Prevent Child Abuse America, 2005). It was found that most sexual abuse offenders are acquainted with their victims; approximately 30% are relatives of the child, such as brothers, fathers, mothers, uncles or cousins; around 60% are other acquaintances such as friends of the family, babysitters, or neighbours; strangers are the offenders in approximately 10% of CSA cases. Based on National Child Abuse Statistic (2011), over 90% of CSA victims knew their perpetrator in some way and 68% of them are abused by family members. In Malaysia, from the cases reported to Social Welfare Department in 2008, 72% from 733 cases were incest cases (UNICEF Malaysia, 2010).

Sexual abuse will impacts negatively to the children during their childhood and future adulthood in physical, emotional, and psychological context (Nor Maizura and Salwana, 2010). Studies found that the victims of the CSA are traumatized greatly in life compared to other types of child abuse. Darshini and Teoh (2009) adds that a child who is abused will be traumatized and will need counselling and professional intervention by a child psychologist or psychiatrist. It is important because when a child is sexually abused they are scared, frightened, and do not know what to do. In addition, children who are sexually abused often do not tell anyone that they have been sexually abused, because they feel ashamed and guilty. Also, they are afraid of causing the perpetrator angering, or because the perpetrators threat them to stay quiet, or convincing the children that no one will believe them. This can often lead for the abuse to last for years.

2.2 *Persuasive technology*

In recent years, technology has been used to persuade users to change their attitudes and behavior. This field of research has been termed "Captology". Fogg (2003) define captology as design, research and analysis of interactive computing product intentionally designed for the purpose of changing people's attitude or behaviors. This new area of study explores the overlapping of persuasion (increasing awareness, influence, motivation, behavior change, etc) and computing technologies. This application of persuasion strategy by means of computing technology is defined as persuasive technology. Persuasive technology focuses on the computers and the persuasion, which are combined together into an assistive tool.

Fogg developed a functional triad for captology, which neatly people organized three different ways people respond to computer technology. Firstly, computer as a tool to persuade people by making some behavior easier or more efficient to do, or leading people through a process of performing calculations or measurements that motivates. Secondly, the computer as a medium can persuade people by allowing people to explore cause- and–effect relationships, or providing people rehearse a behavior. Thirdly, the computer as a social actor can persuade people by rewarding people with positive feedback, or modeling a target behavior or attitude or providing social support.

Currently, there are various researches that apply the persuasive technology approach in designing and developing applications with the intention to change a particular aspect of human behavior in a predefined way. Among the important current targets for persuasive technology is promoting healthy behaviour (Segertstahl and Kukkonen, 2007), environmental conservation and education (IJsselsteijn et al., 2006). Empirical results have shown that persuasive technology can change people's attitude and behavior to some extent (Cheng, 2003). This research extends the application of persuasive technology in influencing children behavior toward child sexual abuse and in increasing their awareness level of any dangerous situation that might impact their personal safety. It is believed that computer applications are now able to persuade as effectively as human (Canaday, 2004).

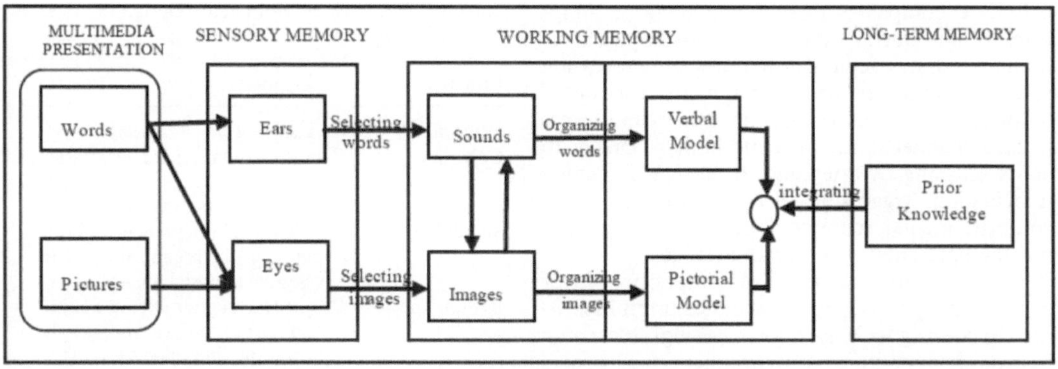

Figure 1. Cognitive theory of multimedia learning by Mayer (2001).

2.3 *Multimedia learning*

Multimedia can utilize the advantage of persuasive technology in connecting the sense of human to inform, persuade, and entertain (McCracker and Wolfe, 2004). The most effective multimedia can provides learning experience that real mirror the real world experiences and allows learners to apply the contents in different contexts (Shank, 2005). She also adds that multimedia certainly has the potential to extend the amount and types of information available to learners.

Multimedia learning refers to learning from words and pictures (Mayer, 2009). It is the process of learning, usually in a classroom or similarly structured environment, through the use of multimedia presentation and teaching methods. This can typically be applied to any subject and generally any sort of learning process can either be achieved or enhanced through a proper application of multimedia applications (Wisegeek, 2010). The process of multimedia learning can be viewed as information acquisition (in which multimedia messages are information delivery vehicles) or as knowledge construction (in which multimedia messages are aids to sense making). According to Mayer (2001) multimedia learning can occur when people build mental model from words (such as spoken text and printed text) and pictures (such as illustrations, photos, animation, or video). Mayer explains how we process information through two basic channels, verbal and visual (Fig. 1).

3 LEARNING THEORIES, PRINCIPLE AND DESIGN GUIDELINES

Next section will discuss about theories, principles and design guidelines used in designing persuasive multimedia learning application for increasing children awareness about child sexual abuse.

3.1 *Persuasive design principle*

The principles involve from persuasive theories are attractiveness, similarity, suggestion and simulation in real world context. The principle of attractiveness suggests that a computer technology that is visually attractive to target users is likely to be more persuasive as well. One of the most powerful persuasion principles in the area of psychological cues is similarity. The principle of similarity suggests that, in most situations, people who we think are similar to us in terms of personality, preferences, or in other attributes, can motivate and persuade us more easily than people who are not similar to us. A computing technology will also have greater persuasive power if it offers suggestions at opportune moments. Suggestion technology is defines as an interactive computing product that suggests a behavior at the most opportune moment. In order to guide children to act appropriately when facing any dangerous situation, the principle of suggestion is significant to be concern. If the children have the knowledge and they already practice the suggestion guideline, they are expected to be more aware and know appropriate action based on the suggestions provided. One category of simulation that is relevant to persuasive multimedia learning in this study is based on principles of simulated in real-world contexts. Example experiences that have provided to the children will design in the real situation. These will help children more understand the concepts of learning and guide them in making decision in their real life.

3.2 *Cognitive Theory of Multimedia Learning (CTML)*

Cognitive Theory of Multimedia Learning (CTML) describe on how people learn from words and pictures (Mayer, 2001). Figure 1 depicts cognitive theory of multimedia learning intended to represent the human information processing system. Mayer's

CTML encompasses three fundamental assumptions for multimedia design. These assumptions are humans possess separate channels for processing visual and audio information, humans are limited in the amount of information that they can process in each channel at one time and humans engage in active learning by attending to relevant incoming information, organizing selected information into coherent mental representations, and integrating mental representation other knowledge.

Referring to the third assumption, Mayer explains his philosophy for the meaningful learning. According to Mayer, for meaningful learning to occur in a multimedia environment, the learners must engage in five cognitive processes. Firstly, the learner selects relevant words for processing in verbal memory. Secondly, the learner selects relevant images for processing in visual working memory. Thirdly, organizing selected words into a verbal mental model. Then, organizing selected images into a visual mental model and lastly, integrating verbal and visual representations. In line with CTML, Mayer (2009) proposed twelve research-based principles for the design of multimedia application. Multimedia Design Principles explain how student learn better from the multimedia elements. Each of the principle is also suggested to be implemented in light of the CTML.

3.3 *Principle of multimedia design*

Various experimental studies had been carried out by Clack and Mayer (2008) and yielded twelve major principles that can serve as a guideline for developers to design a multimedia system capable of reducing the extraneous processing, managing essential processing and fostering generative processing as describes in Table 1. However, in this study, two principles which are personalization and voice principles were adopted for designing and developing the persuasive multimedia learning application. Since this application will involve an instruction and guidance to the children on how to avoid any dangerous situation, the text and narration use must design in a proper way. So, personalization and voice principle considered as the accurate option.

3.3.1 *Personalization principle*

Personalization principle describes that users learn better from multimedia lessons when words are in conversational style rather than formal style. The theoretical rationale for this principle is when learners feel that the author is talking to them; they are more likely to see the author as a conversational partner and therefore will try harder to make sense of what the author is saying. However, this principle may be most effective when it is not overdone and when the learners are beginner.

Table 1. Principle of Multimedia Design (Source: from Mayer, 2009).

Principle	Descriptions
Principles for reducing extraneous processing	
Coherence	People learn better when extraneous words, pictures, and sounds are excluded rather than included.
Signalling	People learn better when cues that highlight the organization of the essential material are added.
Redundancy	People learn better from graphics and narration than from graphics, narration, and on-screen text.
Spatial contiguity	People learn better when corresponding words and pictures are presented near rather than far from each other on the page or screen.
Temporal contiguity	People learn better when corresponding words and pictures are presented simultaneously rather than successively.
Principles for managing essential processing	
Segmenting	People learn better when a multimedia lesson is presented in user-paced segments rather than as a continuous unit.
Pre-training	People learn better from a multimedia lesson when they know the names and characteristics of the main concepts.
Modality	People learn better from graphics and narration than from animation and on-screen text.
Multimedia	People learn better from words and pictures than from words alone.
Principles for fostering generative processing	
Personalization	People learn better from multimedia lessons when words are in conversational style rather than formal style.
Voice	People learn better when the narration in multimedia lessons is spoken in a friendly human voice rather than a machine voice.
Image	People do not necessarily learn better from a multimedia lesson when the speaker's image is added to the screen.

3.3.2 *Voice principle*

In addition to the personalization principle, voice principle also consider as accurate preference in designing instruction to the children in this study. The goal of using this principle is to determine how to use aspects of the instructor's voice to foster a sense of social partnership in human computer interaction. The voice principle explains that learners learn more deeply when the narrations in multimedia lesson are spoken by a friendly human voice rather than by a machine voice.

3.4 *Constructivist learning environment*

In relation to cognitive thinking, effective learning environment is required to develop effective learning instruction. Constructivist Learning Theories holds the belief that knowledge is constructed not transferred from teachers. Therefore, under this environment, the learners are expected to explore the learning material by their own. Learners construct their own knowledge by looking for meaning and order; they interpret what they hear, read, and see based on their previous learning and habits (Driscoll, 2005). The understanding of this perceptible inconsistency is that constructivist learning depends on the learner's cognitive activity, not the learner's behavioral or social activity. The learner constructs his own conceptualisations and finds his own solutions to problems, mastering autonomy and independence. In constructivist thinking, learning is inescapably affected by the context and the beliefs and attitudes of the learner. Here, learners are given more latitude in becoming effective problem solvers, identifying and evaluating problems, as well as deciphering ways in which to transfer their learning to these problems. Learning is successful when students can demonstrate conceptual understanding. From constructivist learning theory, the meaningful learning occurs when learners actively select relevant information, organize it into coherent representations, and integrate it with other knowledge (Mayer, 1996; Mayer, 1999a; Wittrock, 1990 as cited by Mayer & Moreno, 2002). Therefore, accordingly, learning is most effective when learning materials is designed and presented based on their level of cognitive structure.

3.5 *Design guideline for children*

Since the persuasive multimedia learning application will be design for children as a user, the design guideline for children must be considered. In this study a Guidelines for Educational Websites for Children by Meloncon (2010) were chosen as a rule to refer. This guideline contains 10 elements of design guidelines for children and the elements are group in three categories that are navigation, search, appearance and content. The guidelines are target for children age 7 to 9 years old.

4 CONCEPTUAL FRAMEWORK

The formulation of the above learning theories, principle and design guidelines than were guide to implemented of the theoretical framework for this research. The theoretical framework explains the overall structure and expected outcome of the research. To ensure the study is systematically conducted, the theoretical framework of this study is based on a micro and macro design strategies.

Under the micro strategy, the combination of persuasive design principles were applied in order to strengthen researcher's believes on increasing children's awareness. This can be achieved by combining four persuasive principles which are attractiveness, similarity, suggestion and simulation in real-world context. All the persuasive principles applied were representing the functional triad for captology in persuasive strategies. For the first triad, suggestion principle was chosen as a tool for persuading children in making some behavior easier or more efficient to do. Then, simulation in real-world context was chosen as a medium that can persuade children by allowing them to rehearse a behavior. Lastly, attractive and similarity principles were chosen as a social actor that can persuade children by rewarding them with positive feedback, or modeling a target behavior or attitude or providing social support. This functional triad for captology neatly children in organizing three different ways they can respond to computer technology.

4.1 *Perceived motivation*

On the other hand, macro strategies are concerned with the selection, sequence, and the organization of the subject matter are to be presented. In this study, macro strategies absorb in this theoretical framework are merge Mayer's Cognitive Theory of Multimedia Learning (2001) and Constructivist Learning Theories. Additionally, macro strategies also occupy Meyer Design Principle (Mayer, 2009) and Design Guideline for Children (Meloncon et al., 2010) since target user of this study is children. Figure 2 depicted the conceptual framework of this study.

The expected outcome from this research will be an increasing of children awareness about child sexual abuse and perceived of children motivation while using the proposed persuasive multimedia learning application. The implementation of this conceptual framework is important as it will map and integrate all the relevant theories, persuasive principles and multimedia principles together to form a concrete pathway for the development of persuasive multimedia learning application. This could validate the researchers' hypothesis that certain persuasive theories and multimedia learning principles can increase children's awareness in this view.

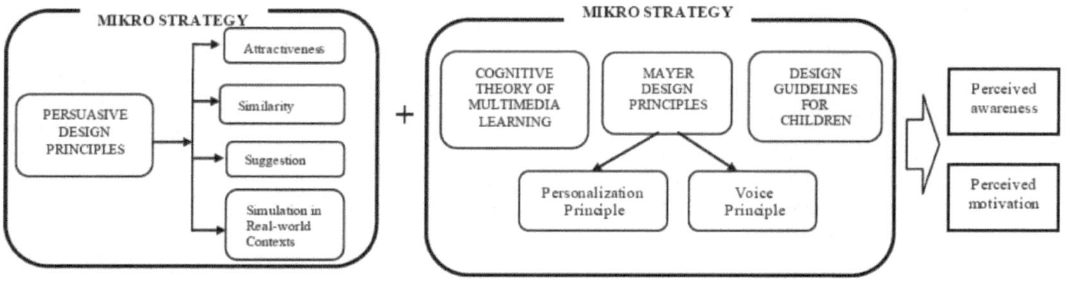

Figure 2. Conceptual framework.

5 CONCLUSION

Realizing the important impact in enhancing children awareness about child sexual abuse, this study tried to design a persuasive multimedia learning application that could help in providing basic knowledge about the dangerous and effects of child sexual abuse and skill to avoid and act appropriately if any dangerous situation happen to them. However, the success of implementing persuasive multimedia learning application will be based on carefully and specific preference of learning theories, multimedia principles and persuasive principles. It is hoped that our proposal of persuasive multimedia learning application will be a viable alternative approach for children to get information about child sexual abuse and prepare them to be aware and know appropriate action if any dangerous situation happened to them.

REFERENCES

Canaday, V.D. Persuasive Technology and Emotional Agents. Technical Report. Norfolk State University, 2004.

Cheng, R. Persuasion strategies for computers as persuasive technologies. Department of Computer Science, University of Saskatchewan. 2003.

Clack, R. and Mayer, R.E. E-learning and the science of instruction (2nd ed). Pearson Merill Prentice Hall. 2008.

Darshini, K. and Teoh E.S. 10,000 child abuse cases in four years. The Malay Mail. Available: http://stage2.mmail.com.my/node/18955. 2009.

Driscoll, M.P. Psychology of learning for instruction. Boston, MA: Allyn & Bacon Publisher. 2005.

Finkelhor, D. and Leatherman, J.D. Victimization Prevention Programs: A National Survey of Children's Exposure and Reaction. Child Abuse & Neglect. Interaction Design: beyond human computer interaction. John Wiley & Sons, Inc. USA, 1995. vol 19(2). 129–139.

Finkelhor, D. The prevention of Childhood Sexual Abuse. The future of Children, 2009. vol. 19(2).

Fogg, B.J. Persuasive Technology: Using Computer to Change What We Think and Do. Morgan Kaufmann Publishers. USA. 2003.

Hitrec. Teaching children to protect themselves from sexual Abuse. Protecting children from sexual violence. 2011.

IJsselsteijn, W.A. Kort, Y. de Midden, C. Eggen, B and Hoven E. van den (Eds). Persuasive Technology for Human Well-Being: Setting the Scene. Persusive 2006. Springer-Verlag Berlin Heidelberg. 2006, pp. 1–5.

Mayer, R.E. & Moreno, R. Aids to computer-based multimedia learning. *Learning and Instruction,* 12, 107–119. 2002.

Mayer, R.E. Multimedia Learning, Cambridge Universiti Press. 2001.

Mayer, R.E. Multimedia Learning. Cambridge Universiti Press. 2009.

McCracker, D.D. and Wolfe, R.J. User-Centered Web Site Development: A Human-Computer Interaction Approach. Prentice Hall Inc. New York, 2004.

Meloncon, L. Haynes, E. Varelmann, M. & Groh, L. Building a Playground: General Guidelines for Creating Educational Web Sites for Children. Technical Communication. 2010. vol. 57(4).

Noor Maizura, M.N. and Salwana, M. Intelligent Interpretation and Analysis of Child Sexual Abuse Forensic Evidence: a Preliminary Study. IEEE Journal, 2010. 624–629.

Prevent Child Abuse America, Prevent Child Sexual Abuse. Available: http://preventchildabuse.org. 2005.

Rebecca, M.B. Child Sexual Abuse: Prevention or Promotion? Social Work, vol 48(2), National Association of Social Workers Inc. 2003.

Renk, K. Lijequist, L. Steinberg, A. Bosco, G and Phares, V. Prevention of Child Sexual Abuse: Are We Doing Enough? Trauma, Violence, & Abuse. 2002. vol. 3, no 1. 68–84.

Segertstahl, K. and Kukkonen, H.O. Distributed User Experience in Persuasive Technology Environments. Persuasive 2007. Springer-Verlag Berlin Heidelberg. pp. 80–91. 2007.

Shank, P. The value of multimedia in learning. Adobe System Incorporated, USA. 2005.

UNICEF Malaysia. Penderaan Kanak-kanak di Malaysia. Available: http://www.uniteagaintabuse.my, 2010.

Wisegeek. What is Multimedia Learning? Available:http://www.wisegeek.com/what-is-multimedia-learning.htm, 2010.

Section 3: Computer science and technology

E-Commerce, E-Business and E-Service – Lee (Ed)

A review on recognition of traffic signs

Zhijia Zhang, Wenqiang Li, Dan Zhang & Wei Zhang
Technology Computer Vision Group, Shenyang University of Technology, Shenyang, China

ABSTRACT: Traffic sign recognition system, as part of the intelligent transportation, has attracted attentions from a growing number of researchers. In this paper, recognition algorithms for traffic signs are reviewed and the current research status on the recognition of traffic signs, both domestically and internationally, is introduced. Additionally, the common methods for the two stages of traffic sign recognition, detection stage and matching stage, are introduced. The advantages and disadvantages of each method are analyzed from multiple perspectives. Finally, the prospect of research on vehicle traffic sign recognition systems is given.

Keywords: traffic signs; detection; recognition

1 INTRODUCTION

Traffic accidents happen frequently nowadays. The number of casualties has been increasing in the recent years. How to improve traffic safety has become an imminent question. In order to give solutions to this question, the driver assistance system was invented. The system collects the real-time video image information of the road using a car camera and concludes usable traffic information through computer processing. The information is then feedbacked to the driver so that traffic accidents caused by drivers' carelessness can be reduced or avoided. In the past couple decades, researchers and scholars developed multiple shape matching algorithms for the recognition of traffic signs. Each algorithm has its own advantages and shortcomings. A summary and classification of these algorithms then become essential. This paper introduces a number of representative shape matching methods, which would be helpful to readers' further research.

2 THE CURRENT RESEARCH STATUS OF TRAFFIC SIGN RECOGNITION SYSTEM

Currently, there are several research groups have successfully applied the traffic sign recognition system to the field of machine vision for intelligent vehicles. They include UBM in Germany, the project group led by Professor Broggi in Italy, and Science Application International Corporation. In terms of practical applications, car manufacturers including BMW, Audi, and Volvo put forward their driverless concept car with intelligent software and driving environment detection system.

There are two basic parts for the TSR assistance system. The first one is the detection (separation) of traffic signs. This part determines whether the collected image involves traffic signs. If detected, the traffic signs are located and separated. The second part is the recognition (matching) of traffic signs. This part is responsible for the characteristic extraction, characteristic analysis, and final recognition of the separated traffic signs.

3 COMMON DETECTION (SEPARATION) ALGORITHMS

The detection stage usually uses colors and shapes of traffic signs to detect areas that might involve traffic signs, which lays the groundwork for the recognition in the next step.

3.1 *Traffic sign recognition based on color features*

Although the color features of traffic signs can be easily affected by factors such as illumination, they are still one of the important clues for the detection of traffic signs. The reason is that all traffic signs share the same external characteristics. Only by colors can they be differentiated. On the other hand, color-based selection can remove much background information, which reduces the complexity in the following processes.

3.1.1 *Detection algorithms based on RGB space*

RGB color space is the base of all kinds of other color spaces. Aryuanto Soetedjo et al. transformed the RGB image to a binary image that corresponds to the red component by setting a threshold on the color ratio for traffic signs with red frames [1]. Additionally, Hu et al. proposed a dynamic threshold traffic sign separation approach based on a three component chromatic aberration method. Another class of color threshold separation methods separates the image by establishing a color lookup table. For example, M.M. Zadeh et al. realized the color separation for typical traffic signs using this kind of methods where the RGB color space is transformed to a color lookup table with $256 \times 256 \times 256$ bits [2]. The advantages and shortcoming of above methods are summarized in Table 1.

Most of the traffic signs have color except the two prohibition traffic signs, 'Unban overtaking' and 'elimination of speed limits'. As a result, color based image separation algorithms cannot process these two signs. To solve this problem, S. Maldonado-Bascon et al. proposed a new algorithm for these kinds of non-colored traffic signs. The equation is [3]:

$$f(R,B,G) = \frac{(|R-G| - |G-B| - |B-R|)}{3D} \quad (1)$$

where D is a constant for the classification of non-color components. When $f(R, G, B) > 1$, the corresponding pixel is colored, otherwise non-colored. It was concluded that the algorithm had the best results when $D = 20$.

3.1.2 *Detection algorithms based on HSI color space*

The representation of color by the HIS color model which is composed by Hue, Intensity, and Saturation better matches the human visual understanding for color. Furthermore, the three components are independent of each other, which completely eliminate the influence caused by illumination. Therefore, the model is favored by researchers.

H.X. Liu et al. proposed to perform traffic sign separation in a normalized HSI color space [4], where a normalized RGB space is transformed to an HIS space and the HIS space is again normalized. Literature [5] removed redundant area by prior knowledge of the traffic signs before the separation using HSI color model. This increases the efficiency of the algorithm. A. de la Escaler et al. designed two lookup tables for blue color and red color in the HSI space for color separation and located traffic signs using genetic algorithm [6, 7]. Kiran C.G. et al. also proposed a traffic sign image separation algorithm based on an enhanced color lookup table. The algorithm first transforms the RGB image to the HIS color space and then locates the image of corresponding traffic signs by enhancing the H values and S values of red, yellow, and blue [8].

3.1.3 *Detection algorithms based on other color space*

Except the above two common color spaces, there are other color spaces used for the detection of traffic sign images as well, including HSV space and YUV space. S. Vitabile et al. transformed RGB images to HSV color space and pointed out that HSV color space can be divided into three regions: non-color region ($S \leq 0.25$ or $V \leq 0.2$ or $V \geq 0.9$), unstable colored region ($0.25 < S < 0.5$ and $0.2 < V \leq 0.9$), and colored region ($S > 0.5$ and $0.2 < V \leq 0.9$) [9]. Shadeed W.G. et al. analyzed the advantages and disadvantages of traffic sign image separation algorithms based on HSV space and YUV space and found that the two separation algorithms can compensate each other's deficiencies. They also stated that better separation results can be obtained by combining these two color spaces by performing an 'and' operation [10].

3.1.4 *Summary of color feature detection*

In summary, among all color spaces, RGB space is the most basic one, but the image information reflected in the space is not intuitive and does not match human visual perception. Additionally, the three variables in the RGB color space are significantly correlated, which make the information extracted from the image very sensitive to illumination. Therefore, it is not commonly used

Table 1. A comparison of RGB color space methods from different literature

Literature	Year published	Detection method	Advantages	Disadvantages
[3]	2006	Transform RGB images to binary images corresponding to the red component	The method is simple	Need manually set the threshold
[4]	2009	Three components chromatic aberration method	Separation by Otsu dynamic threshold binarization method	Only for red and blue traffic signs, fails for yellow traffic signs
[5]	1997	Establish color lookup table	Easy implementation	Low accuracy

in the detection of traffic signs. HSV color model is similar to HSI model. The three components of them have very little correlation. However, low saturation and intensity cause unstability of hue. Under natural conditions, the performance of detection for traffic signs is not ideal. Stable hue can be obtained only when the saturation and intensity are in certain ranges. YUV color space does not have a strong perception, though it is a uniform color space. The disadvantage of these methods is the sensitivity to illumination, which makes them lack of universality in practical use and time consuming when treating complex scenes.

3.2 *Traffic sign detection based on color-shape features*

The color-shape feature based image detection extracts the shape features after extracting the color features in order to further confirm whether the 'region of interest' is really the right region or not.

3.2.1 *Edge detection operator*

There are many basic algorithms for edge detection, including Roberts operator, Sobel operator, La place operator, and Canny operator. Some researchers proposed to use Sobel operator to extract the edges and then use Hough transform to detect circular signs in the image [11]. Literature [12] proposed to apply a contour detection method for binary images and then determined whether there were circular or triangular traffic signs by performing geometric analysis to the detected contours. Literature [13] adopted the modified Canny edge detection operator to detect edges and generated chain codes from the edges to remove regions with non-circular contour. Finally, the location parameters of circular traffic signs are determined by nonlinear least squares regression.

3.2.2 *Hough transform*

Hough transform can detect geometric shapes including straight lines and circles. Therefore, it is a common method for the detection of traffic signs. Literature [14] used Hough transform to detect circular signs, but due to high signal-to-noise ratio caused by noise interference included in imaged collected from natural scenes, the parameters and thresholds of the transform are hard to determine. 'False peaks' and 'missing detections' often happen. Barnes et al. utilized the symmetry of the shape of traffic signs and proposed a symmetry based traffic sign detection approach. The method can be seen as a modification of Hough transform. The computation effort of it is smaller than that of Hough transform. The method solves sign blocking and fading to some extent and is robust to illumination. The advantages of the method are its rotation and scale invariance and high noise immunity. But the method requires heavy computation efforts so that it has a poor real-time performance.

Figure 1. Matching results of image block and codebook.

3.2.3 *Shape features*

Literature [25] calculated geometric features including angular point and eccentricity by setting a color threshold. In literature [16], a codebook with the shape features of traffic signs is constructed. Signs can be extracted by combining the image blocks and the codebook, as shown in Figure 1. Furthermore, experimental results show that the method has high precision and searching efficiency.

There are also other algorithms exist. For example, to extract traffic signs from dynamic or static complex scenes, literature [17] combined log Gabor filter and quaternion frequency domain phase spectrum and proposed a novel multiscale visual saliency traffic sign detection algorithm.

3.2.4 *Summary on color-shape detection methods*

Detection based on the combination of color and shape features has a more comprehensive expression for the image content. Compared with detections based on single feature, it makes best use of the advantages and bypasses the disadvantages, which results in better detection performance. In the shape based detection stage, edge detection operators are commonly used due to short detection time, but its accuracy is in general not high. The detection accuracy of shape based detection methods is better than that of edge detection operators, but the detection time increases correspondingly. Therefore, different detection operator should be selected based on specific cases.

4 COMMON RECOGNITION (MATCHING) ALGORITHMS

The recognition of traffic signs reads the meaning of traffic signs. This is the ultimate goal of sign detection. There are a number of pattern recognition algorithms. The common ones include statistical classification and neural network classification.

4.1 *Statistical pattern recognition*

Statistical pattern recognition is a commonly used classification and recognition method for which a complete theoretical system has been developed. The nature of it is to use all kinds of distribution characteristics to make classification possible. Many researchers used SVM (Support Vector Machine) to classify and recognize traffic signs and got decent results [18, 19]. Literature [20] partitioned the training of SVM into two stages. In the first stage, the contour of the traffic signs is determined. In the second stage, the graphic symbol of the sign is figured out. Static image classification can be done by training and testing through the above two stages. Min et al. discussed using global features of binary images and Zernike moment features respectively for training. They also analyzed and compared the experimental results using two different types of SVM with four different kernels and obtained decent results [21]. The realization of SVM has its own advantages, but it is only suitable for cases that are linearly separable, which indicates that its interference immunity is poor. Literature [22] used reversely propagated MLP (Multiplayer Perception) to accomplish the recognition of the target, but the computational time is too long to be used in real time for complex traffic scenes.

4.2 *Neural network*

Neural network classification is one of the algorithms with the most amounts of development and applications in the current traffic sign recognition field. Some commonly used methods include 3-layer BP neural network and RBF neural network algorithm. Literature [23] used Convolution Neural Network (CNN) so that there is no need to extract individual feature, but a 48×48 pixel normalized color image patch is needed. P. Douville et al. first normalized the size of the image and then performed the classification using a 3-layer perception using the Gabor value of the normalized image as the characteristic. However, this kind of method that performs classification using multilayer perception network has poor interference immunity and robustness. Moreover, only simple comparisons can be done. Literature [24] used a two-stage neural network structure to perform classification first and then recognition. The contour feature of traffic signs and the moment invariant features are extracted respectively to be used as the input characteristic of the classification network and the recognition network. Using a two-stage neural network can reduce the miss recognition rate. Literature [25] simplified the recognition problem. It used neural network groups to form a two-stage intelligent classifer, which realized the crude classification and fine classification of traffic signs. After that, some researchers also proposed to use fuzzy feature training set to train the BP neural network where small scale training set realizes the crude classification, which obtained good results [26]. Other than that, the preliminary research on cellular neural network has already been started due to its advantages including good real-time performance, high speed parallel processing, and easy VLSI realization [27].

4.3 *Algebraic feature*

Edge analysis method detects traffic signs by performing geometric analysis for the edge information of images. In 1962, Hu gave a definition for continuous moment functions [28]. Using the nonlinear combination of moments, seven invariants that are invariant to translation, rotation, and scaling can be obtained. The three features of the moments offer them good recognition properties because the quality of recognition will not be affected due to changes in shooting distance or shooting angle. M.A. Garci a-Garrido et al. used this method. However, when detecting traffic signs in a full size image, this method consumes a huge amount of computational effort, which is not favorable for the real-time operation of the system. N. Barnes et al. utilized the inherent symmetric feature of traffic signs and proposed a radial symmetry detector based on Hough transform for the recognition of traffic signs. The method largely reduced the amount of calculations and also had decent results. In order to overcome the difficulty in the recognition of traffic signs when their shape changes [29, 30], literature [31] used a modified Hu invariant and Zemike moment to extract feature vectors to establish a feature library. Literature [32] used BP neural network to perform classification using the Hu invariant moment of traffic sign images as characteristic, where the number of node of the hidden layer need to be selected based on prior experience. Due to the large size of the training set and slow convergence rate, the optimal number of hidden layers and number of nodes are hard to find. Multiple simulations have to be done in order to determine the parameters of the neural network.

Additionally, invariant feature descriptor is also commonly used. Literature [33] proposed a prohibition traffic sign feature extraction algorithm that combines SURF (Speeded-Up Robust Features) descriptor and LLC (Locality-constrained Linear Coding). Experimental results showed that the proposed approach not only had a high recognition rate, but also required less computational time. A detailed example is shown in Figure 2.

(a) Original image

(b)Local feature pints

Figure 2. Traffic sign detection example using SURF feature points.

4.4 *Summary on recognition algorithms*

Image matching methods are generally simple, but they need to match images to all the images in the library. Therefore, this kind of methods is time consuming and has low accuracy. Comparatively, due to their scientific feasibility, complete theoretical foundation, and decent recognition performance in practical application, neural network and SVM have been widely used by researchers. SIFT, as a newly developed recognition method, is promising for further development.

5 CONCLUSIONS

Although the research work for Traffic signs recognition system already had some achievements, there also exist several technical difficulties. For example, the complexity of real-time traffic condition complicates the recognition the traffic signs. The changes in the natural sunlight will cause significant color distortion of traffic signs. Geometric distortion will also happen due to variations in vehicle velocity and shooting angles. Additionally, it is also a problem how to feedback the recognized sign information to the drivers. The solution to these difficulties requires the recognition method to have both high recognition accuracy and good robustness and stability. Therefore, how to overcome the above obstacles, improve the current detection and recognition algorithms, and increase the stability and robustness of algorithms has become the focus of future research. With the deepening of the research and the continuous improvement of detection and recognition technology, the automatic traffic sign recognition system will finally be applied to real life in the near future.

ACKNOWLEDGEMENTS

This work was financially supported by Shenyang Science and Technology Foundation (Grant No. F13-316-1-38, 'Research on affine invariant shape recognition based on sub-space').

REFERENCES

[1] Aryuanto Soetedjo and Koichi Yamada. An Efficient Algorithm for Traffic Sign Detection [J]. Journal of Advanced Computational Intelligence and Intelligent Informatics, 2006: 409–417.

[2] M.M. Zadeh, T. Kasvand, and C.Y. Suen. Localization and Recognition of Traffic Signs for Automated Vehicle Control Systems [C]. International Conference on Intelligent Transportation Systems, part of SPIE's Intelligent Systems & Automated Manufacturing, Pittsburgh, Pa, USA,1997: 272–282.

[3] C.G. Keller, C. Sprunk, C. Bahlmann, J. Giebel, and G. Baratoff, "Real-time recognition of U.S. speed signs," in IEEE Intelligent VehiclesSymposium (IV 2008), 2008.

[4] H.X. Liu, B. Ran. Vision-based Stop Sign Detection and Recognition System for Intelligent Vehicle [J]. Transportation Research Record, 2001, 1748: 161–166.

[5] Hasan Fleyeh. Shadow and Highlight Invariant Colour Seglnentation Algorit For Traffic Signs, Cybernetics and Intelligent Systems, 2006 IEEE Conference on, 2006.7,1~7.

[6] A. de la Escalera, J.M. Armingol, M. Mata. Traffic Sign Recognition and Analysis for Intelligent Vehicles [J]. Image and Vision Computing, 2003, 21(3): 247–258.

[7] B. Cyganek. Color Image Segmentation with Support Vector Machines for Road Sign Recognition [J]. Technical Report, AGH-University of Science and Technology, 2007.

[8] Kiran C.G, et al. support vector machine learning based traffic sign detection and shape classification using Distance to Borders and Distance from Center features, TENCON2008-2008 IEEE Region 10 Conference, 2008.11, 1~6.

[9] Shadeed, et al. Road traffic sign detection in color image, Electronics, Circuits and Systems, 2003. ICECS 2003. Proceedings of the 2003 10th IEEE International Conference on, 2003.12, 890–893.

[10] Zhangka, Sheng Yehua, Zhao Dongbao, Measurable real visual images automatically detect traffic signs, Journal of Scientific Instrument, 2012.

[11] XuHuaqing, Chen Ruinan, Detection method of road traffic signs, Fuzhou University Journal, 2010.

[12] MaLingjiao, Features based on color and shape—based image retrieval technology and applications, Jilin University, master's degree thesis, 2011.

[13] Chen Zhi, Jiangrui, Image matching research [J], Information technology, 2007.

[14] H. Gomez-Moreno, S. Maldonado-Bascon, P. Gil-Jimenez, and S. Lafuente-Arroyo, "Goal Evaluation of Segmentation Algorithms for Traffic Sign Recognition," Intelligent Transportation Systems, IEEE Transactions on, vol. 11, no. 4, pp. 917–930, Dec. 2010.

[15] R. Kastner, T. Michalke, T. Burbach, J. Fritsch, and C. Goerick, "Attention-based traffic sign recognition with an array of weak classifiers," in Intelligent Vehicles Symposium (IV), 2010 IEEE, June 2010, pp. 333–339.

[16] R. Timofte, V. Prisacariu, L. Van Gool, and I. Reid, "Combining Traffic Sign Detection with 3D Tracking Towards Better Driver Assistance," Emerging topics in computer vision and its applications, 2011.

[17] Traffic Sign Classification Using Invariant Features and Support Vector Machines Hasan Fleyeh and Mark Dougherty 2008 IEEE Intelligent Vehicles Symposium Eindhoven University of Technology Eindhoven, The Netherlands, June 4–6, 2008
[18] Min Shi, Haifeng Wu, and Hasan Fleyeh. Support Vector Machine for Traffic Signs Recognition [C]. IEEE International Joint Conference on Neural Networks, Hongkong, 2008:3820–3827.
[19] Auranuch Lorsakul and Jackrit Suthakorn. Traffic Sign Recognition Using Neural Network on Open CV: Toward Intelligent Vehicle/Driver Assistance System. Mahidol University. Center for Biomedical and Robotics Technology (BART LAB)
[20] D. Ciresan, U. Meier, J. Masci, and J. Schmidhuber, "A committee of neural networks for traffic sign classification," in Neural Networks (IJCNN), The 2011 International Joint Conference on. IEEE, 2011, pp. 1918–1921.
[21] F. Boi and L. Gagliardini, "A Support Vector Machines network for traffic sign recognition," in Neural Networks (IJCNN), The 2011 International Joint Conference on. IEEE, 2011, pp. 2210–2216.
[22] SuGuangmin, WangJing, Detection and recognition of traffic signs in natural background, Beijing University Journal2010.
[23] F. Zaklouta, B. Stanciulescu, and O. Hamdoun, "Traffic sign classification using K-d trees and Random Forests," in Neural Networks (IJCNN), The 2011 International Joint Conference on, 31 2011-aug.5 2011, pp. 2151–2155.
[24] D. Ciresan, U. Meier, J. Masci, and J. Schmidhuber, "A committee of neural networks for traffic sign classification," in Neural Networks (IJCNN), The 2011 International Joint Conference on. IEEE, 2011, pp. 1918–1921.
[25] Ritendra D, Dhiraj J, et al. Image Retrieval: Ideas, Influences, and Trends of the New Age [J]. ACM Transactions on Computer Survey, 2008, 40(2): 1–66.
[26] King Hann Lim, Li-Minn Ang and Kah Phooi Seng. New Hybrid Technique for Traffic Sign Recognition [J]. IEEE International Symposium on Intelligent Signal Processing and Communication Systems, 2009: 1–4.
[27] B. Hoferlin and K. Zimmermann, "Towards Reliable Traffic Sign Recognition", Intelligent Vehicles Symposium, 2009 IEEE, Xi'an, 2009, 6(23):324–329.
[28] H. Gomez-Moreno, S. Maldonado-Bascon, P. Gil-Jimenez, and S. Lafuente-Arroyo, "Goal Evaluation of Segmentation Algorithms for Traffic Sign Recognition," Intelligent Transportation Systems, IEEE Transactions on, vol. 11, no. 4, pp. 917–930, Dec. 2010.
[29] Chang Loufeng, Research traffic sign detection and recognition technology vehicle auxiliary systems, Central South university thesis, 2010.
[30] [N. Barnes, G Loy. L.S. Fletcher. Real-Time Speed Sign Detection Using the Radial Symmetry Detector [J], IEEE Transactions on Intelligent Transportations 2008,9(2): 322–332.
[31] S. Vitabile, G. Pollaccia, G. Pilato and F. Sorbello. Road Signs Recognition Using a Dynamic Pixel Aggregation Technique in the HSV Color Space [C]. The 11th International Conference on Image Analysis and Processing, Palermo, Italy, 2001: 572–577.
[32] Miguel A. Garc'ıa-Garrido. Complete Vision-Based Traffic Sign Recognition Supported by an I2V Communication System. Sensors 2012, 12, 1148–1169.
[33] S. Maldonado-Bascon, et al. Road-sign Detection and Recgnition Based on Support Vector Machines. IEEE Transactions on Intelligent Transportation Systems,2007,8(2):264~278.

Zhijia Zhang received the B.E. and M.E. Degrees from Northeastern University, Shenyang, China in1996 and 2002 respectively. In 2006 graduated from Shenyang Institute of Automation, which attached to the Chinese academy of sciences, received the Ph.D. degree in Pattern Recognition and Intelligent System profession.

Currently, he is a vice president working in Shenyang University of Technology. His present research is focused on image processing and target recognition.

Wenqiang Li is reading in Shenyang University of Technology for the Master Degree, mainly research is focused on image processing and recognition

Chunjing He is reading in Shenyang University of Technology for the Master Degree, mainly research is focused on image processing and recognition

E-Commerce, E-Business and E-Service – Lee (Ed)
 ISBN 978-1-138-02646-9

Research on spatial characteristics of 60 GHz MIMO channel and multiplexing

Chunlei Wu, Xuerong Cui, Zhongwei Li & Peiying Zhang
Department of Computer and Communication Engineering, China University of Petroleum (Huadong), Qingdao, China

ABSTRACT: With the development of information technology, there have been a growing number of pervasive and mobile computing based E-Services. 60 GHz wireless communication technology is considered to be the promising physical layer connectivity solution. However, its channel is very different to that of the low-frequency system with a rich multipath environment, and the adjacent channels produced by antenna array have a larger correlation because they mainly rely on the LOS paths. To solve these problems the channel characteristics of the 60 GHz MIMO structure is studied, and the configuration guidelines of antenna array are given. Then a two-layer antenna array based 60 GHz wireless communication link model is proposed. Finally, a spatial multiplexing example system for multi-receiving devices is simulated. The results show that the proposed two-layer antenna array can not only enhance the link performance against a severe path loss with the use of codebook based beamforming, but also achieve a larger spatial degree of freedom for high level spatial multiplexing capability.

1 INTRODUCTION

Multiplexing is an important way to increase channel capacity and enhance system performance for wireless communication systems. The directional signal propagation enables 60 GHz wireless communication additional multiplexing potentials. Therefore, the study of 60 GHz spatial characteristics can help to improve the data rate, system performance, and ability of multi-user access. However, the millimeter-wave propagation characteristics of 60 GHz are very different to that of the lower-frequency wireless communication systems, especially the reduction of multipath effect leads to greater correlation between adjacent channels, which will damage the multiplexing gain. Therefore, the 60 GHz MIMO channel characteristics need to be further studied, and more reasonable space division multiplexing methods need to be proposed.

The MIMO structure exists in 60 GHz systems naturally, because the millimeter arrays are usually adopted at both ends of 60 GHz communication links for the beamforming. And the antenna element spacing and channel independence are not critical, where the half of the wavelength is used for simplicity. However, the array configuration can't apply to the systems with the purpose of spatial multiplexing, because the MIMO channels are required independent completely in this kind of systems.

The low-frequency MIMO systems have been deeply studied, and the dense multipath effects caused by the rich scattering, reflection environment are considered key factors to the channel independence of the MIMO channels. However, the multipath components are significantly reduced in 60 GHz systems even with complex indoor environment because of the severe propagation loss, reflection loss and penetration loss. Then the LOS path has the absolute dominance in communication [1,2]. Therefore, the space division multiplexing of 60 GHz can no longer depend on multipath effects, but the independence of adjacent channels should be considered by other methods. Fortunately, the wavelength of 60 GHz signals is about 5 mm, and only centimeter-size space is required for the inter-spacing extension of adjacent antenna elements to several wavelengths. The new idea is to maintain an adequate spacing between adjacent array elements to increase the independence of neighboring channels. It shows that when the antenna elements spacing is increased to a certain extent, a high-level space division multiplexing can also be achieved even in the only LOS environments [3].

In [4], the authors demonstrate that only when the inter-spacing of array elements reaches 2λ, the desired capacity of MIMO system can be obtained. Furthermore, the spacing is required to be $\sqrt{l\lambda}$ when there is only LOS path in the communication environment to achieve the ideal space multiplexing capability, where l where represents the distance between transmitter and receiver. How to configure 60 GHz linear antenna arrays to obtain better spatial characteristics, and how to develop the space division multiplexing based on the beamforming for a higher antenna gain are key issues in this paper.

The remainder of this paper is organized as follows. In Section II, the characteristics of 60 GHz MIMO channel is discussed. Section III presents 60 GHz space division multiplexing model. The simulation results are given in Section IV, and Section V concludes the paper.

2 CHARACTERISTICS OF 60 GHZ MIMO CHANNEL

2.1 *System model*

A typical 60 GHz HDMI transmission system is considered in this paper. It is assumed to include a high-definition video source transmit device, e.g. a DVD Player, and several receiving devices, e.g. TV Display, Laptop, and sound box, etc. We assume that all devices are equipped with MMW antenna arrays for directional communication between Dev1 and other devices. Without loss of generality, we focus on the "downlink", the communication from the video source device, denoted as Dev1, to the video display device, denoted as Dev2. The proposed scheme is also applicable to the "uplink".

The ray-tracing method is used to find and compute the important paths between the two devices in a predefined indoor environment, whose dimensional model with size $L_x \times L_y$ is illustrated in Figure 1. As shown in the figure, Dev1 with the coordinates (x_1, y_1) is assumed to have a uniform linear antenna array with N_t elements, whose length is L_t, Dev2 (x_2, y_2) has an array with N_r elements, whose length is L_r.

Denoting the $N_t \times 1$ signal vector as s, whose component signals are emitted from each array element of Dev1 respectively, the received signal vector $\boldsymbol{y}$ at Dev2 can be written as

$$y = Hs + n, \tag{1}$$

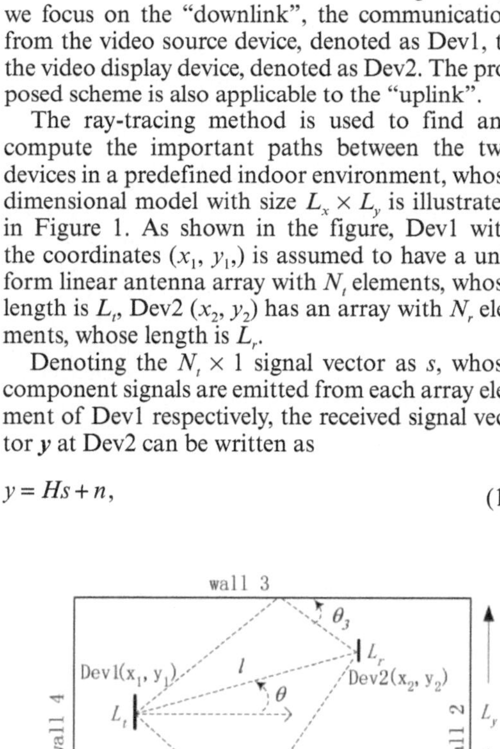

Figure 1. Indoor wireless communication model based on ray-tracing.

where H represents the MIMO channel matrix with dimensions of $N_r \times N_t$, and n is a AWGN vector, whose components are all independent and identically distributed Gaussian random variable with zero mean and variance of σ^2. In order to analyze effects of angle, location and other factors on the spatial characteristics on the channel, the ray tracing method is adopted to model the LOS channel, which is proved to be consistent with IEEE 802.15.3c channel for simplified model CM1 [5]. Then any element $h(n_r,n_t)$ of H can be denoted as [6]

$$h(n_r,n_t) = \frac{l(1,1)}{l(n_r,n_t)} e^{-j\frac{2\pi}{\lambda}l(n_r,n_t)} \approx e^{-j\frac{2\pi}{\lambda}l(n_r,n_t)}, \tag{2}$$

where $h(n_r, n_t)$ represents the channel complex gain from the n_t-th ($1 \ll n_t \ll N_t$) transmit antenna of Dev1 to the n_r-th ($1 \ll n_r \ll N_r$) receive antenna of Dev2, and $l(n_r,n_t)$ is the corresponding distance.

2.2 *Spatial freedom of MIMO channel*

The spatial degrees of freedom of MIMO channel reflect the spatial multiplexing capability, i.e. the degree value of freedom determines the link number for independent and parallel data transmission. Through the singular value decomposition of the channel matrix H, the MIMO channel can be converted into a plurality of parallel, independent SISO (Single Input Single Output) channels. Therefore, the distribution of singular values of H, denoted as $\sigma_i (i = 1, 2, \ldots \min(N_t,N_r))$, reflect the spatial degrees of freedom.

For the linear uniform antenna array with its element spacing being the half a wavelength, the maximum element number can be quickly calculated and they are $N_t = [2L_t/\lambda]$, $N_r = [2L_r/\lambda]$ for transmitter and receiver, respectively, where $[x]$ represents the bottom operation. According to common sizes of home appliances, e.g. $L_t = 40$ cm, $L_r = 70$ cm, and assuming $\lambda = 5$ mm for 60 GHz, we can get that $N_t = 120$, $N_r = 280$. Without loss of generality, Dev1 is assumed to be placed at the coordinates of $(0, L_y/2)$, and Dev2 is placed at $(x_2, L_y/2)$ parallel to Dev1, where the value of x_2 varies from 1 to L_x. Because of the various link distances between Dev1 and Dev2, we can get the distributions of spatial degrees of freedom by singular value decomposition of different H. Assuming $L_x = 10$ m and $L_y = 10$ m, the distributions are obtained and shown in Figure 2.

In the figure, the asterisk legend represents the primary singular values by decomposition of H, while the other legend is obtained by Rayleigh criterion estimation method [3]. It shows that the

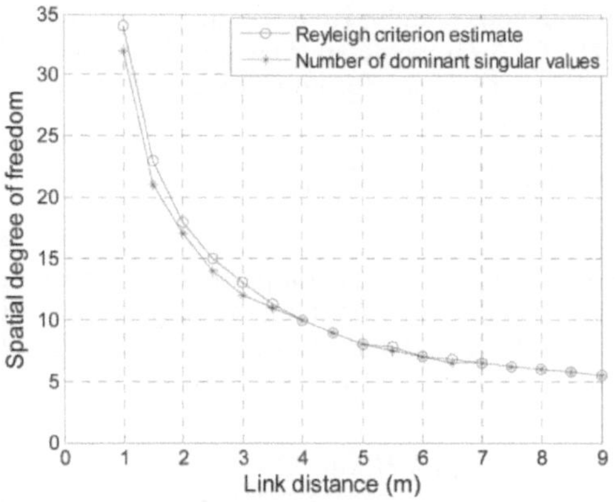

Figure 2. Distributions of spatial degrees of freedom in LOS environment.

two kinds of values are very consistent, and both of them show that the freedom degree of channel reduce significantly as the link distance increases. However, we can get a degree no less than 12 when the distance is in the range of 3 m, which indicates 60 GHz system has the great potential for the space division multiplexing even in the LOS environment.

2.3 *Optimal configuration of antenna array*

Cost and complexity also need to be considered in addition to the size limitation of devices. In fact, a device will not be equipped with a very big antenna array, while it should retain a very limited number of array elements. Then, for a given-size device, the optimal number of antenna elements and element inter-spacing should be investigated for the ideal channel degree of freedom. We propose that the columns of the MIMO channel matrix H should become orthogonal, then the redundant elements can be removed, and we can get a simplified structure and efficient antenna array. Assume that both transmitter and receiver have N elements, and the LOS link distance is l, the inter-spacing of elements can be given by Rayleigh spacing criterion as

$$d_t d_r = \frac{l\lambda}{N'} \tag{3}$$

where d_t and d_r represent the transmit and receive antenna element spacing, respectively. In addition, the array length, element number and spacing have arithmetic relationship as follows

$$d_t = \frac{L_t}{N-1}, \quad d_r = \frac{L_r}{N-1}. \tag{4}$$

Table 1. Optimal configuration of commonly used 60 GHz linear antenna array.

Number of array elements	Element spacing: Value (cm)	Element spacing: Multiples of wavelengths	Array length (cm)
2	8.66	17.32	8.66
4	6.12	12.25	18.37
8	4.33	8.66	30.31
16	3.06	6.12	45.93

Then the optimal element number can be solved as

$$N = \left\lfloor 2 + \frac{L_t L_r}{l\lambda} + \sqrt{\left(2 + \frac{L_t L_r}{l\lambda}\right)^2 - 4} \right\rfloor. \tag{5}$$

As an example, assuming that the exactly same antenna arrays are adopted at receiver and transmitter, i.e. $L_t = L_r$, $d_t = d_r$, the physical length of each array element is ignored, and the link distance is 3 m, the array length, element spacing can be given by (4) and (5). Table 1 lists some of the commonly used linear antenna array and its recommended configuration. It shows that devices with the size similar to DVD players and televisions can contain a millimeter array with no less than 8 elements. That means this size of 60 GHz wireless communication system has a very high level of space division multiplexing capability.

3 60 GHZ SPACE DIVISION MULTIPLEXING MODEL

As is well known, the path loss of 60 GHz signal is severe. If the antenna array is used only for the spatial multiplexing and not for beamforming any longer, it may cause the small coverage and poor performance for wireless communication. Therefore, it is an important challenge for us to ensure the coverage and performance while develop the space division multiplexing capability simultaneously. A structural model for millimeter-wave communication systems is proposed by Ecric Torkildson in [6]. It has a two-layer architectural style to provide the necessary beamforming gain while keeping the channel independence. Because the adaptive beamforming scheme based on the channel state information is used, the model does not apply to those small devices in 60 GHz systems. In this paper, we propose to build a codebook beamforming based two-layer array model, as shown in Figure 3.

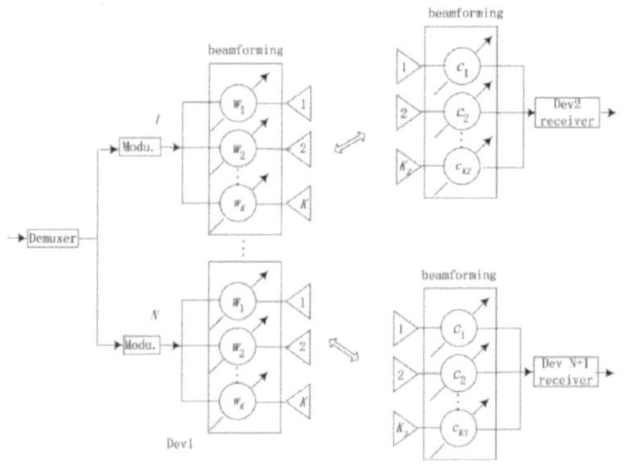

Figure 3. 60 GHz space division multiplexing model for multi-receiver devices.

At Dev1, the outer layer contains an N-element antenna array, and the inter-spacing between adjacent elements is set to $d(d \gg \lambda)$ according to the proposed optimal configuration. So the channels can keep independent for spatial multiplexing. From the inner layer perspective, each antenna of the N ones is a K-element smaller antenna array, which is designed to form a codebook based beam pattern for the better link performance. The inner inter-spacing can be set to $\lambda/2$. Each receive device is assumed to be equipped with only a monolayer millimeter-wave antenna array for beamforming. Before the communication process from Dev1 to other receive devices, one or more outer antenna of the N-element array will form the right beams to point to corresponding target devices, and the receive devices will also change their beams pointing to Dev1. Therefore, several independent and parallel communication links are built based on beamforming around Dev1.

Without loss of generality, the communication from Dev1 to Dev2 is considered to investigate the Signal to Interference Noise Ratio (SINR) and its influencing factors. We assume that the first antenna of Dev1 is used to build link with Dev2, and the emitted signal from this antenna is denoted as s_1. The signals emitted from the other N-1 antennas are denoted as s_i $(i = 2,\ldots,N)$, then the signals received by Dev2 can be written as

$$y = H_1 w_1 s_1 + \sum_{i=2}^{N} H_i w_i s_i + n, \tag{6}$$

where $H_i\,(i = 1, \ldots, N)$ represents the channel matrix generated by the i-th inner antenna array of Dev1 and the array of Dev2. Its size is $K_2 \times K$, where K_2 is the element number of Dev2' array. The parameter w_i represents the beamforming weight defined in codebooks, and n is an AWGN vector with the size of K_2. The signal in (6) will be weighted again by the receive beamforming at Dev2, written as

$$r = c_2^H H_1 w_1 s_1 + \sum_{i=2}^{N} c_2^H H_i w_i s_i + c_2^H n, \tag{7}$$

where c_2^H represents the complex conjugate vector of c. We assume that the total transmit power of Dev1 is P, and each antenna element has the same power, i.e. $p = P/NK$, the SINR at Dev2 can be given as

$$\gamma_2 = \frac{p\left|c_2^H H_1 w_1\right|^2}{p\sum_{i\neq 1}\left|c_2^H H_i w_i\right|^2 + N_0 c_2^H K_2}. \tag{8}$$

According to the Shannon formula, the channel capacity for Dev1 is given by

$$C = \sum_{i=1}^{N} \log_2\left(1+\gamma_i\right). \tag{9}$$

The conclusion can be draw by (9) that the proposed two-layer antenna array model can greatly increase the capacity of communication systems through the channel division, and it benefits from the independence of channels caused by a larger inter-spacing of the outer antenna array. In addition, the SINR is also improved by the codebook based beamforming gain, which is generated by the inner layer millimeter wave array.

4 NUMERICAL RESULTS

In this part, following the guidelines of IEEE 802.15.3c specifications, we present some simulation results to illustrate performances of the proposed model for 60 GHz spatial multiplexing. As an example, we assume that there are two outer antenna elements, which are used to build communication link with two receivers, e.g. Dev2 and Dev3, respectively. All the antennas are borne by millimeter antenna arrays with K elements for each one. Further, the signal transmission power is fixed to 40 dBm, and the AWGN is assumed to be the thermal noise generated at the receivers.

Several different sizes of codebooks are adopted here to provide various beam resolution and coverage. The SINR results via angles between the two communication links are simulated, and the sample result is shown in Figure 4 when the link distance is 5 m.

From the simulation results, we can draw some important conclusions: 1) For the given 4 kinds of

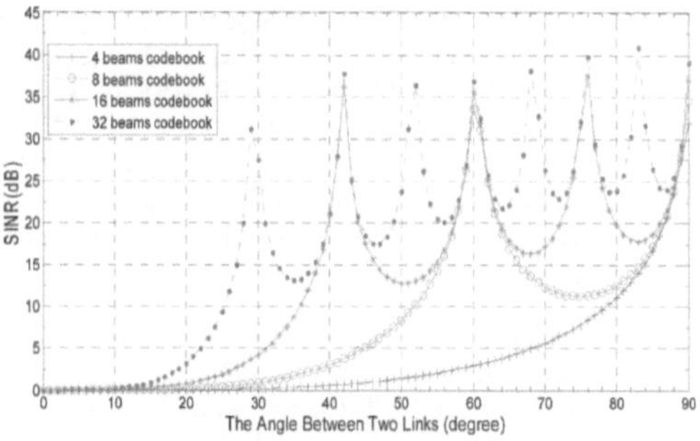

Figure 4. SINR values via angles between two links at Dev2 ($l = 5$ m).

codebooks, all of the SINR values show increasing trends along with the increasing angels. It indicates that the link performance of the proposed model depends on the angle between links, because the beams generated by codebooks cover a certain space, which will interfere with the adjacent links. 2) For same angles, the codebook with 32 beams has the best performance; the codebook with 4 beams has the worst performance. The reason is that the beam width becomes small when the number of beams in codebook increases. Therefore, the size of the antenna array and codebook should be improved as much as possible for better link budget and less interference. 3) The SINR have some maximum and minimum values in the figure, and conduct regular alternation. That's because the 4-phase codebook design mechanism is adopted here, which leads to a high sidelobe level of beams. Therefore, the improvement of codebook design to suppress the sidelobe level is also very important for 60 GHz spatial multiplexing [8,9]. 4) For beams with more than 16 beams, the proposed model has the ideal spatial multiplexing capability in LOS environment at 5 m distance.

Capacity is another very important indicator to measure wireless communication systems [10]. Assuming both the width and length of the room are 10 m, and the coordinates of Dev1 and Dev2 are (0,5) and (5,5), respectively, Dev3 (x,y) is not fixed. The channel capacity and its influencing factors of the proposed model are investigated. Different codebooks are used for simulations and the results are shown in the Figure 5.

As shown in the figure, the relative positions of devices, codebook size, and the generation mechanism of beams are key factors for the proposed spatial multiplexing model. As an example, when the codebooks with more than 16 beams are used, the channel capacity can reach 8 bps /Hz in 70 percent of the indoor area.

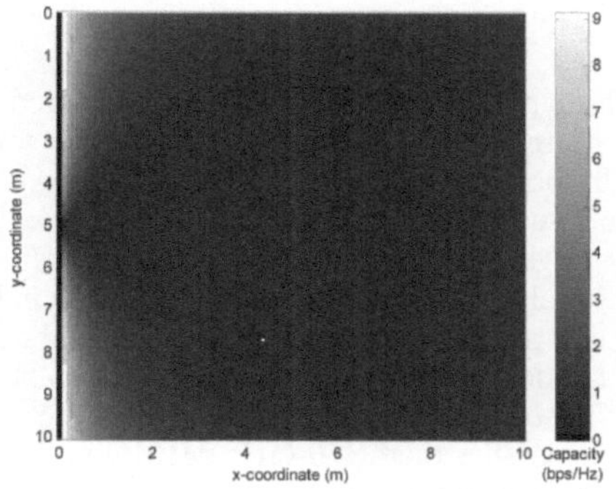

(a) codebook with 4 beams

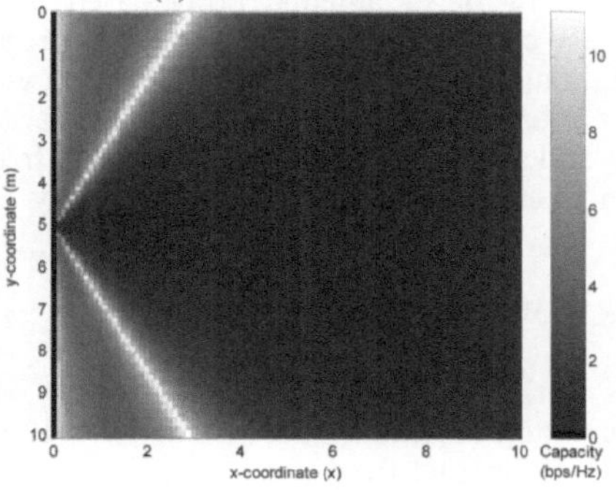

(b) codebook with 8 beams

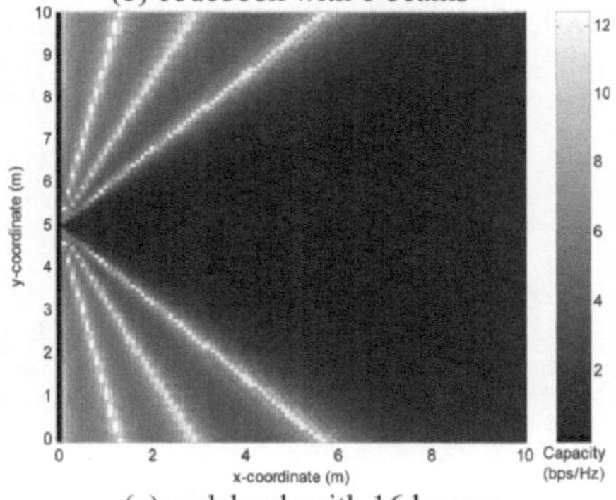

(c) codebook with 16 beams

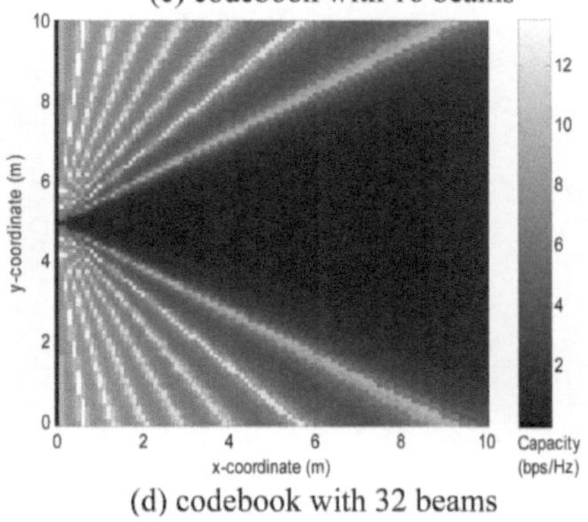

(d) codebook with 32 beams

Figure 5. Capacity distributions of channel between Dev1 and Dev2 via Dev3's position.

5 CONCLUSION

Although 60 GHz technology is considered to be the promising physical layer connectivity solution for indoor high-speed wireless applications, its millimeter-wave characteristics lead to the great path loss, larger reflection and transmission attenuation, and weaker scattering. Therefore, how to enhance the link performance, extend coverage, and develop the spatial multiplexing are key issues to be studied. To solve these problems, we studied the channel characteristics of the 60 GHz MIMO system, and propose the optimal array configuration guidelines. Then a two-layer antenna array based 60 GHz wireless communication link model is proposed. The simulation results show that the proposed two-layer antenna array can enhance the link performance and achieve a larger spatial degree of freedom for high level spatial multiplexing capability.

ACKNOWLEDGMENT

This work was supported by the Nature Science Foundation of China under grant No. 61301139, the Fundamental Research Funds for the Central Universities under grant No. 14CX02139A, No. 13CX02026A and No. 13CX02031A, and the Key Technology Development Projects in Qingdao Development Zone under grant No. 2013-1-52.

REFERENCES

[1] Manojna, D.S., Kirthiga, S., Jayakumar, M. Study of 2 × 2 spatial multiplexed System in 60 GHz indoor environment. In: 2011 International Conference on Process Automation, Control and Computing (PACC), 2011. 1–5.

[2] Xiaoyi Zhu, Doufexi, A., Kocak, T. A performance evaluation of 60 GHz MIMO systems for IEEE 802.11ad WPANs. In: PIMRC2011, 2011. 950–954.

[3] F. Bohagen, P. Orten, G. Oien. Design of optimal high-rank line of-sight MIMO channels. IEEE Transactions on Wireless Communications, 2007, 6(4): 1420–1425.

[4] Z. Genc, B.L. Dang, J. Wang, *et al.* Home networking at 60 GHz challenges and research issues. Ann. Telecommun., 2008, 63: 501–509.

[5] "IEEE Standard for Information technology—Telecommunications and information exchange between systems—Local and metropolitan area networks—Specific requirements. Part 15.3: Wireless Medium Access Control (MAC) and Physical Layer (PHY) Specifications for High Rate Wireless Personal Area Networks (WPANs) Amendment 2: Millimeter wave-based Alternative Physical Layer Extension," IEEE Std 802.15.3c- 2009 (Amendment to IEEE Std 802.15.3–2003), pp. c1–187, Dec. 2009.

[6] Eric Torkildson, Colin Sheldon, Upamanyu Madhow, *et al.* Millimeter-Wave spatial multiplexing in an indoor environment. In: 2009 IEEE Globecom Workshops, 2009. 1–6.

[7] Katsuyuki Haneda, Afroza Khatun, Marzieh Dashti, *et al.* Measurement-Based analysis of spatial degrees of freedom in multipath propagation channels. IEEE Transactions on Antennas and Propagation, 2013, 61(2): 890–900.

[8] Wei Feng, Zhenyu Xiao, Depeng Jin, et al. Circular-Antenna-Array-Based codebook design and training method for 60GHz beamforming. In: 2013 IEEE Wireless Communications and Networking Conference, 2013. 4140–4145.

[9] Bin Li, Zheng Zhou, Weixia Zou, et al. On the efficient Beam-Forming training for 60 GHz wireless personal area networks. IEEE Trans. on Wireless Communications, 2013, 12(2): 504–515.

[10] Seung Joon Lee, Wooyong Lee. Capacity of multiple beamformed spatial stream transmission in Millimeter-Wave communication channels. IET Communications, 2013, 7(12): 1263–1268.

E-Commerce, E-Business and E-Service – Lee (Ed)
© 2014 Taylor & Francis Group, London, ISBN 978-1-138-02646-9

An improved semantic similarity approach based on Cilin

Peiying Zhang & Weishan Zhang
School of Computer and Communication Engineering, China University of Petroleum (East China) Qingdao, Shandong, China

Zhanshan Zhang
International School Tongji University, Shanghai, China

Chunlei Wu
School of Computer and Communication Engineering, China University of Petroleum (East China) Qingdao, Shandong, China

Youkai Sun
Geophysical Research Institute of Shengli Oilfield Branch, Donging, Shandong, China

ABSTRACT: Semantic similarity metric is very important for many applications of computational linguistics and artificial intelligence. In the past, many similarity measurements have been proposed, exploiting explicit knowledge-such as the structure of a taxonomy-or implicit knowledge-such as information content distribution. This paper proposed a semantic similarity approach that based on Cilin. One of the benchmarks is Miller and Charles' list of 30 noun pairs which had been manually designated similarity measurements. We correlate our experiments with those computed by several other methods. Experiments on Chinese word pairs show that our approach is the closet to human similarity judgments.

1 INTRODUCTION

The measurement of the semantic similarity between words contributes to the better understanding of textual resources. As a result, it has been applied in many different tasks such as FAQ system [1], document classification, and automatic language translation. According to Liu [2], the approach of semantic similarity can be roughly divided into two categories, including corpus-based measurements and ontology-based measurements. The latter measurements mainly use taxonomies as ontology to calculate its similarity. Such as [3] utilize WordNet to calculate the English words semantic similarity, Li feng [4] uses HowNet as a Chinese semantic dictionary to measure semantic similarity between words and concepts.

Semantic similarity is understood as the degree of taxonomic proximity between terms. Similarity measurements assess a numerical score that quantifies this proximity as a function of the semantic evidence observed in one or several knowledge sources.

2 RELATED WORKS

Semantic similarity measurement methods supporting multiple ontologies are based on the edge-counting and feature-based paradigms. Petrakis et al. [8] proposes an approach relying on the matching between synonym sets and concept glosses. They considered that two terms are similar if their synonyms and glosses and those of the concepts in their neighborhood are lexically similar. Al-mubaid and Nguyen [9] combines the features path length and common specificity of the compared concepts. Authors quantity the common specificity of two concepts by subtracting the depth of their Least Common Subsumer (LCS) from the depth of the taxonomic branch to which they belong.

Resnik [10] was the first to consider the use of formula as bellow for the purpose of semantic similarity measurements:

$$ic_{res}(c) = -\log p(c) \tag{1}$$

The ic value is then obtained by considering the negative log likelihood, where c is some concept in WordNet and p(c) is the probability of encounting c in a given corpus. To evaluate semantic similarity between two concepts, formally, the formula is as below:

$$sim_{res}(c_1, c_2) = \max_{c \in s(c_1, c_2)} ic_{res}(c) \tag{2}$$

According to Resnik, semantic similarity depends on the amount of information two concepts have in common, this shared information is given by the Most Specific Common Abstraction (MSCA) that subsumes both concepts, here $S(c_1,c_2)$ are the set of concepts that subsume c_1 and c_2.

3 INTRODUCTION OF CILIN

Tongyici Cilin [5] is developed by Harbin Institute of Technology Center for Information Retrieval, which organizes words in a hierarchy structure. The vocabulary is divided into large, medium and small categories. There are 12 large categories, 97 medium categories, and 1400 small categories. Each small category contains a lot of words, which are divided into a number of paragraphs according to the proximity of meaning and relativity. The words in each paragraph are divided into a number of rows. The words in the same row have the same meaning or have a strong correlation. The paragraphs in small categories can be considered as the fourth level classification and the row of the paragraph can be considered as the fifth-level classification. In this way, there is a five-layer structure in Tongyici Cilin.

3.1 *Cilin classification methods*

Tongyici Cilin organizes all the indexed words into the hierarchy tree of taxonomy. Its dictionary utilizes hierarchical classification system, with five-layer structures, as shown in Figure 1.

The deeper its layer is, the more detailed its meaning is. In the fifth layer, the number of words in each category is limited, many have only one word, and it can not be divided which called atom words. Different levels of classification may provide different services for natural language, for instance, the fourth and fifth layers of classification are applied in information retrieval, text classification, automatic question answering and other research areas. Studies have shown that an effective extension of meanings or synonyms for words to do the replacement can significantly improve information retrieval, text classification performance and automatic question answering systems.

3.2 *The coding methods of Tongyici Cilin*

Tongyici Cilin provides five layer coding. The first layer is represented by capital letters, the second layer is represented by lowercase letters, the third layer is represented by two decimal numbers, the fourth layer is represented by capital letters, and the fifth layer is represented by two decimal numbers. Such as: “Aa01 A01 = 人 士 人物 人士 人氏 人选”, “Aa01 A01 = ” is the code, and “人 士 人物 人士 人氏 人选” is its words.

3.3 *Semantic similarity measurement based on Cilin*

Chinese words often expressed a lot of meanings, i.e. every word can be expressed as one or several concepts. Such as “things” in the thesaurus dictionary have three meanings to explain it. Their five layer codes are: people (Aa01 A06), materials (Ba01 A01), and things (Da28 A01).

3.4 *The similarity based on path length*

The path length of two words depends on their common ancestor’s layer level, such as: the code of word “体制” is “Di09D01 = ”, and the code of word “阶级” is “Di06 A01 = ”, the common ancestor “Di” in the second layer, the third layer is different, so the path length of two words is 6. One word maybe have some different codes, we take the shortest path length as the final path length.

Although the path length is more intuitive, but simply use it as semantic similarity metric is not appropriate. Study show that semantic similarity between words is not a linear relationship with

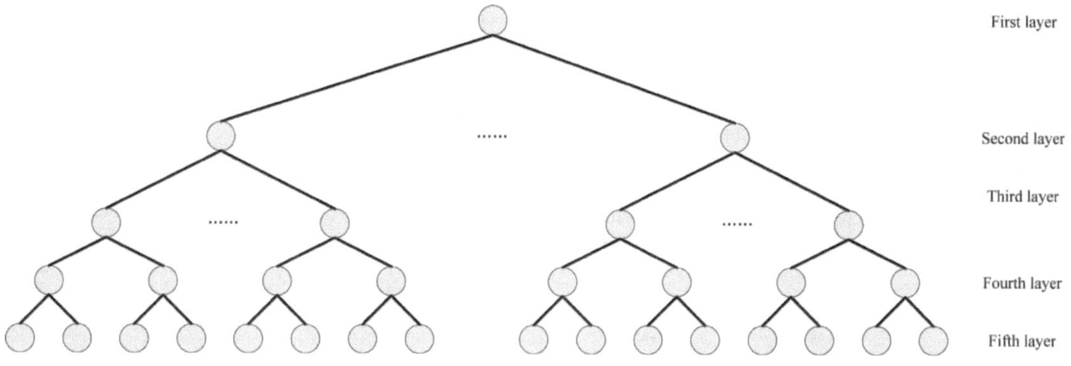

Figure 1. Five-layer structure of Tongyici Cilin.

their path length, but an exponential relationship, so we use this formula to measure similarity:

$$sim_1(c_1,c_2) = e^{-\alpha l} \tag{3}$$

3.5 *The similarity based on word node density*

The density information between the words need calculating the number of words separated between two words. The general approach is to calculate the number of words included in the common ancestor in Cilin or HowNet. The density of two words is described as follows:

$$d = -\log\frac{\text{nums(ancestor)}}{N} \tag{4}$$

where nums (ancestor) denotes the number of their ancestor, N denotes the number of words in Cilin. To limit the density value between 0 and 1, we use a nonlinear hyperbolic tangent function to denote the density similarity between two words.

$$sim_2(c_1,c_2) = \frac{e^{\beta d} - e^{-\beta d}}{e^{\beta d} + e^{-\beta d}} \tag{5}$$

3.6 *The similarity between concepts*

We use the formula below to measure the semantic similarity between two concepts:

$$sim(c_1,c_2) = \sigma \times sim_1(c_1,c_2) + (1-\sigma) \times sim_2(c_1,c_2)$$

where σ and $1-\sigma$ is the weight of sim_1 and sim_2, $0 \le \sigma \le 1$.

3.7 *The similarity between words*

Chinese words may have one or more concepts. We take the maximum similarity between the concepts as the similarity between the two words. We define the similarity as follows:

$$sim(w_1,w_2) = \max sim(c_{1i},c_{2j})$$

Table 1. The table of code of word.

Code bit	1	2	3	4	5	6	7	8
Instance	A	a	0	1	A	0	1	=\#@
Attribute	Large category	Medium category	Small category		Word group	Atomic word group		
Level	First	second	third		Fourth	Fifth		

Table 2. Comparison of semantic similarity measurements.

Noun pair	WordNet edges	Hirst St. Onge	Jiang conrath	Leacock chodorow	Propsed method
汽车 机动车	30.000	16.000	1.000	3.466	0.847
男孩 少年	29.000	5.000	0.231	2.773	0.837
工具 农具	29.000	4.000	0.546	2.773	0.694
兄弟 和尚	29.000	4.000	0.294	2.773	0.691
少年 兄弟	26.000	3.000	0.071	1.856	0.691
旅程 轿车	17.000	0.000	0.075	0.827	0.134
修士 圣人	23.000	0.000	0.058	1.386	0.691
公墓 森林	21.000	0.000	0.049	1.163	0.176
海岸 丘陵	26.000	2.000	0.148	1.856	0.068
森林 墓地	21.000	0.000	0.050	1.163	0.146
相关度	0.732	0.689	0.695	0.821	0.847

where c_{1i}, c_{2j} is the i-th and j-th concept associated with w_1 and w_2 respectively. $Sim(c_{1i},c_{2j})$ is the similarity between c_{1i} and c_{2j} that can be computed by Eq. 4.

4 EXPERIMENT AND ANALYSIS

4.1 *Dataset and measurements*

Rubenstein and Goodenough [6] established synonymy judgments for 65 pairs of nouns. They invited 51 people to assign every pairs a score between 0.0 and 4.0 to indicate semantic similarity. Miller and Charles [7] followed this idea and restricted themselves to 30 pairs of nouns selected from Rubenstein and Goodenough's list, divided equally amongst words with high, intermediate and low similarity.

4.2 *Experimental results*

We use the human measurements of RG's experiment and MC's experiment as the baseline. We choose some nouns from the MC's list of 30 nouns since some English words can not translate into Chinese properly. We compute the correlation coefficient between the human judgments and the measures achieved by our approach.

5 CONCLUSION AND FUTURE WORKS

This paper proposes a semantic similarity approach that based on Cilin. This approach mainly considers the shortest path length, the depth of their word node, and the density of their ancestor node. Experiment results show that our algorithm is comparative with other classic similarity algorithms, with the results closet to human similarity judgments.

In the future, we will utilize some other ontology taxonomy to calculate the semantic similarity. The Cilin only has 77343 words items, some words can not in the taxonomy. We will concentrate on the multi-ontology based approach.

ACKNOWLEDGEMENT

This work is supported by "the Fundamental Research Funds for the Central Universities" of China University of Petroleum (East China) (Grant No.13CX02031A and Grant No. 14CX02139A). The authors also gratefully acknowledge the helpful comments and suggestions of the reviewers, which have improved the presentation.

REFERENCES

[1] Pei-ying Zhang. Sentence similarity metric and its application in FAQ system. Advanced Materials Research Vols. 718–720 (2013) pp. 2248–2251.

[2] Liu Qun, Su Jian. "HowNet" lexical similarity calculation. Proc of Chinese lexical semantics workshop third Technology, Taipei, pp. 59–76, 2002.

[3] Agirre E, Rigau G. A proposal for word sense disambiguation using conceptual distance [C] // Proceedings of the First International Conference on Recent Advanced in NLP. 1995.

[4] Li feng, Li fang. A new approach measuring semantic similarity in hownet 2000, Journal of Chinese information processing. 2007, 21(3), pp. 99–105.

[5] Tongyici Cilin (Extended) [online], available: http://ir.hit.edu.cn/demo/ltp/sharing plan.htm, October 26, 2009.

[6] H. Rubenstein, J.B. Goodenough. Contextual correlates of synonymy. Communications of the ACM, 1965, 8(10), pp. 627–633.

[7] G.A. Miller, W.G. Charles. Contextual correlates of semantic similarity. Language and Cognitive Processes, 1991, 6(1):1–28.

[8] Petrakis, E.G.M., Varelas, G. et al. X-similarity: Computing semantic similarity between concepts from different ontologies. Journal of digital Information management, 2006, 4: 233–237.

[9] Al-Mubaid, H., Nguyen, H.A. A cluster-based approach for semantic similarity in the biomedical domain. In 28th An nual international conference of the IEEE engineering in medicine and biology society, EMBS 2006, pp. 2713–2717.

[10] P. Resnik. Information content to evaluate semantic similarity in a taxonomy. In: Proceedings of IJCAI, 1995, pp 448–453.

E-Commerce, E-Business and E-Service – Lee (Ed)

The application of real-time data remote transmission for LWD/MWD in oilfield

Qifeng Sun
China University of Petroleum, China

Youkai Sun
Shengli Oil Field Geophysical Research Institute, Shengli Oilfield Branch, Donging, Shandong, China

ABSTRACT: Real-time data transmission technique is one of the key technologies to build LWD remote real-time control center. In this study, we designed LWD real-time decoding and remote transmission devices, builded remote real-time interpretation center based on the cluster server hardware and remote receiver software components on the server side, designed and implemented a web-based information released system to share of real-time data. The system can meet the need of real-time Geosteering while drilling.

1 INTRODUCTION

Geosteering while drilling is the the most promising new drilling technology in 21st century. It is the combination of the drilling engineering, reservoir geology, logging technology, automatic control and computer technology.

Geosteering is more mature in foreign countries at present. It has built up a complete technical solution which contains Measurement While Drilling (MWD), real-time data transmission, processing and interpretation. Besides, it has been applied successfully in many drilling projects of different oil fields. In 2004, Parthasent LWD (Logging While Drilling) information to Geosteering Remote Assistance Center which is based on virtual reality technology in real time by satellites, reduced the risk and budget of drilling by forward predicting bit information technology. Based on geosteering diagnosis system, Thevoux-Chabuel [3] proposed a new approach to control drilling process. By comparing between seismic information and LWD information, designed drilling trajectory and drilling trajectory, it can be used for risk warning and improvement of drilling quality. In addition, it also can be used to monitor the location of each drilled hole through a 3D geological model. In 2009, Salem Al-Jaberi [4] put forward a geosteering system based on real-Time 3D modeling calibration and applied it in carbonate reservoir exploitation. It has successfully solved the problems in thin hydrocarbon zones' exploitation. Ignacio Gorgone has explicitly point out that geosteering is changing from "Remote decision support" to "Remote drilling control" under current technological change and application fundament. Besides, Ignacio Gorgone proposed solutions to staff collaboration, real-time data transmission, virtual reality, remote control and other key technologies.

In this study, based on realities of situation and researches of MWD/LWD logging tools which are widely used in domestic market of China, we designed LWD real-time decoding and remote transmission devices, builded remote real-time interpretation center based on the cluster server hardware and remote receiver software components on the server side, designed and implemented a web-based information released system to share of real-time data. This system is an important technical support to the transforming of geosteering while drilling from "Remote decision support center" to "Remote Drilling Control center".

2 SYSTEM REQUIREMENTS AND DESIGN

Real-time LWD/MWD data remote transmission system is primarily designed for the transmission of borehole trajectory and geological parameters while drilling to data center in the rear.

As shown in Figure 1, This system is composed of three parts: well site function module, remote transmission module and real-time data processing and release module of data center. Well site function module is mainly for the decoding and validation of real-time data while drilling and the real-time display of LWD engineering parameters and geological parameters. Remote transmission module is responsible for the storage of received

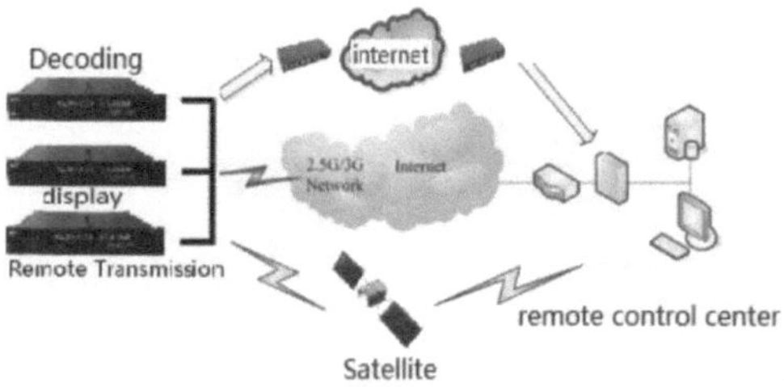

Figure 1. System requirement and structure.

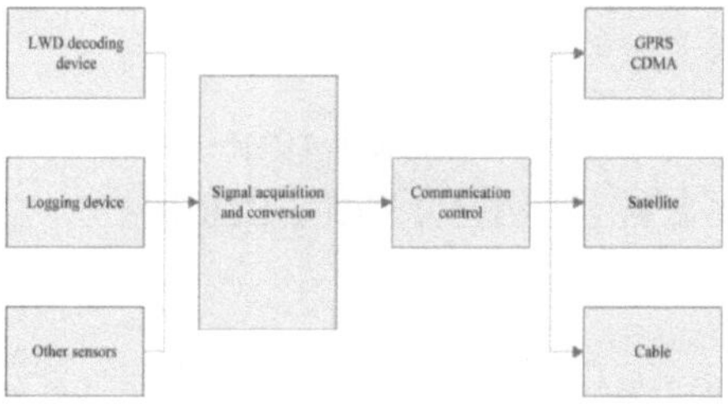

Figure 2. Decoding and remote transmission devices.

real-time data and the supply of communication modes such as network cable, GPRS, satellite and so on according to the network environment and status. It is able to change data transmission mode when the quality of network connections declining, make sure that LWD real-time data can be sent to data center as quickly as possible. Data center is made up of cluster servers, real-time receiving and processing softwares and web-based real-time display programs. It is used for the processing and display of received well site data, it can meets dozens of wells' work requirements at the same time.

3 KEY TECHNOLOGY AND METHOD

3.1 *Real-time decoding and remotetransmission devices*

In the process of drilling, engineers need to be aware of the working condition of subsurface equipments, well trajectory, geological survey and other information. The acquisition of this information is one of the keys in wireless MWD system. Currently, the transmission technology which is based on the use of the pulse of drilling fluid is mature and wide in applications. In this study, we designed LWD real-time decoding and remote transmission devices based on the research of primary devices in domestic market and the systematic research and analysis of the decoding technology of some major MWD and LWD devices.

System architecture is shown in Figure 2. The data from real-time collection and other devices will be decoded and converted at client-side. After the conversion, the data will be sent to database of data analysis center in a relatively stable communication channel which is selected by computer. The data can be automatic buffered on client-side due to the data transmission interruption or transmission errors which are caused by communication link problems. In order to maintain the consistency of field data and back data, clients can automatically resume the data transmission after the recovery of communication channel.

3.2 *Real-time transmission and channel selection*

The information level of oilfields in China is increasing rapidly at present as some well sites are able to have broadband access. Therefore, LWD real-time remote transmission device is designed to be able to make use of different link methods such as network cable, mobile wireless and satellite and so on. But the main transmission method is dominated by wireless transmission due to the situation that most well sites are located at remote places where endowed with bad traffic condition. There are three principal means to achieve wireless data transmission: satellite, microwave and 3G/EDGE/GPRS network.

Satellite transmission has many great advantages such as great network coverage, fast data transmission speed, strong real-time, high security and reliability. But in practical application, it charges expensive and the movement is ordinary, the use of the spectrum should be applied in advance, the cost is too high. Therefore, satellite transmission is not suitable for large-scale use in actual applications.

Microwave transmission can guarantee data's real-time property, but microwave requires point-to-point transmission, antennas must be set up, the cost of station construction and maintenance is really high and there is a certain distance limitations. So the limitation of microwave transmission is relatively large.

3G/EDGE/GPRS/CDMA network is a voice communication platform that built up by China Mobile, China Telecom and China Unicom. It does not need to set up any other antennas and purchase other expensive ancillary facilities and can be used flexibly. It supports real-time online and has low operating cost. EDGE/GPRS/CDMA wireless network has already formed a complete network structure now. It has large network coverage, faster data transmission speed and stable operations. For example, the peak rate of GPRS is 115.2 kbit/s, CDMA 1X system is 153.6 kbit/s, the theoretical maximum data transfer rate of EDGE is 473.6 Kbps. 3G network speed is in the theoretical speed 3 Mb/s or so, speed testes at this stage are not very stable and the network coverage is still not perfect.

Taking the situation of well sites, features of LWD data, coverage of network, strength and stability of signals and other factors into account, system provides two kinds of data transmission methods for selection: web-based network transmission and GPRS wireless data transmission.

3.3 *Design of reliable UDP transport protocol*

When listing facts use either the style tag List signs or the style tag List numbers. According to the work conditions and the distribution of well sites, this system is an application which scatters widely, has a small amount of data, high real-time requirements and a large number of terminals. System's transport layer has adopted UDP Transport Protocol. In order to improve the transmission efficiency of data, we conducted optimization design on UDP Transport Protocol. We designed the frame structure of data that sent by application layer and on this basis, we defined a communication protocol.

Reliable UDP Transport Protocols use a resizable sliding window to send and receive message synchronously and control flow. The sequence numbers delivered by protocol use two-byte unsigned integer and range from 0 to 65535, they can repeat counted in the data transmission process. The value written in the window fields of a UDP packet header segment is the maximum limit value of the send window. Send window sets up initial values when it is establishing the connection. However, in the communication process, the receiving end can dynamically adjust (increase or decrease) each other's maximum limit value of the send window at any time according to its own resources condition [12]. In order to guarantee the transmission efficiency of real-time data, both of the server and the client have a larger buffer queue. We establish and maintain two timers for each connection due to the instability of wireless network: send timer and receive timer. These two timers can be reused by different state of each connection. In this reliable UDP Transport Protocol, accumulated confirmation and timeout confirmation are combined to achieve the confirmation so that we can have better control of the flow and maximize the use of network resources.

The actual testes that have been taken on the way which goes through GPRS DTU -> GPRS network -> Internet -> User Data center in GPRS network when the network under normal circumstances shows that the UDP transport effectiveness is greater than 99% and the TCP transport effectiveness is approximately equal to 100%. Practical applications prove that reliable UDP not only has high transfer efficiency but also can meet the accuracy requirements of the system at the same time.

3.4 *Cluster server and real-time processing software*

LWD real-time interpretation cluster system is composed of two members, member server is a DELL PowerEdge R710 server. We configured AHA 2944 SCSI channel adapter on each server, connected it to RAID Array 3000 array controller card on the RA3000 disk cabinet. There are four Ultra2 SCSI-3 hard disks placed in RA3000 disk cabinet. Each server was configured with two Intel 10/100 adaptive network cards, one is used for internal communication and the other one The other one is used for providing services to the client. Every member server is running with Microsoft Windows 2008 Server Enterprise Edition, Microsoft Cluster Server and Microsoft SQL Server Enterprise Edition client. The system is stable and has high reliability.

The real-time processing software on server side is developed on the basis of language C # 2.0 and .NET Framework. Besides, by using multi-threaded programming techniques it can meets the service needs of multi-well field at the same time and it is able to analyze the data that meet WITS (Wellsite Information Transfer Standard)/ WITSML (Wellsite Information Transfer Standard Markup Language)—the international standard protocol for petroleum industry data transmission, able to achieve the real-time data format conversion technology for the wellsite data that comes from various sources (such as LWD, mud logging, drilling, directional drilling, etc.), improve system's compatibility.

3.5 *Ajax-based real-time web publishing system*

Real-time drilling data dissemination system uses multi-B/S mode software architecture, based on Microsoft. NetFrameWork4.0 and SQL Server 2005 database, developed with ASP.NET, lots of Ajax technologies have been adopted to improve system's real-time property and interactivity. Ajax technology is an integrated application of multiple technologies in which XHTML and CSS are used for standardization rendering, DOM for dynamic display and interaction, XML and XSTL for data exchange and processing, XMLHttpRequest objects for asynchronous data reading, Javascript for binding and processing of all data.

During the running, system releases Web Method through Web Services platform and exposes real-time data to the Ajax client at first. Then, Ajax client pages uses XML HttpRequest objects to send SOAP requests to Web server, invokes Web Method that published by Web server through Web Services platform. At last, Web Method makes deals for Ajax requests, the data sequence that the browser client needs will be transformed into a XML document and returned in SOAP response messages. Finally, the browser uses DOM objects to parse the XML documents and get the data, refresh the page at last.

Based on Ajax technologies, system can implements the real-time display of drilling engineering parameters and Logging curves. There are several differences between Ajax applications and traditional Web applications. First, the response capability is more rapid if the communication with the server is in an asynchronous way. Second, it is not necessary to refresh the entire page when the data is updated, communicate with the server within the page. Third, the display effect is richer and the interactive capability is strong.

4 CONCLUSIONS

In this paper, the key technology in LWD real-time data transmission system has been researched and analysed. A perfect real-time transmission system with the following features has been established based on existing technologies: Good Compatibility, High Transmission Stability, Strong Processing Ability, Real-Time. This program and the equipment have been applied in Shengli oilfield and other oilfields' LWD real-time data transmission up to now. It has proved that this system has strong real-time, high stability, good compatibility and accurate data display.

The research was supported by supported by "the Fundamental Research Funds for the Central Universities".

REFERENCES

[1] Su Yinao. "The technology Survey and research of Geosteering while drilling in our country. "Petroleum exploration and development (32) 2005: 92–95.

[2] Partha P. Mitra, T.R. Joshi, Hugues Thevoux-Chabuel, Paradigm Geophysical. "Real Time Geosteering of High Tech Well in Virtual Reality and Prediction Ahead of Drill Bit for Cost Optimization and Risk Reduction in Mumbai High L-III Reservoir." Paper presented at SPE Asia Pacific Oil and Gas Conference, Perth, Australia, 18–20 October 2004.

[3] Thevoux-Chabuel, Fejerskov, Norsk."Geosteering Diagnosis: A New Approach to Monitor the Well Position Within a 3D Geological Model." Paper presented at SPE Annual Technical Conference, Antonio, Texas, 24–27 September 2006.

[4] Salem Al-Jaberi, Saeed Dama, Jasim Al-Rayaisah, Saad A.H. Said. "Real-Time 3D Modeling and Geosteering in a Heterogeneous Carbonate Reservoir" Paper presented at SPE/IADC Middle East Drilling Technology Conference & Exhibition, Manama, Bahrain, 26–28 October 2009.

[5] Ignacio Gorgone, Juan Gomez, Gary Uddenberg. "Remote Operations, the Future of Drilling: Firectional Drilling Remote Control" Paper presented at IADC/SPE Drilling Conference, Orlando, Florida, USA, 4–6 March 2008.

[6] Jeffrey Kok, PathFinder, Jeff DeJarnett, Daniel Geary. "Successful Geosteering in Low Resistivity Contrast Reservoirs of the Permian Basin" Paper presented at SPE Eastern Regional Meeting, Columbus, Ohio USA, 17–19 August 2011.

[7] S.K. Anand, J. Kumbhar, U. Singh, MSHM Jeelani, J. Singh. "Evolution of Geosteering Technology Helped in Successive Development of a Brown Field—Case Study from Mumbai High" Paper presented at the International Petroleum Technology Conference, Bangkok, Thailand, 7–9 February 2012.

[8] Martinez, E., Corpoven, S.A. "Directional Drilling Expert System" Paper presented at the second Latin American Petroleum Engineering Conference, Bangkok, Thailand, 8–11 March 2002.

[9] Tagir R. Salakhov, Vil U. Yamaliev, Vladimir Dubinsky. "An Expert System Approach to Real-Time Diagnosis of Drill Bit Condition and Preventing Its Damage" Paper presented at SPE Oil and Gas India Conference, Mumbai, India, 20–22 January 2010.

[10] A.S. Al-Yami, J. Schubert, G. Beck. "Expert System for the Optimal Design and Execution of Successful Completion Practices Using Artificial Bayesian Intelligence" Paper presented at Brasil Offshore Conference, Macae, Brazil,14–17 June 2011.

[11] Andrew S.T. "Computer Networks." Prentice Hall PTR, 2008.

E-Commerce, E-Business and E-Service – Lee (Ed)
© 2014 Taylor & Francis Group, London, ISBN 978-1-138-02646-9

Application of AF-SVR model for the endpoint prediction of converter steelmaking

Leifu Gao, Xiuli Zhang, Shijie Zhao & Dongmei Yu
Institute of Mathematics and System Science, Liaoning Technical University, Fuxin, China

ABSTRACT: The Endpoint control of converter steelmaking has the great significance for steelmaking and the neural network prediction model has the disadvantages of "over-fitting study" and "under-fitting study". The AF-SVR prediction model of the artificial fish swarm algorithm and the supporting vector regression's combination is proposed. And it is used in the prediction analysis of the endpoint temperature and carbon content of the converter steelmaking. This model has combined the better parallel and searching performance of the fish swarm algorithm and the better nonlinear fitting ability of the supporting vector regression and has overcome neural network existed over-fitting study problems. This model puts the estimation sample deviation sum minimized as the optimality principle. Then the prediction analysis is conducted for the molten steel's endpoint temperature and carbon content. Experiment simulation is carried out for the measured data of the convertor steelmaking by using the AF-SVR prediction model. The result shows that the better prediction accuracy and the higher hit rate can satisfy the error requirement produced by the steelmaking.

1 INTRODUCTION

Converter steelmaking is a very complex process of cyclical temperature, carbon reduction, decontamination, during the steel smelting process occurs with a series of very complex diverse, heterogeneous, high temperature physical and chemical reaction, at the same time there are many variable of time-varying and nonlinear making influence, and there are strong complex coupling relationship between variables, it bring great difficulty to the accurate endpoint prediction of converter steelmaking. The stand or fall of the converter steelmaking endpoint control has a close relationship with the rates of steelmaking production, metal yield, production cost and the quality of the steel and other indexes,[1] and the purpose of the converter steelmaking is to achieve the requirements about endpoint temperature and carbon content index of the steel grade at the end of the melting process of the molten steel, so the prediction analysis and research of converter steelmaking end-point control has been an important research direction. At present, the prediction models of converter steelmaking end-point include the prediction model based on the samples of variable selection,[2] the prediction model based on statistical regression analysis,[3] the prediction model based on image analysis[4] and the prediction model based on artificial neural network,[5–7] etc.

With the development of intelligent algorithms, more and more intelligent algorithms are applied to the prediction analysis of converter steelmaking endpoint control, mainly include: grey model,[8] case-based reasoning analysis,[9] relevance vector machine,[10] and the prediction model with the fusion of many kinds of intelligent algorithms.[11–13]

Since the establishment of the statistical regression model needs a large number of data samples, and is aimed at a specific converter making its reproducibility poor, using it in the endpoint prediction is able, but there is still a certain gap between forecast accuracy with index requirements if using it in the analysis of the real-time monitoring, although the artificial neural network has better fitting performance for nonlinear problems, and has a certain effect on converter steelmaking endpoint control, but it has disadvantages of "over-fitting study", "under-fitting study" and "dimension disaster" problem in the training process. Considering the above problems, there proposed AF-SVR prediction model (Artificial Fish-Support Vector Regression prediction model), and used it in the prediction analysis of converter steelmaking endpoint, it combines the supporting vector regression machine with better nonlinear fitting ability and at the same time overcomes the disadvantages of over-fitting study or under-fitting study of artificial neural network with the artificial fish swarm algorithm of great parallel search performance. Artificial fish swarm algorithm[14–16] is a new bionic algorithm with strong ability of parallel processing and high optimization speed, it's not sensitive to the initial value, and has the advantages of global optimization ability. The model optimizes the parameters selection of support vector regression machine by artificial fish swarm algorithm, it can converge to the global extreme value point faster by using the parallel performance of artificial fish swarm algorithm, and establishes the objective function by using minimizing the sample deviation sum as the optimization principle, so as to realize the optimized selection of kernel parameters and penalty parameter of support vector regression machine, and to enhance the accuracy of data prediction and fitting effect of support vector regression machine. THE THEORY OF SUPPORT VECTOR REGRESSION MACHINE

If the training sample data set is $\mathbf{A} = \{(x_1,y_1),(x_2,y_2),\ldots,(x_N,y_N)\}$, and $x_i \in \mathbf{R}^n$ correspond to the influence factors of the endpoint temperature and carbon content in converter steelmaking endpoint control; $y_i \in \mathbf{R}$ correspond to the value of the endpoint temperature and carbon content in converter steelmaking, $i = 1, 2, \ldots, N$. For linear ε-support vector regression machine,[17] the convex quadratic programming of original optimization problem is

$$\begin{aligned}&\min_{\omega,b,\xi^{(*)}} \frac{1}{2}\omega^T\cdot\omega + C\sum_{i=1}^{N}(\xi_i+\xi_i^*)\\ &s.t.\ ((\omega\cdot x_i)+b)-y_i \le \varepsilon+\xi_i,\ i=1,2,\ldots,N\\ &\quad y_i-((\omega\cdot x_i)+b)\le \varepsilon+\xi_i^*,\ i=1,2,\ldots,N\\ &\quad \xi_i^{(*)}\ge 0, i=1,2,\ldots,N\end{aligned} \tag{1}$$

where $\omega \in \mathbf{R}^n$, $b \in \mathbf{R}(*)$ is the simple notation of vectors with * or not.

Use Lagrange function to solve problem (1), the corresponding convex quadratic programming problem is

$$\begin{aligned}&\min_{\lambda^{(*)}\in R^{2N}} \frac{1}{2}\sum_{i,j=1}^{N}(\lambda_i^*-\lambda_i)(\lambda_j^*-\lambda_j)(x_i^*\cdot x_j)\\ &\qquad +\varepsilon\sum_{i=1}^{N}(\lambda_i^*-\lambda_j)-\sum_{i=1}^{N}y_i(\lambda_i^*-\lambda_j)\\ &s.t.\sum_{i=1}^{N}(\lambda_j^*-\lambda_i)=0, 0\le\lambda_i^*\le C, i=1,2,\ldots,N\end{aligned} \tag{2}$$

where λ_i are lagrange multipliers.

Solve the dual problem can get the optimal solution $\bar{\lambda}^{(*)} = (\bar{\lambda}_1,\bar{\lambda}_1^*,\ldots,\bar{\lambda}_N,\bar{\lambda}_N^*)^T$. For $\bar{\lambda}$, select the positive component as $\bar{\lambda} > 0$, then calculate

$$\bar{b} = y_i - \sum_{i=1}^{N}(\bar{\lambda}_i^* - \bar{\lambda}_i)K(x_i\cdot x_j) - \varepsilon \tag{3}$$

and get the decision function of linear ε-support vector regression machine

$$f(x) = \sum_{i=1}^{N}(\bar{\lambda}_i^* - \bar{\lambda}_i)K(x_i\cdot x) + \bar{b} \tag{4}$$

For linear inseparable problems, need to replace linear partition with nonlinear partition, ordinary the data are mapped to high-dimensional feature space from low-dimensional space by nonlinear mapping (selecting the proper kernel function $K(x_i,x_j)$) to make it linear separable. Common kernel functions of SVR are mainly Poly kernel, Gaussian kernel (RBF kernel), Sigmoid kernel, and it's able to construct the corresponding kernel function according to the specific issues. The decision function of nonlinear ε-support vector regression machine is

$$f(x,\lambda) = \sum_{i=1}^{N}(\bar{\lambda}_i^* - \bar{\lambda}_i)K(x_i,x_j) + \bar{b} \tag{5}$$

where $K(\cdot)$ is kernel function.

2 THE PREDICTION MODEL OF MOLTEN STEEL ENDPOINT BASED ON AF-SVR ALGORITHM

Converter steelmaking process is a complex nonlinear process, the molten steel endpoint of molten pool is influenced by many nonlinear factors, mainly including the quality of molten iron and scrap steel, the temperature and carbon content of molten steel in sublance, the content or time of blowing oxygen, the quality of the coolants and additives (such as massive limestone, light burning white lime, the sphere of magnesite, massive dolomite, etc) etc. While support vector regression machine has great processing performance, it's beneficial to improve the endpoint hit rate of converter steelmaking, at the same time, due to the performance of SVR is affected by the kernel function, namely the need to optimize the kernel parameter selection, in order to improve the prediction performance of support vector regression machine, so AF-SVR prediction model is put forward, the model uses the parallel optimization performance of artificial fish swarm algorithm to optimize the kernel parameters of SVR model, and uses the optimized kernel parameters for SVR model in prediction analysis of the temperature and carbon content of converter steelmaking endpoint. The parameters of AF-SVR prediction model for predicting converter steelmaking endpoint temperature and carbon content are mainly penalty parameter C of SVR and kernel parameter σ, and the optimization goal is to determine the optimal combination of parameters (C,σ) so as to make the model get the minimum deviation between prediction value and actual value of the samples, the specific operation steps of the model are:

Step 1: Set the model parameters. Artificial fish swarm algorithm parameter setting includes fish scale *Fish_N*, maximum iterations number Max_G of fish swarm, maximum test number *Try_N* of foraging, percept distance *Visual*, moving step length $Step\_L$ and crowded degree factor δ of fish; Support vector regression machine parameter setting is to set the upper and lower limit values of penalty parameter C and kernel parameter σ;

Step 2: Initialize the artificial fish. Each of the artificial fish is SVR parameter combination (C, σ) to optimize; Initialize the artificial fish random according to the value ranges of C and σ in Step1, the whole fish swarm is a matrix of $2 \times Fish\_N$, each operation makes the *Fish_N* artificial fish parallel optimize, so it improves the optimization performance;

Step 3: Selecting the converter steelmaking measured data set by support vector regression machine, mainly includes the training set M_Train, training tag set L_Train, prediction set P_Test and prediction tag set L_Test;

Step 4: Calculate the initial food density of fish. Select appropriate kernel function of support vector regression machine, train SVR according to the training set M_Train and training tag set L_Train, and use the trained model in the regression prediction of prediction set P_Test and prediction tag set L_Test, where SVR regards the sum of all samples' prediction error as artificial fish food density and compares the densities, selects the minimum as the optimal value of the current fish swarm, and saves the parameter combination (C, σ) of artificial fish corresponding to current optimal value and the prediction values of prediction sample set;

Step 5: Operate the behavior of artificial fish in swarm. Artificial fish simulate foraging behavior, rear-end behavior and group behavior, and carry out the behavior with smallest food density. Missing behavior is random behavior, according to the percept distance *Visual* and moving step length $Step\_L$ of artificial fish random walk;

Step 6: Select the optimal food density. Each time perform a behavior operation, will calculate the lowest food density of current fish: if there is an artificial fish with lower food density than current optimal density, use it to replace the current optimal food density, and save the parameter combination (C, σ) of this artificial fish and the prediction values of prediction sample got from the model, otherwise save the original density and its corresponding parameter combination (C, σ) of artificial fish and the prediction values of prediction sample;

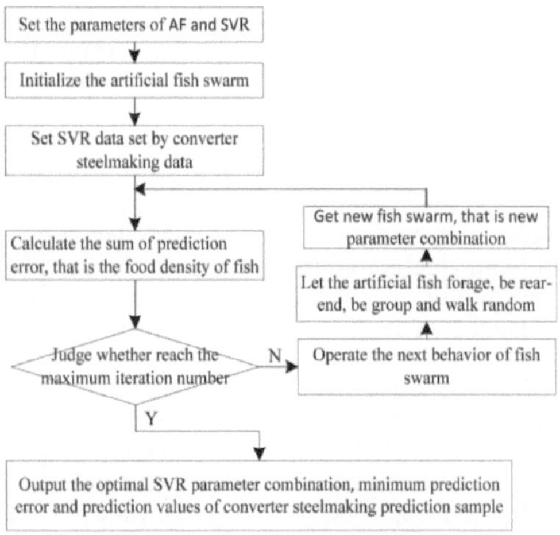

Figure 1. The flow chart of the prediction of converter steelmaking endpoint based on AF-SVR algorithm.

Step 7: Judge whether meet the termination condition of the algorithm: judge whether reach the maximum iteration number Max_G of fish, if reach then output the optimal food density value of fish and its corresponding parameter combination (C, σ) of artificial fish and the prediction values of prediction sample, otherwise the number of iteration add 1, and jump to operate Step 5.

The flow chart of the prediction model of converter steelmaking endpoint temperature and carbon content by based on AF-SVR algorithm as in Figure 1.

3 EXPERIMENTAL RESEARCH AND ANALYSIS

Take 60 converters measured data for 180 tons converter from some steelmaking plant as the analysis and research object, predict the converter steelmaking data by AF-SVR, take the prediction error of converter steelmaking endpoint by the model and hit rate as evaluation standard, and in the form of visual display that AF-SVR prediction model has good fitting effect, to verify the effectiveness and feasibility of AF-SVR in predicting converter steelmaking measured.

3.1 *Data processing and data set selection*

The influence factors of converter steelmaking are multidimensional nonlinear factors, the data from different factors have different dimension and order of magnitude, in order to eliminate the dimension difference of various factors, the data need to be standardized. Standardization of its formula is:

$$y_{ij} = \frac{x_{ij} - \overline{x_i}}{\sigma_i} \tag{6}$$

where $\overline{x_i}$ and σ_i respectively represents the arithmetic average value and standard deviation of influence factor index columns of the converter steelmaking measured data, x_{ij} is the actual measured value, y_{ij} is the standardized measured value.

For 60 converters measured data, select 30 converters data through the method of random selection to build SVR model, the remaining data are used to test the prediction effect of the model.

Random selection is defined as: random selection of data sets: according to the number of the converter steelmaking data N, arrange the N integers of integer set $\{1, 2, \ldots, N\}$ random, then according to the number n of training set M_Train in SVR model's need, select the front n positions of the random set, select the samples in same positions from original data set and input training set, finally get training set M_Train composed of n samples; the remaining N-n samples are classified as prediction data set P_Test (Here $N = 60$, $n = 30$).

3.2 *Model parameters set*

In training of support vector regression machine model, use Gaussian kernel function, the expression is

$$K(x_i, x) = exp\{\gamma | x_i - x |^2\}$$

Table 1. Comparison table of prediction analysis of the two models for converter steelmaking data.

	Converter steelmaking endpoint temperature T (°C)					Converter steelmaking endpoint carbon content C ($\times 10^{-2}$%)				
Number	Actual value	BP value	Deviation	AF-SVR value	Deviation	Actual value	BP value	Deviation	AF-SVR value	Deviation
1	1630	1623	7	1616	−14	9	10	−1	11	2
2	1625	1643	−18	1622	−3	8	23	−15	16	8
3	1640	1644	−4	1635	−5	13	24	−11	12	−1
4	1640	1634	6	1654	14	12	9	3	11	−1
5	1650	1626	24	1645	−5	22	25	−3	20	−2
6	1665	1677	−12	1657	−8	10	10	−0	18	8
7	1615	1616	−1	1617	2	13	12	1	13	0
8	1610	1628	−18	1617	7	13	10	3	13	0
9	1635	1642	−7	1638	3	14	10	4	13	−1
10	1630	1638	−8	1641	11	47	27	20	33	−14
11	1630	1628	2	1622	−8	17	25	−8	19	2
12	1590	1595	−5	1621	31	14	9	5	14	0
13	1645	1627	18	1636	−9	12	24	−12	12	0
14	1645	1643	2	1656	11	17	19	−2	17	0
15	1625	1613	12	1616	−9	16	20	−4	16	0
16	1650	1683	−33	1649	−1	12	21	−9	18	6
17	1625	1611	14	1620	−5	57	46	11	33	−24
18	1625	1609	16	1627	2	15	11	4	15	0
19	1630	1636	−6	1646	16	13	24	−11	18	5
20	1620	1675	−55	1616	−4	13	12	1	15	2
21	1620	1629	−9	1635	15	11	8	3	16	5
22	1680	1607	73	1691	11	17	10	7	13	−4
23	1630	1605	25	1637	7	8	8	0	9	1
24	1635	1697	−62	1612	−23	15	24	−9	17	2
25	1655	1630	25	1661	6	6	4	2	6	0
26	1640	1615	25	1617	−23	10	6	4	10	0
27	1640	1637	3	1633	−7	21	24	−3	19	−2
28	1640	1630	10	1634	−6	10	16	−6	18	8
29	1605	1613	−8	1617	12	15	16	−1	14	−1
30	1610	1605	5	1612	2	15	19	−4	19	4

where $\gamma = -1/2\sigma^2$. In the parameters to optimize of SVR, the value range of kernel parameter γ is [0,1], the value range of penalty parameter C is [0,50]; the setting parameters of artificial fish swarm algorithm are: evolution number *Max_G* is 50, fish scale *Fish_N* is 50, maximum test number *Try_N* of foraging is 100, crowded degree factor δ of fish is 0.618, percept distance *Visual* is 1 and moving step length *Step_L* is 0.1. The setting parameters of BP neural network are: node number *R_num* of input layer is 30, node number *Y_num* of hidden layer is 7, node number *C_num* of output layer is 30, number *Train_num* of network training is 10000, error δ of network training is 0.0001 and learning rate *lr* of network training is 0.1.

3.3 *Model evaluation and results analysis*

For 60 converters measured data, select 30 converters data through the method of random selection to train model, the remaining data are used to predict converter steelmaking endpoint by AF-SVR model, put the estimation sample deviation sum of the 30 prediction data minimizing as the optimality principle, analyze and judge the hit rate of converter steelmaking endpoint by the model, prediction analysis for converter steelmaking data of AF-SVR model and BP neural network are shown in Table 1. Where the hit rate of converter steelmaking endpoint prediction by of model is defined as:

hit rate = the number of converter meeting the error limit/actual total number of converter predicted.

Table 1 shows that: analyze the prediction of the 30 measured converter steelmaking by using AF-SVR prediction model, when the limited condition of endpoint temperature is $|\Delta| \leq 15$, the hit rate of endpoint temperature by AF-SVR model is 86.7%, and is higher than that of BP neural network 60.0%; when the limited condition of endpoint carbon content is $|\Delta C| \leq 0.05\%$, the hit rate of endpoint carbon content by AF-SVR model is 80.0%, and is higher than that of BP neural network 63.3%. Accordingly, in the prediction analysis of the 60 converters measured data for 180 tons converter from the steelmaking plant, AF-SVR model has higher hit ratio and relative smaller prediction deviation than BP neural network.

For further analyze the intuitional relations between prediction value and actual value of AF-SVR model and BP neural network for the prediction samples, the contrast diagrams for prediction values and actual values of the two models in molten steel endpoint temperature and carbon content are as shown in Figures 2 and 3.

Figures 2 and 3 show that: in general, in the prediction analysis of the 60 converters measured data for 180 tons converter from the steelmaking plant,

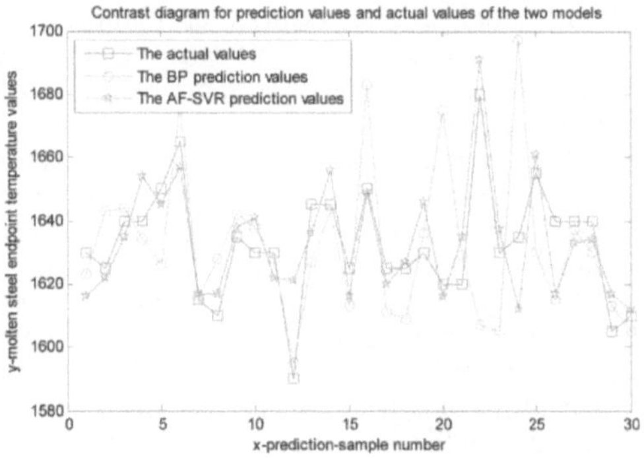

Figure 2. Contrast diagram for prediction values and actual values of the two models in molten steel endpoint temperature.

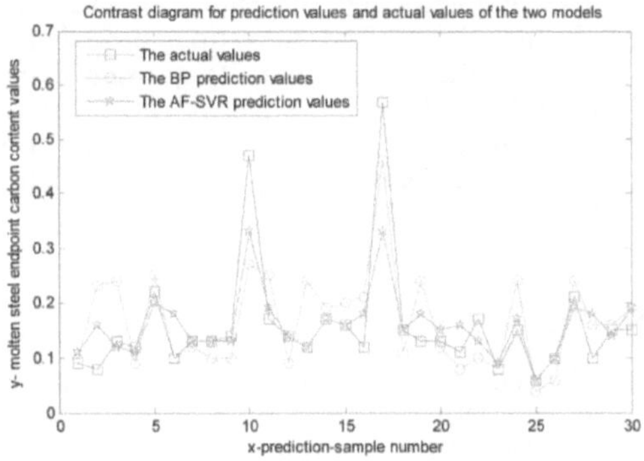

Figure 3. Contrast diagram for prediction values and actual values of the two models in molten steel endpoint carbon content.

AF-SVR model has better fitting performance for the measured samples of converter steelmaking prediction, as well as higher prediction accuracy and smaller prediction deviation, it can meet the deviation limit requirements of industrial production better, so AF-SVR model is effective and feasible for the prediction analysis of converter steelmaking endpoint.

4 CONCLUSION

1. Converter steelmaking process is influenced by many nonlinear factors, simple linear model is difficult to carry on the accurate prediction control, and traditional regression model has poor reproducibility, here use support vector regression machine with better nonlinear fitting ability in the prediction analysis of converter steelmaking endpoint, use artificial fish swarm algorithm to realize the optimized selection of

SVR model parameters, in order to improve the prediction accuracy and hit ratio of converter steelmaking endpoint by support vector regression machine.
2. Support vector regression machine prediction model (AF-SVR) based on artificial fish swarm algorithm and put forward here, puts the estimation sample deviation sum minimized as the optimality principle, by using the strong parallel search performance and the feature of being insensitive to initial value of artificial fish swarm algorithm, enhances the performance of AF-SVR model for converter steelmaking endpoint prediction in nonlinear processing and parallel search, etc., and it's in favour of meeting prediction precision better and faster.
3. In the contrast analysis of experimental prediction of converter steelmaking endpoint between AF-SVR model and BP neural network, AF-SVR model has higher hit ratio and prediction accuracy, and it has better fitting performance for each prediction sample, it can meet the deviation requirements of production better, it's a kind of feasible and effective prediction model.

REFERENCES

[1] China metallurgical encyclopedia. Iron and steel metallurgy/metallurgy encyclopedia editor committee of China iron and steel metallurgy, editorial board, metallurgical industry press, the China metallurgical encyclopedia newsroom. Beijing: metallurgical industry press, 2001.3.

[2] Wang Xin-zhe, Han Min. Variable selection based BOF endpoint prediction model[J]. Control and Decision, 2010, 25(10):1589–1592.

[3] Zhang Hui-yi, Zhou Qi-long, Yuan Zhi-xiang, Liu Zhi-ming, Zhou Yun. Application of the Regression Algorithm Based on Sample-Self-Selection in BOF Steelmaking[J]. Journal of Iron and Steel Research, 2011, 23(12):5–8.

[4] Yang You-liang, Zhang Yan, Meng Fan-wei. Prediction model of converter steelmaking endpoint based on spectral information and temperature[J]. Metallurgical Analysis, 2012, 32(11):1–5.

[5] Xie Shu-ming, Tao jun, Chai Tian-you. BOF steelmaking endpoint control based on neural network[J]. Control Theory & Applications, 2003, 20(6):903–907.

[6] Fileti A.M.F., Pacianotto T.A., Cunha A.P. Neural modeling helps the BOS process to achieve aimed end-point conditions in liquid steel[J]. *Engineering Applications of Artificial Intelligence*, 2006, 19(1):9–17.

[7] Han M., Wang X.Z., Wang Y.J. Applying ICA on neural network to simplify BOF endpoint predicting model[J]. IEEE World Congress on Computational Intelligence. Hong Kong, 2008:772–777.

[8] Xie Shu-ming, Gao Xian-wen, Chai Tian-you. BOF Endpoint Prediction Based on Grey Model[J]. Journal of Iron and Steel Research, 1999, 11(4):9–12.

[9] Wang Xin-zhe, Han Min. Causality-based CBR model for static control of converter steelmaking[J]. Journal of Dalian University of Technology, 2011, 51(4):593–598.

[10] Han Min, Zhao Yao, Yang Xi-lin, Lin Dong. Endpoint prediction model of basic oxygen furnace steelmaking based on robust relevance-vector-machines[J]. Control Theory & Applications, 2011, 28(3):343–349.

[11] Sriram, M.V.V.N., Singh, N.K., Rajaraman, G. Neuro fuzzy modelling of Basic Oxygen Furnace and its comparison with Neural Network and GRNN models[C]// Computational Intelligence and Computing Research. Coimbatore: IEEE, 2010:1–8.

[12] Han Min, Jiang Li-wen, Zhao Yao. Endpoint Prediction Model of Basic Oxygen Furnace Steelmaking Based on PSO-ICA and RBF Neural Network[J]. Information and Control, 2010, 39(1):83–87.

[13] Zhou Guo-xiong, Li Lin, Shen Xue-jie. Integrated Prediction for Carbon Content in Fly Ash Based on Online Support Vector Machine and Grey Prediction[J]. Journal of System Simulation, 2013, 25(4):727–731.

[14] Li Xiao-lei. A New Intelligent Optimization Method-Artificial Fish School Algorithm[D]. Hangzhou: Zhejiang University, 2003.

[15] Li Xiao-lei, Shao Zhi-jiang, Qian Ji-xin. An optimizing method based on autonomous animats: fish-swarm algorithm[J]. Systems Engineering-Theory & Practice, 2002, 22(11):32–38.

[16] Li Xiao-lei, Qian Ji-xin. Studies on artificial fish swarm optimization algorithm based on decomposition and coordination techniques[J]. Journal of Circuits and Systems, 2003, 8(1):1–6.

[17] Deng nai-yang, Tian ying-jie. Support Vector Machine—methory, algorithm and development[M]. Beijing:Science Press, 2009.

E-Commerce, E-Business and E-Service – Lee (Ed)
 ISBN 978-1-138-02646-9

Feasibility analysis of spectra partition under a monopolization and sharing pattern

P. Zhang & L.N. Wu
School of Information Science and Engineering, Southeast University, Nanjing, China

ABSTRACT: The limited spectral resource is increasingly shortage under classical rectangular spectra partition pattern which can not make full use of spectral resource. To improve spectral efficiency, a non-rectangular spectra partition pattern (namely monopolization and sharing spectra partition pattern) is proposed in view of the excellent spectral efficiency performance of Extended Binary Phase Shift Keying (EBPSK) modulations. The channel bandwidth and capacity of EBPSK is analyzed, and then the feasibility of this monopolization and sharing spectra partition pattern is studied from the perspective of information theory. Results indicate that, this new spectra partition pattern is feasible, and is expected to become a more advanced spectra partition pattern.

1 INTRODUCTION

Classical rectangular spectra partition pattern allocates exclusive spectral resource to authorized users, and moreover the ideal rectangular frequency domain filter can not be realized, so there are many unused spectrum "holes" and underused spectrum blocks. Obviously, traditional wireless communication systems employ wasteful static spectrum allocation which significantly decreases the efficient usage of spectrum.

With the advent of Cognitive Radio (CR), people can use the portion of radio spectrum which is unused or underused to transmit information. By spectrum sensing and spectrum sharing, spectral efficiency can be improved. A typical example is the Ultra Wide Band (UWB), which has a very wide bandwidth and very low Power Spectral Density (PSD). This extremely low spectral density of UWB transmission avoids significant interference to existing narrowband communication systems operating in the range of the UWB transmission (Wu & Natarajan 2007). Another modulation technique opposite to UWB is the one called Ultra Narrow Band (UNB) (Walker 1997). Since Walker proposed this idea, a series of UNB modulation methods have been put forward. The typical characteristic of UNB in frequency domain is an ultra narrow "spectrum tree (discrete spectrum)" plus broad "spectrum grass (continuous spectrum)". The ultra narrow "spectrum tree" contains the most part of signal energy, while the PSD of "spectrum grass" may be lower than UWB, so UNB can get an excellent spectral efficiency.

The PSD of "spectrum grass" is extremely low, even submerged in background noise, therefore can coexist with other users with little interference to them. The researches on UNB for many years by Wu Lenan team in Southeast University have made EBPSK (Zhang & Wu 2007) tend to be mature, not only bring about a great improvement of spectral efficiency, but also make people begin to think about the shortcomings of classical rectangular spectra partition pattern and explore the high-capacity non-rectangular spectra partition pattern. Due to the more excellent spectral efficiency of UNB compared with traditional modulations, it is expectable that the intensive multi-carrier UNB transmission mode will be a better solution for spectral efficiency problem, which can form a non-rectangular pattern of spectra partition with a monopolized ultra narrow spectrum tree plus shared broad spectrum grass. In this paper, the feasibility of this monopolization and sharing spectra partition pattern is studied from the perspective of information theory. The UNB method used in the article is EBPSK.

2 CHANNEL BANDWIDTH OF EBPSK

EBPSK modulation method (Wu et al. 2011):
Assuming that:

① The symbol width of data (namely the temporal length of a code) T lasts $N \geq 1$ cycles of the carrier, $T = N/f_c$ where f_c is the carrier frequency.
② The temporal length $\tau = K/f_c$ of a hopping waveform lasts $K \leq N$ cycles of the carrier.
③ For logical "0", waveform $g_0(t)$ is transmitted in the symbol period T, and for logical "1", $g_1(t)$ is transmitted. The modulation angle is θ.

Then the expression of EBPSK is defined as below.

$$g_0(t) = A\sin 2\pi f_c t, \qquad 0 \le t \le T \tag{1}$$

$$g_1(t) = \begin{cases} B\sin(2\pi f_c t + \theta), & 0 \le t \le \tau, 0 \le \theta \le \pi \\ A\sin 2\pi f_c t, & \tau \le t \le T \end{cases} \tag{2}$$

The PSD of EBPSK modulation has been analyzed (Feng et al. 2008). The information is contained by continuous spectrum, so we focus on the bandwidth of continuous spectrum. According to the PSD of EBPSK, we can work out the main lobe bandwidth of EBPSK continuous spectrum, namely the null-to-null bandwidth:

$$W = 2/\tau \tag{3}$$

The modulation duty cycle is $\tau/T = K/N$, and $T = 1/f_s$, where f_s is the symbol rate, so we can get

$$\tau = \frac{K}{N}T = \frac{K}{Nf_s} = \frac{K}{f_c} \tag{4}$$

Hence the null-to-null bandwidth of EBPSK signal is

$$W = \frac{2f_c}{K} \tag{5}$$

3 CHANNEL CAPACITY OF EBPSK

The channel capacity of EBPSK can be written as below

$$C = W\log_2\left(1 + \frac{S}{n_0 W}\right) \quad (bit/s) \tag{6}$$

where n_0 is the single-side PSD of noise, and S is the signal power of continuous spectrum. Assuming the average PSD of continuous spectrum is Φ_p, namely $S = \Phi_P W$, then

$$C = W\log_2\left(1 + \frac{\Phi_P}{n_0}\right) \tag{7}$$

It is can be seen from above expression that the channel capacity of EBPSK depends on W and Φ_P/n_0. Due to the sparse characteristic in time domain of UNB, the EBPSK modulation parameter K is very small, so $W = 2f_c/K$ is considerably big. Hence although the extremely low power spectral density of continuous spectrum makes Φ_P/n_0 be small, the channel capacity C still can be very large.

An example: If carrier frequency is $f_c = 2.4\,GHz$, and modulation parameters are $K = 2$, $N = 1200$, then the null-to-null bandwidth of EBPSK is 2.4 GHz, even if the power of continuous spectrum to noise power ratio is only $-10\ dB$, namely $\Phi_P/n_0 = 0.1$, the channel capacity still is considerable as below.

$$C = 2.4\times 10^9 \log_2(1+0.1) = 330\,Mbps \tag{8}$$

But due to the big modulation parameter $N = 1200$, the transmission rate is relatively smaller than the channel capacity, which is merely $R_b = f_c/N = 2\ Mbps$. According to Shannon information theory (Shannon 2001) we know that, error-free transmission can be achieved when the transmission rate is less than the channel capacity.

4 FEASIBILITY ANALYSIS OF THE PROPOSED SPECTRA PARTITION PATTERN

4.1 *Monopolization and sharing spectra partition pattern*

Shannon channel capacity equation indicates that the maximum amount of information can theoretically be transmitted in unit time when the power of signal and Additive White Gaussian Noise (AWGN) in a channel with certain bandwidth is given. Only if the transmission rate is less than or equal to channel capacity, theoretically we can always find a channel coding method to achieve error-free transmission. Else if the transmission rate is greater than channel capacity, it is impossible to achieve error-free transmission. The channel capacity of the example in Section 3 is 330 *Mbps*, much more than the 2 *Mbps* transmission rate. It is thus clear that the channel capacity is not fully used and still has a lot of surplus; this will be the theoretical basis for the monopolization and sharing spectra partition pattern. By partitioning spectra under this pattern, building an intensive multi-carrier UNB transmission system, and making each carrier share spectrum, in theory, we can make full use of the channel capacity and improve spectral efficiency considerably. So it is expected to become a more advanced spectra partition pattern than existing rectangular spectra partition pattern.

The sketch map for spectrum sharing of the intensive multi-carrier UNB transmission system is shown as Figure 1, in frequency domain any two neighboring carriers are placed like that. It can be seen that in intensive multi-carrier UNB transmission system, for each carrier channel there is not only AWGN, but also the interference from many other carrier channels. According to the Central Limit Theorem (CLT), the distribution of many independent EBPSK signals superposed tends to

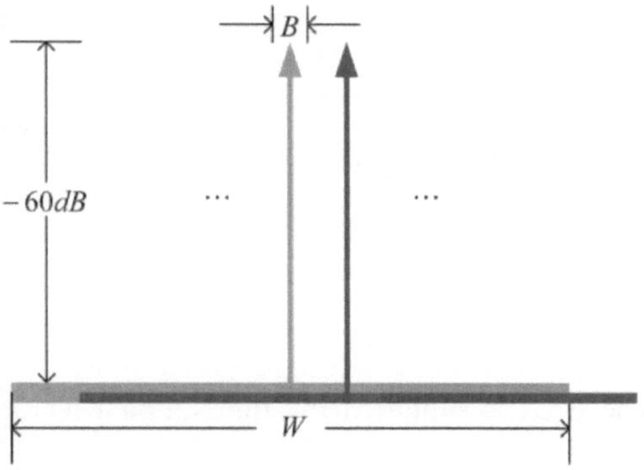

Figure 1. The sketch map for spectrum of intensive multi-carrier UNB.

be the normal distribution. So we can use Signal To Interference and Noise Ratio (SINR) to substitute for Signal to Noise Ratio (SNR).

The channel capacity of any carrier channel in intensive multi-carrier EBPSK can be represented as follow:

$$C = W\log_2\left(1+\frac{\Phi_p W}{n_0 W + \sum_{i=1}^{n}\Phi_{pi}W_{pi}}\right) \tag{9}$$

where W is the bandwidth of the carrier channel, Φ_p is the average PSD of continuous spectrum of signal, n is the total interference number from other carrier channels within the bandwidth W, Φ_{pi} is the average PSD of ith interference (only continuous spectrum is considered as interference), and W_{pi} is the corresponding bandwidth of ith interference.

Notice that each EBPSK carrier channel suffers interference from many other carrier channels, which is characterized by $\sum_{i=1}^{n}\Phi_{pi}W_{pi}$. The SINR is significantly decreased, so for error-free transmission, the allowable interference should be limited; namely the channel capacity with interference can not be less than the transmission rate, otherwise the information will not be transmitted accurately. The allowable amount of interference determines the number of carrier channels that can share spectrum with each other.

Still taking 2.4 *GHz* carrier frequency for example, even if the SINR is decreased to –30 *dB*, namely the signal power of continuous spectrum is merely one thousandth of the interference and noise power, the channel capacity is still greater than transmission rate. If the SNR is 0 *dB* when interference is not taken into account and the SINR is –30 *dB*, to measure the allowable amount of interference, we can assume the power of interference is x times that of signal. So if S is the signal power of continuous spectrum, N_0 is noise power and xS is interference power, an equation can be established as follow:

$$\begin{cases} \dfrac{S}{N_0} = 1 \\ \dfrac{S}{N_0 + xS} = \dfrac{1}{1000} \end{cases} \tag{10}$$

The solution to this equation is $x = 999$. That means in the situation of the SINR decreased from 0 *dB* to –30 *dB*, the allowable interference is 999 times as large as the signal power of continuous spectrum. This is a considerable number. Actually, W_{pi} in (9) is usually less than the bandwidth W, namely $W_{pi} \le W$, especially when the other carrier frequency is far away, W_{pi} is much less than W. So it is expected that the number of carrier channels sharing spectrum can be bigger, and the interval between carrier frequencies can be smaller. Thus an intensive multi-carrier UNB transmission system can be established.

4.2 *Feasibility analysis*

The discussion above with an example has just preliminarily shown the feasibility of the monopolization and sharing spectra partition pattern. In order to be more general and convincible, the following gives the constraint inequality in channel capacity aspect and analysis.

In essence, the monopolization and sharing spectra partition pattern is the spectrum sharing pattern of multi-carrier EBPSK. For any carrier channel, its channel capacity has been given in (9). According to the viewpoint of Shannon information theory, to ensure error-free transmission of information, we must make the channel capacity no less than the transmission rate. Hence, the key of the monopolization and sharing spectra partition pattern is to ensure that the channel capacity of any carrier channel is greater than or equal to the transmission rate when this carrier channel is interfered by noise and other carrier channels. That is:

$$C = W\log_2\left(1+\frac{\Phi_p W}{n_0 W + \sum_{i=1}^{n}\Phi_{pi}W_{pi}}\right) \ge R_b \tag{11}$$

Assuming that $S = \Phi_p W$, $N_0 = n_0 W$ and $xS = \sum_{i=1}^{n}\Phi_{pi}W_{pi}$, because $W = 2f_c/K$ and $R_b = f_c/N$, so the inequality (11) can now be rewritten as below:

$$2\frac{f_c}{K}\log_2\left(1+\frac{S}{N_0 + xS}\right) \ge \frac{f_c}{N} \tag{12}$$

It can simplified as

$$\log_2\left(1+\frac{S}{N_0+xS}\right)\geq\frac{K}{2N} \tag{13}$$

The purpose of doing like this is still to measure the allowable amount of interference by comparing interference with signal power. The solution to inequality (13) is

$$x\leq\frac{1}{2^{K/2N}-1}-\frac{N_0}{S} \tag{14}$$

It is thus clear that the maximum value of x depends on K/N and N_0/S, namely depends on the modulation duty cycle and the signal to noise ratio.

Substituting $K=2$, $N=1200$ into inequality (14), and assuming that the signal of continuous spectrum to noise ratio is -10 dB, then the solution to the inequality is $x \leq 1720.7$. The maximum value of x is considerable, it means that a large number of interference about 1720.7 times the power of continuous spectrum is allowable. In the bandwidth W, a large number of EBPSK signals can share spectrum and the error-free transmission can be also achieved.

5 CONCLUSIONS

From the previous analysis we can see that, the monopolization and sharing spectra partition pattern is feasible in the perspective of information theory. Theoretically, an intensive multi-carrier EBPSK transmission system is expected to be established.

But from the perspective of demodulation, when the SINR is decreased to some extent, the demodulation performance will be worse. As a new spectra partition pattern, and to solve the problems of demodulation, there are still some key issues such as channel coding and multiple-access techniques need to be further studied.

ACKNOWLEDGMENTS

The work is supported by the National Natural Science Foundation of China under the grant 61271204 and the National Key Technology R&D Program under the grant 2012BAH15B02.

REFERENCES

Feng, M. et al. 2008. Analysis and optimization of power spectrum on EBPSK modulation in throughput-efficient wireless system. *Journal of Southeast University (English Edition)* 24(2): 143–148.

Shannon, C.E. 2001. A mathematical theory of communication. *ACM SIGMOBILE Mobile Computing and Communications Review* 5: 3–55.

Walker, H.R. 1997. Understanding ultra narrow band modulation. *Microwaves & RF* 12: 173–186.

Wu, L.N. et al. 2011. Recent patents on ultra narrow band modulations. *Recent Patents on Signal Processing* 1(1): 36–47.

Wu, Z. & Natarajan, B. 2000. Interference tolerant agile cognitive radio: Maximize channel capacity of cognitive radio. *Consumer Communications and Networking Conference* CCNC 2007: 1027–1031.

Zhang, S.K. & Wu, L.N. 2007. Raise spectrum efficiency with extended BPSK modulation. *11th World Multi-conference on Systemics, Cybernetics and Information* 2007: 242–246.

E-Commerce, E-Business and E-Service – Lee (Ed)
© 2014 Taylor & Francis Group, London, ISBN 978-1-138-02646-9

An implementation of hardware-based TCP for TS transmitting

T. Tan & X. Wang
Shenzhen Engineering Laboratory of IS&DRM, Department of Electronic Engineering, Graduate School at Shenzhen, Tsinghua University, Shenzhen, China

G. Sun & A. Ji
Integrated Circuit Department, Unitend Technologies Inc., Beijing, China

ABSTRACT: Transport Stream (TS) transmitting over TCP can bring heavy burden to CPU on the network because the CPU must filter the TS packets, generate TCP frames in RAM, and maintain TCP connection information which would cause frequent interrupts and occupy a lot of computation resource. In this paper, we propose an implementation of hardware-based TCP for TS data transmitting. It could convert the filtered TS packets to TCP frames directly and send them to the client in LAN. And it could also maintain TCP states and process ACK frames automatically after a TCP connection is established by CPU. It employs improved flow control and congestion control strategy to assure transmission stability in wired or wireless environment. In the actual test, we evaluated it on a FPGA board and input a 20 Mbps TS data flow under wireless environment, the client could play the program well as expected.

1 INTRODUCTION

Transport Stream (TS) is commonly used in digital television broadcast system[1]. With the development of internet and information technology, it is a trend to reuse the TS over the Internet or home network. Although TCP needs the sender and receiver interact with each other during data transmission which consumes extra bandwidth, choosing TCP as transmission protocal for TS has the follow benefits[2]: (a) the data's order and correctness can be assured, especially under wireless environment; (b) TCP is friendly to firewall which means it has a wide applicable range. TS packets could be encapsulated into HTTP stream which most media players can support. TS transmission over TCP is very fit to employ special hardware because the conversion of TS packets to TCP frame has fixed steps and it could also reduce lots of interrupts during ACK frames receiving.

Related technologies have been proposed and implemented. Liu proposed a FPGA-based TS over IP module which encapsulated TS into UDP-based Real Time Protocal (RTP)[3]. Yang designed and implemented multi-channel bidirectional TS over IP gateway on FPGA which could support four TS input channels, four output channels and one RTP output channel[4]. Hardware-based TCP could be used as TCP offload engine in backbone network. A TCP offload accelerator for 10 Gb/s Ethernet was developed by Y. Hoskote, Bradley A. Bloechel and so on[5]. And T. Uchida implemented a SiTCP which could process one TCP connection[6]. Some embedded systems also employed hardware-based TCP devices. W5100 series devices produced by WIZnet could maintain 4–8 sockets with full hardware TCP[7].

This paper implements TSoverHTCP which can execute the conversion from TS packets to TCP frame and hardware TCP data transmission. This paper is arranged as follows. Section II describes two basic data structures in TSoverHTCP; Section III introduces the inner architecture of TSoverHTCP. Transmission control implementation is discussed in Section IV. The concrete work process is described in Section V. Section VI and VII shows the test result and conclusion.

2 RELATED DATA STRUCTURE

2.1 *Frame structure*

To make every frame's data can be easily parsed as TS packets, TCP frame must contain integral number of TS packets, and to improve the transmission efficiency, TCP frame's data field must approximate to its maximum length. According to RFC894[8], the maximum length of a IP packet over Ethernet is 1500 bytes, and a TS packet has a fixed size of 188 bytes[1]. So the frame length transmitted in TSoverHTCP is 1370 bytes including 14-byte MAC header, 20-byte IP header, 20-byte TCP header and 1316-byte TS data (7 TS packets) as shown in Figure 1.

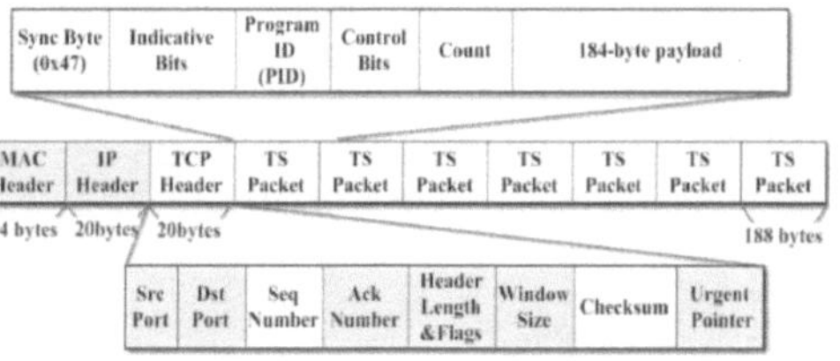

Figure 1. TCP frame structure.

des_valid	reserved	ack_times	Tx Status
Control Bits		Buffer1 Size	Buffer2 Size
Buffer1 Address			
Buffer2 Address			

(a) TX descriptor

des_valid	Rx Status	
Control Bits	Buffer1 Size	Buffer2 Size
Buffer1 Address		
Buffer2 Address		

(b) RX descriptor

Figure 2. TX and RX descriptor structure.

Some fields are not changed after TCP connection established which can be put into a common buffer to improve utilization of external RAM. These fieids include MAC header, IP header and part of TCP header which are marked in gray in Figure 1. To keep frame's continuity, only MAC header, IP header, source port and destination port (38 bytes) are saved in a common buffer called "frame header" and the rest of TCP frame called "frame data" (1332 bytes) is saved in different buffers for different frames.

2.2 *Descriptor*

Descriptors are introduced for different modules to communicate with each other. For example, CPU can set up a TX descriptor to inform frame transmission block to send a frame, and when a frame is received, CPU can also get the frame through reading the RX descriptor.

Figure 2 shows structures of the TX descriptor and RX descriptor. A integral TX descriptor has 16 bytes including 1-bit descriptor valid flag, 2-bit ack_times, 16-bit TX status, 6-bit control bits and two 11-bit buffer size fields, and two 32-bit buffer address fields. Des_valid flag indicates buffer1 or buffer2 has valid data. Ack_times is used to record the times of being acknowledged. The RX descriptor structure is similar to the TX descriptor except the ack_times field. When TSoverHTCP receives a frame, it will inform CPU to process it and CPU should clear the RX descriptor's des_valid bit. The status field is the MAC block processing status and will be filled by TSoverHTCP after sending or receiving a frame. Control bits are used to tell frame transmission block how to process the frame. Every descriptor contains two buffers to adapt the frame structure. Buffer 1 is used to point to the unchanged "frame header", and the other points to the "frame data". TSoverHTCP will check these two buffers before sending.

When TSoverHTCP works, CPU needs to use one software TX descriptor circle FIFO and one RX descriptor circle FIFO to send and receive frames through TSoverHTCP to establish TCP connection. And frame generation block also needs a hardware TX descriptor circle FIFO to inform frame transmission block to transmit frames.

3 TSoverHTCP ARCHITECTURE

3.1 *Frame generation block*

This block generates valid TCP frames for TCP data communication. It is responsible for filtering input TS packets, saving valid TS packets and updating TCP headers. Figure 3 shows its basic architecture. External RAM, CPU and TSoverHTCP (including frame generation block and frame transmission block) are connected together through the AHB bus bridge for CPU and TSoverHTCP must access the external RAM. CPU is used to initiate RAM data structure and configure the inner registers of TSoverHTCP.

Frame generation block maintains an pointer which points to a TX descriptor in the hardware TX descriptor circle FIFO. Before writing any data into a frame, it will check the descriptor des_valid field. If the current descriptor is invalid which means the corresponding frame is sent and acknowledged, this block starts to fill TS packet data and generate a valid frame.

Frame generation block filters TS packets because not all the TS packets are needed by users in an input TS data flow. Every TS packet header contains a Program ID (PID) field which is used to identify what content it carries[1]. To implement this function, a PID table is needed. The PID table is an array of PIDs whose maximum size is 64. We choose 64 as the maximum size because a too large PID table will cause a large lookup time delay, and 64 PIDs are enough to contains a complete program. CPU can put the PIDs needed into a continuous RAM space, and set the PID table base address and its length into frame generation block's registers before TS inputs. When a TS packet is buffered into TS input fifo, TS packet filter will read out the TS packet header and look up the PID table. If the PID in the header is matched in the table, then TS packet filter block will notify

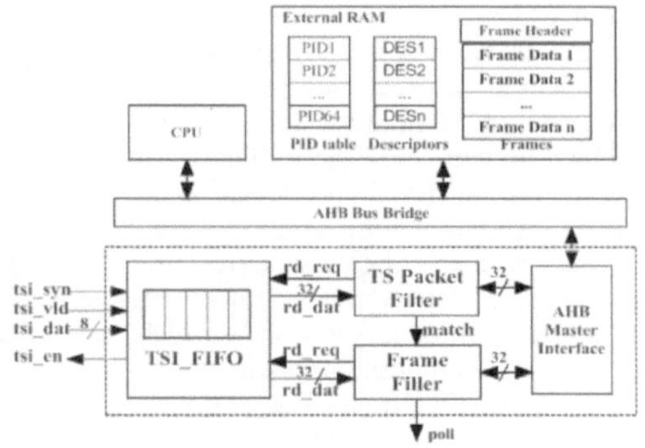

Figure 3. Frame generation block and external RAM content.

the frame filler module to read the integral TS packet out of the TS input fifo, to put into frame data buffer.

The frame filler module is used to fill the PID-matched TS packets into frame data and update TCP headers. When a TCP connection is established, the sequence number field and checksum field in TCP header are always different. Sequence number indicates the position of transmitted data by byte, so the sequence number of the current frame can be calculated by adding 1316(7*188) to the sequence number of the previous frame which is saved in frame filler block. The checksum in the TCP header is used to check the correctness of transmitted data. It is calculated during TS packet data storing into the external RAM and updated into the TCP header after 7 TS packets are stored.

After a valid frame is generated, this block needs to write its frame header address and size, frame data address and size into the current descriptor and set the descriptor valid. Then it signals a "poll" pulse to activate frame transmission block to prepare transmission and increase the current descriptor pointer for next frame generation.

3.2 *Frame transmission block*

Frame transmission block realizes all the sending and receiving function. Normal data transmission over TCP involves frame sending, ACK frame parsing and frame receiving as shown in Figure 4. The AHB master interface is connected to the AHB bus bridge for external RAM access and registers (not shown in Fig. 4) offers an AHB slave interface for CPU to configure. The MAC module offers a MAC Transmitting Interface (MTI), a MAC Receiving Interface (MRI) and Reduced Media Independent Interface (RMII) to interact with TX block, RX block and PHY chip. The key point is the implementation of TX block, RX block and ACK parser which will be discussed in the following part.

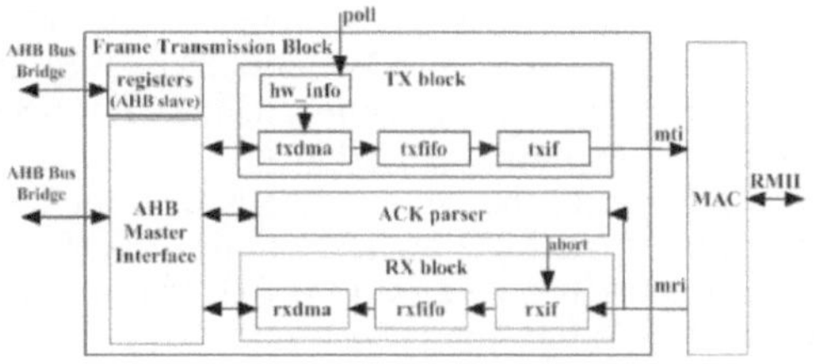

Figure 4. Frame transmission block architecture.

TX block is made of hw_info, txdma, txfifo and txif to implement "software frame" and "hardware frame" sent. A software frame is a valid TCP frame prepared by CPU which is used to establish a TCP and HTTP connection, while a hardware frame is filled by frame generation block which is used to transmit TS data. Hw_info module maintains the current TCP sending state and some other control information, and provides the pointer of hardware frames to txdma module. All the frames are sent through the txdma-txfifo-txif-MAC path to the client. Txdma is responsible for moving frame data from external RAM to txfifo and writing back sending status to corresponding descriptors. When hardware or software frames are prepared, txdma module will select a software descriptor or a hardware descriptor by valid state turns. If the descriptor is valid, txdma module begins to move frame data to txfifo, or emit a descriptor invalid signal. There is a little difference in processing software TX descriptors and hardware TX descriptors after sending frames. After a software frame is sent, txdma module will invalidate the corresponding descriptor and then increase the pointer of software TX descriptor circle FIFO. But for hardware frames, the txdma module just writes the sent status back into the corresponding descriptor and does not affect its validity which is affected by ACK parser. When txfifo module buffers enough data (256 bytes) to transmit data continuously, txif module will signal time sequence for MTI interface to MAC.

ACK parser and RX block use the data received by MAC simultaneously. The ACK parser is used to process ACK frames and maintains some TCP receiving control information. It justifies a frame is a valid ACK frame to hardware frames by comparing MAC address, IP address and TCP port with those CPU has configured, and checking the ACK flag in the TCP header. When ACK parser processes an ACK frame without any (TCP or IP) header option and data field is empty, it will emit an abort signal to drop the frame buffered by rxfifo. Also when it receives an ACK frame with FIN or RST flag which usually means TCP is disconnected, it will stop TX block from transmitting hardware frames. In an ACK frame, Acknowledgment Number (AN) and Window Size (WS) are the information needed to implement TCP congestion and flow control in TX block.

In RX block, the rxif module buffers the data into rxfifo from MRI interface until ACK parser emits a abort signal or rxfifo is full. If a frame is buffered by rxfifo, the rxdma module will read it out of rxfifo, write it into external RAM, then set a RX descriptor valid and send an interrupt signal to CPU at last. This block is designed for CPU to establish TCP/HTTP connection and to receive some other data such like ARP, DHCP frame during TS transmission.

4 TRANSMISSION CONTROL

4.1 *Flow control*

Flow control is realized by tracking the buffer size of the client to calcaulte the number of bytes that could be sent. In a received ACK frame, the Window Size (WS) represents the client buffer size after receiving data and the Acknowlegment Number (AN) represents how many bytes received relative to the Sequence Number (SN) in the oldest sent but not acknowledged frame. In the state shown in Figure 5, when the ACK parser receives a ACK frame, it will read SN from the frame ptr_sack pointed to, and compare AN and SN by the following rules:

1. If AN-SN > 1316, invalidate the current hardware TX descriptor and increase ptr_sack to read next SN until AN-SN= =1316.
2. If AN-SN= =1316, invalidate the current hardware TX descriptor, increase ptr_sack and update ptr_win and RWS in hw_info module.
3. If AN-SN= =0, increase the ack_times value in the current hardware TX descriptor and update RWS. And if the ack_times equals to 3, then send fast retransmission signal to hw_info.
4. If AN-SN < 0, just ignore the frame.
5. If 0 < AN-SN < 1316, transmission error occurs.

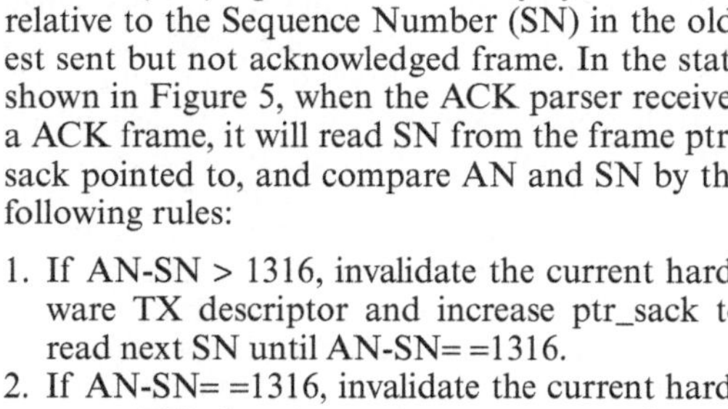

Figure 5. The information used in TCP control.

For example, in Figure 5 state, when an ACK frame with WS = w + 1316 and AN = n + 1316*2, ptr_ack and ptr_win will be at (a) position and RWS = w + 1316 finally.

Hw_info can track the buffer size of the client by ptr_win and RWS and decide whether sending frames forward. Before sending a frame forward, hw_info will calculate the bytes the client can receive at one once which is named free_size. There is another parameter win_shift which can affect the calculation of free_size. win_shift is the scalar factor of WS by left-shifting, and always initiated at the SYN frame received. Let N be the number of frames between ptr_htx-1 and ptr_win, the free_size could be calculated by Equation (1):

$$free\_size = WS * 2^{win\_shift} - 1316 * N \quad (1)$$

If free_size is less than or equal to 1316, then stop sending frames forward and wait for ACK frames to update window size until free_size more than 1316.

4.2 *Congestion control*

Congestion control here is used to avoid TSoverHTCP occupying the network too much to receive ACK frames, especially under wireless network. TSoverHTCP introduces an millisecond timer to control timeout waiting and a Congestion Window (CWND) to control the number of frames sent. And the hw_info module maintains the current sending state to determine which frame to be sent.

The CWND is the maximum value for the number of frames sent but not acknowledged. It is changed between minimum and maximum value with sending state transition. Its minimum and maximum value in TSoverHTCP is set to 2 and 64 respectively. When CWND frames are sent and acknowledged without any retransmission, CWND value will be doubled until it reaches the maximum value. When retransmission happens, CWND value will be set to half of it until it reaches its minimum value. There are two different points from the standard TCP[9]. At first, the unit of CWND is frame while the standard TCP is byte. Secondly, CWND value can only changed by multiplying or dividing by two with an upper limit.

The Retransmission Timeout (RTO) value is employed to control retransmission. The calculation of RTO is based on the Round-Trip Time (RTT). RTT is measured during the normal sending state (no retransmission happens). When TSoverHTCP sends a new frame out, the current time and the position of descriptor is recorded in the sent_time and sent_pos registers. After ACK parser receives the ACK frame to acknowledge the frame send_pos pointed to, RTT is calculated

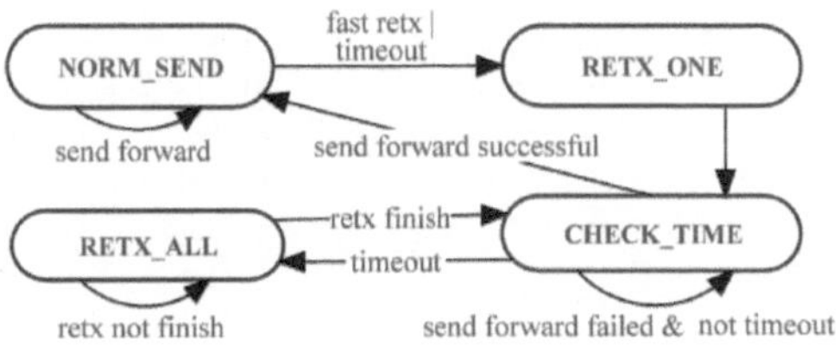

Figure 6. Sending state transition graph.

by current time minus send_time. And RTO is updated as follows[8]:

$$
\begin{aligned}
&RTT_s = (1-a) * RTT_s + a * RTT \\
&RTT_d = (1-b) * RTT_d + b * | RTT_s - RTT | \\
&RTO = \max(RTT_s + c * RTT_d, \min\_RTO) \qquad (2) \\
&a = 1/8, b = 1/4, c = 4
\end{aligned}
$$

In equation (2), RTT_s is the smoothed RTT and RTT_d is the deviation of RTT. And different from the original RTO formula, min_RTO is introduced to avoid that a small RTO value causes frequent retransmission. Waiting time is the current time minus the time last sending frame. When waiting time is more than RTO, retransmission occurs.

Figure 6 is sending state transition graph. After TSoverHTCP is reset, the sending state is initiated as NORM_SEND, when frame generation block emits the "poll" signal to hw_info, it begins to send new frames until current TX descriptor is invalid, or free_size is not more than 1316, or the number of frames sent but not acknowledged is more than CWND. If it receives a fast retransmission signal from ACK parser, or waiting time is more than the RTO, sending state is changed to RETX_ONE to retransmit the oldest frame sent but not acknowledged. After the retransmission, sending state enters into CHECK_TIME state unconditionally to check the waiting time and wait for next sending forward. When "poll" comes or free_size is more than 1316, or frames sent less than CWND, hw_info will try to send a new frame. The sending state will be restored to NORM_SEND if the sending succeeds. And If timeout happens, sending state is changed to RETX_ALL to retransmit all the frames not acknowledged.

A little difference with the standard TCP stipulation can be seen from the state transition, that is, in NORM_SEND state TSoverHTCP just retransmits one frame when timeout happens while the standard TCP stipulates retransmitting all the frame not acknowledged[9].

5 WORK PROCESS

An integral TCP communication consists of three phases—TCP and HTTP connnection, TCP data transmission and TCP disconnecton. TCP connection and disconnecton phases are just performed once, while TCP data transmission phase could last for a long time, especially for TS transmission. TSoverHTCP, as described before, just implements TCP data transmission phase, so the other two phases must be established by CPU.

Figure 7 shows how TSoverHTCP works. When a client wants to request TS data over TCP, the client will start TCP connection by sending a SYN frame to TSoverHTCP. And it will pass the frame to CPU. Then a SYN-ACK frame is sent to the client by CPU through TSoverHTCP. After CPU receives an ACK frame from the client, TCP connection is established.

The client begins to send its http request to tell CPU what program is wanted. Then CPU sends OK response to the client and waits for a ACK frame. Mark 1 means CPU should also do some initial work including preparing a PID table, initiating TX descriptor and RX decriptor circle FIFO, and setting up frame header and frame data in the external RAM. After receiving an ACK frame, CPU starts to configure TSoverHTCP registers such as related date structure's base address and to enable TSoverHTCP to process TS packets and send data.

During data transmission, CPU is under idle state or do some other work, while TSoverHTCP processes ACK frame and sends TS data which is marked as 2 in Figure 7. When client wants to close the current connection, as TCP stipulated, it must send a FIN frame which will be parsed by TSoverHTCP and pass it to CPU. Then TSoverHTCP stops transmitting data and CPU performs the disconnection flow. In some cases, TSoverHTCP may lost connection because of power failure or

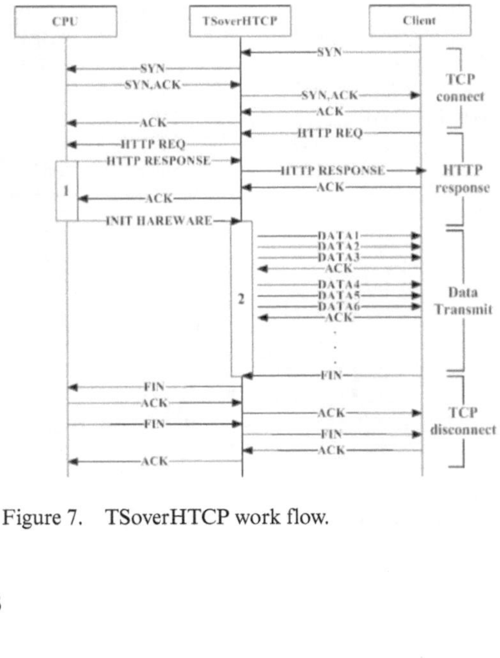

Figure 7. TSoverHTCP work flow.

system crash which could cause no ACK frame received. After retransmitting for several times, TSoverHTCP will send CPU an interrupt to clear the current connnection state.

6 TEST AND RESULT

To evaluate TSoverHTCP, it was implemented on an FPGA using Stratix IV as an FPGA test board. Table 1 shows the resource utilization of TSoverHTCP. TSI_FIFO block uses a 4 K-bit inner RAM to buffer input TS packets and TX_BLOCK and RX_BLOCK use a 16 K-bit inner RAM to buffer frames sent and received. TX_BLOCK and ACK_PARSER cost a lot of resources because TCP control logic.

The entire TSoverHTCP test environment is shown in Figure 8. A CPU (LH79520), a 8 M-byte SDRAM and TS input was connected with FPGA in the test board. The test board also offered a RJ45 interface which was connected with a wireless router whose maximum rate was 150 Mbps. The wired network was always stable and TSoverHTCP worked well in that condition. So the test was under a wireless environment. A laptop was connected with the wireless router through wireless channel. A VLC Player was installed on the laptop which was used to play the program and to observe the transmission rate. The wireless router was fixed at one end of 1.5 m-wide aisle and the laptop was moved from near to far to test the transmission rate as the distance changed.

Table 1. TSoverHTCP resource utilization.

Module name	LC combinations	LC registers	Memory bits
TSI_FIFO	172	190	4096
PKT_FILTER	57	23	0
FRM_FILLER	1331	429	0
AHB_MST	457	246	0
REGISTERS	1139	751	0
TX_BLOCK	2081	960	16384
RX_BLOCK	381	351	16384
ACK_PARSER	763	431	0
AHB_BRIDGE	1731	1195	0
MAC	1309	1071	0
Total	9421	5647	36864

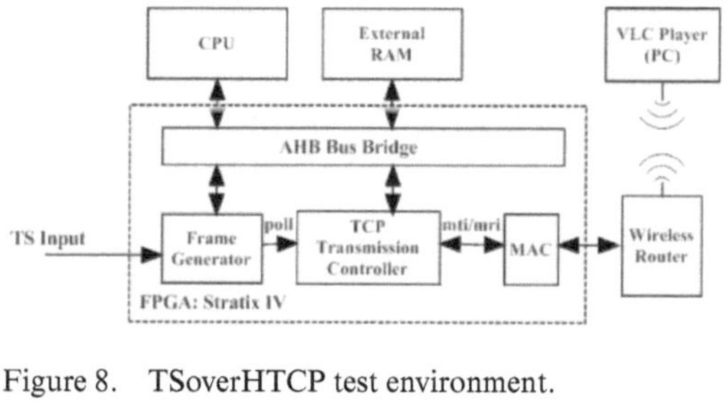

Figure 8. TSoverHTCP test environment.

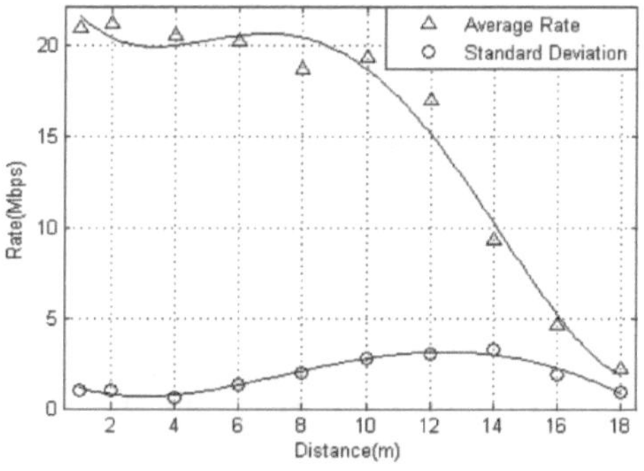

Figure 9. Relationship of transmission rate and distance.

The input TS was a high definition program which had a 20 Mbps bitrate. At each position, multiple transmitting rate values were sampled and their average rate and standand deviation were calculated as a measure of real transmission rate and stability.

Figure 9 has shown the test result. When the distance from the wireless router was less than 10 meters, the transmission rate almost stayed at 20 Mbps and although its standard deviation increased, the program displayed by VLC player was fluent. But when the distance was between 10 meters and 12 meters, the transmission rate decreased below 18 Mbps, and lost some stability, as a result, the program often stoped and picture displayed was not complete. As distance increased from 12 meters, the program almost stoped though there was a non—zero transmission rate which because decoder got too little data to decode a complete picture.

7 CONCLUSION

TSoverHTCP has implemented hardware-based TCP for TS transmission. TSoverHTCP is mainly divided into frame generation block and frame transmission block. Frame generation block can filter TS packets according PID value and convert TS packets to valid TCP frames. Frame transmission block can support CPU and hardware sending frames. Improved congestion control and flow control are employed to assure the stability of TCP transmission.

At last, it is evaluated on the FPGA test board. The result shows that a 20 Mbps high definition TS program can be played well in a reasonable range in WLAN. So it can be used in the home network for TS data sharing.

The defect of TSoverHTCP is that it can just support one TS channel and one TCP connection. So our future work is to extend the TSoverHTCP to support multiple TS channels and TCP connections.

ACKNOWLEDGEMENT

The authors would like to thank Ketuan Zhan, Xianglei Xue for their guidance and Wenrui Chen and Xiaoming Shan for help with the test environment build.

REFERENCES

[1] ISO/IEC-13818-1:2000 MPEG2 Systems.
[2] Marfia, G. and M. Roccetti. 2010. TCP at last: reconsidering TCP's role for wireless entertainment centers at home. Consumer Electronics, IEEE Transactions, 2010. 56(4): p. 2233–2240.
[3] Liu, Y. 2010. The implementation of TS over IP module based on FPGA. China Digital Cable TV, 2010. 290(05): p. 618–620.
[4] Liu, J. 2011. The FPGA design and implementation of multi-channel bidirectional TS over IP gateway. Cable TV Technology, 2011. 264(12): p. 40–42, 59.
[5] Hoskote, Y., et al. 2003. A TCP offload accelerator for 10 Gb/s Ethernet in 90-nm CMOS. Solid-State Circuits, IEEE Journal of, 2003. 38(11): p. 1866–1875.
[6] Uchida, T. 2008. Hardware-Based TCP Processor for Gigabit Ethernet. Nuclear Science, IEEE Transactions on, 2008. 55(3): p. 1631–1637.
[7] WIZnet. 2008. W5100 data-sheet.
[8] RFC894: A Standard for the Transmission of IP Datagrams over Ethernet Networks.
[9] Richard Stevens, W. 2002. TCP/IP Illustrated, Volume 1. New York. Addison Wesley.

E-Commerce, E-Business and E-Service – Lee (Ed)
© 2014 Taylor & Francis Group, London, ISBN 978-1-138-02646-9

Model checking and analysis of systems dependent on CPU speed

Takahisa Mizuno & Shin-Ya Nishizaki
Department of Computer Science, Tokyo Institute of Technology, Tokyo, Japan

ABSTRACT: System behavior is strongly related to the CPU's execution speed in the situation where you have very limited computational resources. Such a situation is difficult to analyze exhaustively. Model checking is one of the automated verification techniques for concurrent systems, and is widely applied not only in academic research but also in industrial practical cases. We have formalized an embedded system to be analyzed as a finite automaton and its required behavior as temporal logic formulas. A model checker automatically checks that the finite automaton satisfies the required behavior by traversing all the possible execution paths. In this paper, we propose a methodology for how to apply model checking to an embedded system strongly related to CPU speed. We show a case study of a smart savings bank in which a microprocessor is embedded in order to count coins. We investigate the smart savings bank by applying a SPIN model checker. Our experience shows that our approach is effective for dealing with problems and their causes even in such a complicated situation strongly related to CPU speed.

1 INTRODUCTION

Many embedded systems interact with their surrounding environments through inputs and outputs. The period of such interaction is sometimes defined clearly, but is sometimes undefined. This kind of embedded system is called a reactive system. There are various examples of reactive systems in the real world: digital watches, computer-controlled brake systems in cars, plant control systems. One of the important features of a reactive system is that a response to an input is expected within a certain period specified in advance as a specification.

Model checking is a major breakthrough in formal verification of software systems. It improves not only coding but also design in software development. A model checker verifies whether a model of a target system fulfills a required specification. Models are described as finite automata; that is, state-transition diagrams, or as program-like codes in modeling languages. System requirements are often formalized as formulas of temporal logic. Many model checkers have been proposed: UPPAAL [2], NuSMV [1,3], and the SPIN model checker [5]. We have studied the application of the model checking technique to various domains [7,8].

As a target system for analysis via the model checker, we used a savings bank, to which we attached an electronic controller, a photo-reflector, and an LED number display. The specifications of the savings bank are explained in the following section.

We first explain the savings bank, or moneybox, which is the basis of the research described in this paper. The bank is designed for use with Japanese coins. When a coin is dropped into the slot, it slides down a slope into the savings bank. There are six holes in the slope, corresponding to the six denominations of Japanese coins: 500 yen, 100 yen, 50 yen, 10 yen, 5 yen, and 1 yen.

Each hole is fitted with a photo-interrupter. The photo interrupter consists of an LED and a photo-transistor, and detects the passage of objects between the LED and the phototransistor. The controller calculates the total quantity of coins in the savings box, based on data obtained from the photo interrupter (Fig. 1).

In the remainder of this paper, we sometimes refer to the phototransistors as photo sensors.

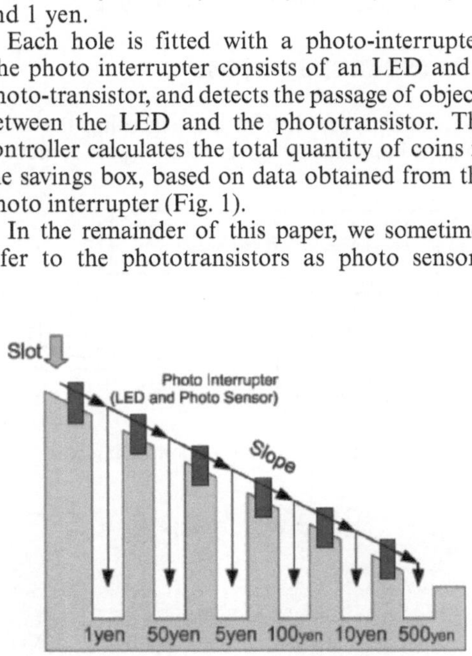

Figure 1. Smart savings bank.

The controlling part of the savings box can be divided into four components:

- CPU
- LED number display
- Photo sensors at the six holes
- Reset switches

Figure 2 shows an overview of these four components, and Figure 3 gives an overview of the controller shown in Figure 2. Figure 3 is a picture of the smart savings bank.

If a coin goes into a slot and passes a photo sensor (photo interrupter), then a signal from the photo sensor is emitted. If the coin goes beyond the photo sensor, another signal is emitted. We therefore know that the coin passes the photo sensor by watching for the first signal (more precisely, the rising edge of a signal from the photo sensor). We describe a detection algorithm in C-like pseudo-code as follows.

```
int currentvalue, previousvalue;
while(1){
  cuurrentvalue = PhotoSensor;
  if(!currentvalue && previousvalue){
  sending a signal to the other processes
  }
  previousvalue = currentvalue;
}
```

The variable PhotoSensor in this pseudo-code keeps the current value of the photo sensor, and we identify the second signal from the photo sensor by watching the variable in the while loop.

2 RESEARCH PURPOSE

We found a problem in implementing the smart savings bank in that the controller part sometimes

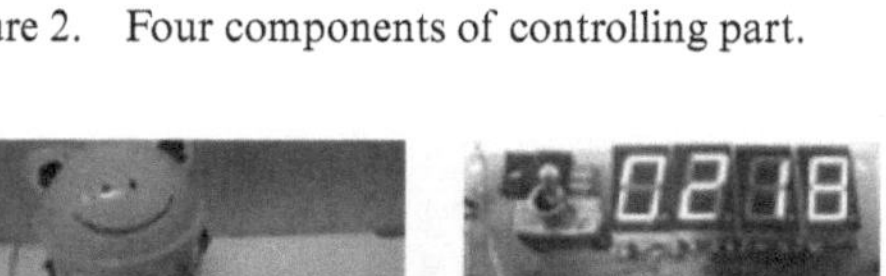

Figure 2. Four components of controlling part.

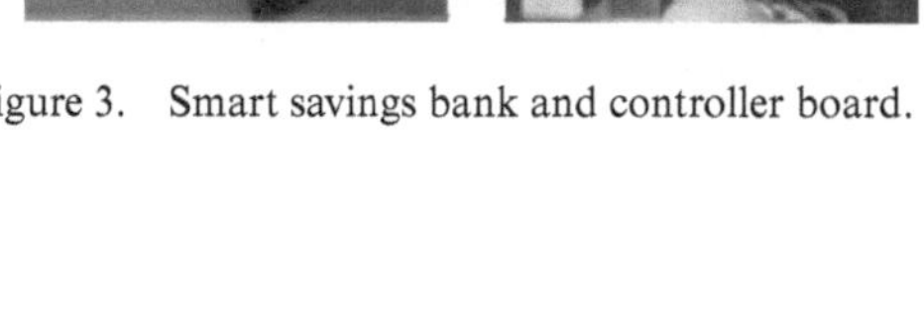

Figure 3. Smart savings bank and controller board.

failed to detect passing coins. Accordingly, in implementing the smart saving bank, we made the following two trials.

1. We changed the timing at which the controller gets a value of the photo sensor.
2. We increased the CPU clock frequency (more precisely, we changed 1 MHz to 8 MHz).

By applying (1) to the control program, we forestalled a malfunction possibly caused by swift fluctuation of the photo sensor's value. We rewrote the controller's program so that the controller observed the photo sensor's value for each loop. Moreover, we guessed that the controller did not observe frequently enough, and therefore we made the trial (2). Thanks to these two trials, we successfully solved the problem. However, we stumbled across the necessity for these two trials, which are very ad hoc. We therefore should consider a more systematic method to improve this erroneous aspect of the system.

3 FORMALIZATION OF TIMING AND EXECUTION TIME DEPENDENCY

We analyzed the smart savings bank using the SPIN model checker, in which we described models in the Promela modelling language. The following are distinguishing characteristics of Promela:

- It uses data types such as integers, Booleans, arrays, etc.
- Distributed programming using message passing between processes through communication channels.
- It has imperative control structures.
- It has deterministic and non-deterministic branching.

The distinctive feature of model checking is that it allows an exhaustive search of all possible execution paths, which is a very powerful and practical method for automatic verification of non-deterministic hardware and software systems. Model checkers use heuristics in order to avoid useless and/or inessential possibilities of execution paths. However, such automatic improvement using heuristics is limited, and therefore users have to remove non-essential execution paths manually before applying it to a model checker.

Another point in formalizing the smart savings bank is weak fairness, which we can assume in a model and SPIN model checker can check given queries under the assumption of weak fairness. However, the weak fairness is actually strong enough to formulate correct behavior of the smart savings bank. Hence, we imposed and formulated a stronger property in its model: *all processes must be executed at the same pace, and no process can proceed exclusively.*

In the following, we show a process scheduler whose behavior is as follows:

while(1){
  if (one of *the other processes stops*){
    make the process restart
  }
}

This code makes it possible to control the pace of each process.

The process scheduler obtains the progress states of the other processes by sharing a variable with them.

4 MODELING THE SMART SAVINGS BANK

If you describe a model in the same manner as program coding, then its search space becomes exponentially large and causes a computational explosion during model checking and the model checker fails to verify it. We therefore reduced the time and memory space required by traversing all the possible execution paths by introducing simplification and abstraction as follows.

Time: Each process does not share a clock with the other processes but maintains its own clock. When an action in a process is completed, a clock in the process is incremented. The clock is initialized to zero if the main loop in the process is finished. Our time model can be considered as cyclic rather than linear.

Number of photo sensors: We reduced the number of photo sensors from six to two in describing a model.

Timing of coin insertion: In reality, coins are entered in arbitrary timing, which makes the search space larger. Accordingly, we assume that coins are entered in a certain punctual timing in our model. From the viewpoint of reality, we can suppose that coins are not inserted consecutively; each next coin is entered after a certain period.

Interrupt: In the real program of the controller, we used a timer interrupt in order to refresh the LED number display. For simplicity's sake, we did not formulate such a timer interrupt in our model. In the following, we give an overview of the model for the SPIN model checker.

Our Promela model for the SPIN model checker consisted of six processes:

- **mainprocess**, modeling the main controlling process,
- **sensor1**, modeling the first photo sensor,
- **sensor2**, modeling the second photo sensor,
- **coin,** modeling behavior of an inserted coin, and
- **scheduler,** modeling the pace controller of the other processes.

Each process except the **scheduler** maintained clock variables which keep progress states. These clock variables are shared with **scheduler**: s1state in **sensor1**, s2state in **sensor2**, cstate in **coin**, and mpstate in **mainprocess.**

The clock variables keep non-negative integers. Every time an action occurs in a process, each clock variable in the process is incremented. If one loop body in a process is finished, the clock variable is initialized to zero. Therefore, these variables take cyclic values. In Table 1, we show the length of the time cycles of the clock variables.

We gave a termination variable to each process. It is initialized to a Boolean value of false. If the process finishes its loop body once, then we put a Boolean value true into the variable. The termination variable takes the two Boolean values alternately.

If a Boolean value of true is stored in all the termination variables, then the pace controlling process **scheduler** sets the variables as false. A process whose termination variable has a value of true waits for the time when the other processes' termination variables also obtain the value of true.

Table 1. Length of time cycle of processes.

Processes	Length of cycle
mainprocess	MP_INTERVAL+3
sensor1	SENSE_END
sensor2	SENSE_END
coin	COIN_INTERVAL

Table 2. Actions in process **mainprocess**.

mpstate	Action
0, …, MP_INTERVAL-1	Making variable mpstate mcrement
MP_INTERVAL	Values of s1value and s2value are stored in curs1value and curs2value. respectively
MP_INTERVAL+1	If the first sensor finds a coin, variable coin1count is incremented.
MP_INTERVAL+2	If the second sensor finds a coin, variable coin2count is incremented.
MP_INTERVAL+3	Values s1value and s2value are stored in pres1value and pres2value, respectively.

Table 3. Actions in process **sensor1**.

s1state	Action
0, …, SENSE_ON-1	Nothing is done
SENSE_ON-	Variable s1state is stored in variable store and true is stored in s2value
SENSE_END	s2state is initialized, false is stored in variable s2value, and coin2count is incremented.
MP_INTERVAL+2	If the second sensor find a coin, coin2count is incremented.
MP_INTERVAL+3	Values s1value and s2value are stored in pres1value and pres2value, respectively.

The termination variables of **mainprocess, sensor1, sensor2,** and **coin** are mpdone, s1done, s2done, and cdone, respectively.

5 MODEL CHECKING OF THE SYSTEM

5.1 *Model checking by SPIN*

We verified the property related to timing:

> *"The difference between the numbers of inserted coins and of counted coins should be less than 1."*

For simplicity's sake, the numbers of coins were formulated not as integer values, but as elements of Z/3Z, in other words, 3-cyclic values. The property to be checked is described as a Linear Temporal Logic formula:

$$\neg\Diamond((p1 \wedge q1) \vee \ldots \vee (p6 \wedge q6))$$

where
(coin1 insert == 2) ↔ p1,
(coin1 insert == 0) ↔ p2,
(coin1 insert == 1) ↔ p3,
(coin2 insert == 2) ↔ p4,
(coin2 insert == 0) ↔ p5,
(coin2 insert == 1) ↔ p6,
(coin1 insert == 0) ↔ q1,
(coin1 insert == 1) ↔ q2,
(coin1 insert == 2) ↔ q3,
(coin2 insert == 0) ↔ q4,
(coin2 insert == 1) ↔ q5, and
(coin2 insert == 2) ↔ q6,

Next, we abbreviated MP_INTERVAL to c_1, COIN_INTERVAL to c_2, and SENSE_ON to c_3, and SENSE_END to c_4.

The result of model checking by SPIN was as follows

Name	c_1	c_2	c_3	c_4	Result
1–1	1	15	6	11	False
1–2	1	15	6	12	True
1–3	1	15	6	13	True
1–4	1	15	6	14	True
1–5	1	15	6	14	True
1–6	1	15	7	12	False
1–7	1	15	7	13	True
1–8	1	15	7	14	True
1–9	1	15	8	13	False
1–10	1	15	8	14	True
2–1	0	15	7	11	False
2–2	0	15	7	12	True
2–3	0	15	7	13	True
2–4	0	15	7	14	True
2–5	2	15	7	13	False
2–6	2	15	7	14	v

5.2 *Analysis of the result*

The smallness of the constant MP_INTERVAL corresponds to the rapidity of execution, which means the rapidity of the loop body's execution in **mainprocess**. In the result from 1–1 to 1–10, the values of SENSE_ON and SENSE_END are changed variously. We know that if the difference between SENSE_ON and SENSE_END is not more than 7, then the value of the photo sensor is considered to be low in comparison with such a short period. We may conclude that the main process fails in 1–1, 1–6, and 1–9, of this result.

We found that the value of MP_INTERVAL was changed in the results from 2–1 to 2–6. If MP_INTERVAL is a small value, then the loop body of **mainprocess** is considered to be executed in a short period. Therefore, in 2–2, the difference between SENSE_ON and SENSE_END is regarded as only 5. On the other hand, if MP_INTERVAL is assumed to be 2, then the period is considered to be longer. Though the difference between SENSE_ON and SENSE_END equals 6, the checking of case 2–5 fails since it is less than MP_INTERVAL+4 and is considered to be small.

We know from this reasoning that we can conduct analysis of the mutual relationship among the

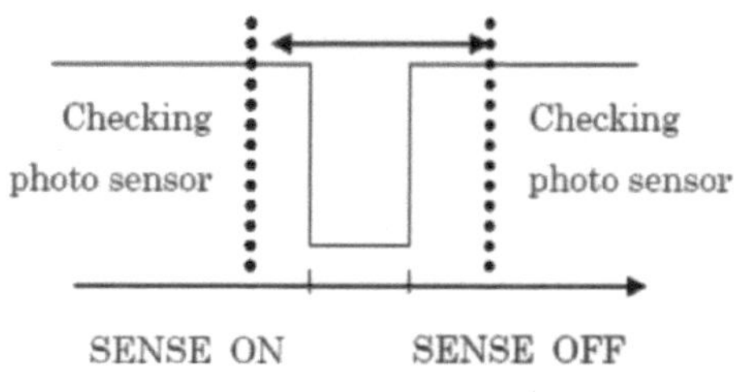

Figure 4. Sensing of photo sensor.

clock variables, and obtain a deep understanding of the target system thanks to model checking.

6 CONCLUDING REMARKS

In this paper, we studied a formalization technique for systems, depending on execution speed and applied model checking to analysis of the mutual relationship among time constraints.

When one analyze timing constraints using the SPIN model checker, we have to simplify and abstract a model of a target system in order to avoid the excessive computational costs of an exhaustive search. In our work, much simplification was implemented. Actually, such simplification is an unavoidable limitation of model checking.

Many researchers have studied model checking for real-time reactive systems. For example, Tripakis studied it based on timed Buchi automata [4]; UPPAAL [2] makes it possible to check timing constraints on clock variables. The probabilistic model checking is considered to be suitable for this kind of verification [9].

ACKNOWLEDGEMENTS

This work was supported by Grants-in-Aid for Scientific Research (C) (24500009).

REFERENCES

[1] Cimatti A, Clarke E, Giunchiglia F, Roveri M. NuSMV: a new symbolic model verifier. In: *Proceedings of International Conference on Computer-Aided Verification, Lecture Notes in Computer Science* 1999; **1633**: 495–9.

[2] Petterson P, Larsen KG. UPPAAL2k. *Bulletin of the European Association for Theoretical Computer Science* 2000; **70**: 40–4.

[3] Cimatti A, Clarke EM, Giunchiglia E, Giunchiglia F, Pistore M, Roveri M, Sebastiani R, Tacchella A. NuSMV 2: An opensource tool for symbolic model checking. In: *Proceedings of International Conference on Computer-Aided Verification (CAV 2002)*, 2002.

[4] Tripakis S, Courcoubetis C. Extending Promela and SPIN for real time. In: *Tools and Algorithms for the Construction and Analysis of Systems, Lecture Notes in Computer Science* 1996; **1055**: 329–48.

[5] Holzmann GJ. *The SPIN Model Checker: Primer and Reference Manual.* Addison-Wesley Professional, 2003.

[6] Kumamoto H, Mizuho T, Narita K, Nishizaki S. Destructive Testing of Software Systems by Model Checking. In: *Proceedings of International Symposium on Communications and Information Technology,* (ICSIT), 2010, IEEE, 26–9.

[7] Ikeda R, Narita K, Nishizaki S. Cooperative Model Checking and Network Simulation for Cost Analysis of Distributed Systems. *International Journal of Computers and Applications* 2011; **33** (4): 323–9.

[8] Tomioka D, Nishizaki S, Ikeda R. A Cost Estimation Calculus for Analyzing the Resisitance to Denial-of-Service Attack. In: *Software Security—Theories and Systems, Lecture Notes in Computer Science* 2004; **3233**: 25–44.

[9] Kwiatkowska M, Norman G, Parker D. PRISM 4.0: Verification of Probabilistic Real-time Systems. In: *Proceedings of 23rd International Conference on Computer-Aided Verification (CAV'11), Lecture Notes in Computer Science* 2011; **6806**: 585–91.

E-Commerce, E-Business and E-Service – Lee (Ed)

A proposal of softness display using honycomb

R. Ishibashi
Faculty of System Design, Tokyo Metropolitan University, Tokyo, Japan

Muhammad Aliff Bin Rosly
Panasonic Co., Ltd., Japan

A. Sano & H. Fujimoto
Graduate School of Engineering, Nagoya Institute of Technology, Nagoya, Japan

ABSTRACT: This paper proposes a new elastic display mechanism for rendering softness. The proposed display mechanism consists of a flexible material that includes a honeycomb structure. The display can be pressed by an actuator, at which point, the flexible form is stretched, and its apparent stiffness will be changed. The internal honeycomb structure ensures that the display has the same stiffness value everywhere on its surface. Thus, we can control the stiffness of the display only with one actuator. Some experiments are conducted to verify the effectiveness of the proposed display.

1 INTRODUCTION

Recently, haptics technology has been applied to a various systems like a medical simulator. For these applications, it becomes important to accurately reproduce and to present the tissue that is being targeted it to improve the sense of being present on location. The tissue has a soft and particular feel, and it is known that normal tissue presents a slightly different feel from that of diseased tissue. This difference in tactile forms is applied during palpation at the medical treatment site, and there is interest in developing a presentation technique capable of displaying the delicate tactile feel of tissue.

Conventional devices take many variegated forms, depending on the applications. For example, methods exist using the end point of the manipulator, such as the PHANToM (Tachi et al. 1994, Hoshino et al. 1999, Papadopoulos et al. 2002), or using a wearable device (Kawai et al. 2002). Here, we propose a device with a tactile display. Because this method presents immediately, without mechanical force transfer, the dynamics of the mechanism of the presentation in the device exert little influence on the presented information. In addition, the operator can experience the tactile characteristic of tissue by directly touching the surface of the display.

Several methods of softness display have been proposed (Fujita et al. 2000 & 2001, Kawamura et al. 2000, Bianchi et al. 2009). In one method, an array of pins arranged under a flexible sheet is rapidly adjusted soon by actuator motion, and the elasticity distribution is expressed on a flexible sheet (Kawamura et al. 2000). To express spatially broad distribution of very minute elasticity, the number of pins and the actuator area correspond to the resolving power. In another method, the rigidity in the normal direction is adjusted by the bias tension of a flexible sheet (Bianchi et al. 2009, Inoue et al. 2003).

The rigidity in the normal direction of the sheet can be adjusted only with the actuator to express the tension. However, when the point of contact moves, the device cannot respond well because the rigidity in the normal direction will have a different value in each part of the sheet under a constant bias tension.

A new softness display is therefore examined, which proposes control based on the contact area of the tip of a finger, which provides changeability in the method (Fujita et al. 2000 & 2001).

In the material approach, a method of adjusting both shape and the elasticity of the display is proposed by using inverted pseudo plastic fluid (Saga et al. 2009). Related study provided effective the display method (Tachi et al. 1994, Hoshino et al. 1999, Sakaguchi et al. 2000, Colgate: 1996, Papadopoulos et al. 2002, Srinivasan et al. 1995, Mitsuda et al. 2002, Kawamura et al. 2002, Kawai et al. 2002, Fujita et al. 2000 & 2001, Kawamura et al. 2000, Inoue et al. 2003, Bianchi et al. 2009, Saga et al. 2009, Iwata et al. 2001 & 2002, Wagner et al. 2003). If we realize these effects by using more simple structure, we can realize useful display (such as touch screen) to show the visual and elastic information.

Viscoelastic material provides the ability to express particular textures not easily evaluated by a single index. However, to provide a sufficiently large display area while considering its ease of use, the driving force for the response must be minimized. The same dimensions of the display area as those of the displayed tissue will be needed to improve the presentation in the future. Moreover, it is necessary to be able to mechanically construct it in a simple structure.

In this research, we propose a new method to attain a sufficiently large display area to represent desired tissue. Figure 1 shows the proposed display material. The proposed display has the feature that the core material of the mesh is viscoelastic. The actuator changes its horizontal special direction, and the internally caused stress is changed because the fluid is able to expand and contract. The rigidity in the vertical special direction is adjusted by changing the residual stress. Thus, the tactile feel can be adjusted. In this principle, the driving force to adjust the sense of touch affects only the actuator regarding the expansion and contraction of the display. The stiffness is the same at each part of the display area due to the core material of the mesh being evenly distributed in the display material. Moreover, the material is not physically restrained, and because a passive element is used for the presentation device, operator safety is easily ensured. In addition, if a transparent resin is used, image projection from the rear can also be enabled. Section II and III introduce an active tactile display of soft tissues using a viscoelastic material, and we explain how the stiffness can be controlled. Section IV shows the experimental setup and some experimental results. Section V provides conclusions for this paper.

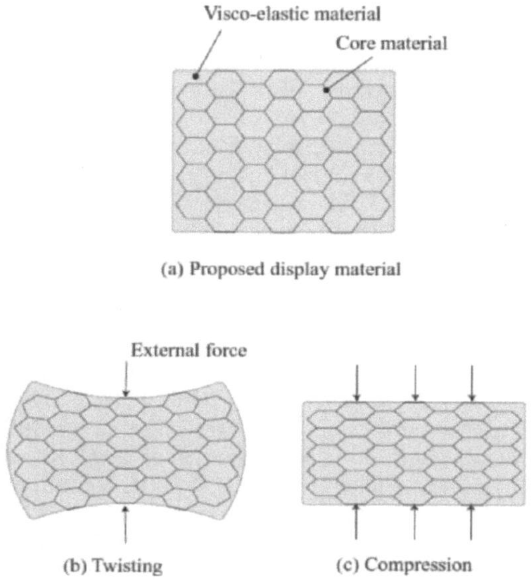

Figure 1. Concept of a proposed display material.

2 BASIC PRINCIPAL OF THE HAPTIC DISPLAY

In this section, constitution and basic principal of the proposed haptic display is shown as follows.

Figure 2 shows a sketch of the proposed haptic display panel. Figure 3 shows a side view of the proposed material. This display controls the residual stress by the expansion and contraction of the direction of the x axis shown in Figure 2. Then, the sense of touch when touching from the direction of the z axis as face of a panel is adjusted.

2.1 *Shape of the internal cell*

In general, it is difficult to change the residual stress of a flexible resin of plane externals thoroughly

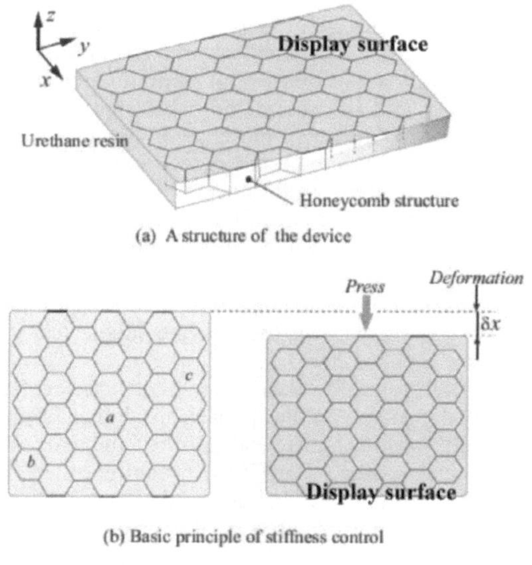

Figure 2. Sketch of the proposed material.

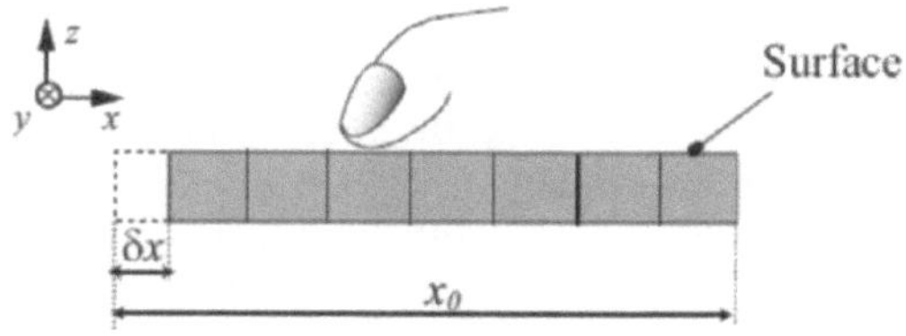

Figure 3. Side view of the proposed material.

by expansion and contraction from the side. Additionally compressing it in excess too much causes the upheaval of the center part. Therefore, it is necessary to stabilize making of the stress distribution and uniform the structure by some methods.

For the practical use, it should be better to save energy at the stress control strategy.

In the proposed procedure, the problem is solved by putting the core material of the mesh in the resin.

At this time, it is preferable that each hole of the reticulation is a similar figure. Moreover, the fewer the surface area of the reticulation seen on the presentation side is, the better.

Then, we think about the problem as the tessellation problem of the hexagon cell geometry. Here, we assume that the cell geometry be graph G, then neighborhood of the Ni of the i th node point becomes as follows.

$$N_i = \{j \mid j \in v(G) : (i, j) \in \varepsilon(G)\} \tag{1}$$

where, $v(G)$ and $\varepsilon(G)$ are finite set. $v(G)$ is vertex set and $\varepsilon(G)$ is edge set of the graph G. Here, we assume that edge connectivity of the i th node point as Zi. Zi be the average number of sides meeting at a vertex. Let n be the average number of sides of a regular polygon, n becomes

$$n = \frac{2Z_i}{Z_i - 2} \tag{2}$$

Here we think about the problem is the tessellation of the regular polygon, Zi and n must be natural number. Then we obtain the combinations $\{n, Zi\}$ as $\{3, 6\}$, $\{4, 4\}$, $\{6, 3\}$, for the tilings in the display. The edge of the honeycomb cell should be better not to affect softness perception with the finger tip contact area. Then, $\{6, 3\}$ as the honeycomb of the hexagon minimizes the display surface area viewed along the z axis. In addition, hexagon shape is the fewest the surface area on the display side, and it is suitable for the image presentation.

2.2 *Deformation of the haptic display*

Next, we think about the transformation by external force which acts to the structure. It is preferable that the number of actuators for expansion and contraction is little. In addition, it is preferable that the actuator to give the external force doesn't disturb the presentation of the visual information with the display. In this research, the external force is given from the side of the structure.

Here, regular polygon is assumed to be a closed linkage structure that has the resilient joint. Let m be the degrees of freedom and we obtain the following equation.

$$m = 3(n - j - 1)\sum_{i=1}^{j} f_i . \tag{3}$$

Here, n is a number of links, j is number of joints, and fi is degrees of freedom of each joint. All the culminations are assumed to be a rotary joint of one DOF (Degrees of Freedom). Then, deformation degrees of the hexagon as combinations $\{6, 3\}$ becomes three degrees of freedom. The compressive deformation that uses one vector becomes possible as shown in Figure 2(b). In addition, the transformation into various shapes becomes possible by assuming the three degrees of freedom. It is possible to expand and contract in the direction of the x axis and the direction of the y axis when the core material is put internally like Figure 2.

Moreover, it has a high stiffness and it stabilizes structurally for the external force from the direction of the z axis. Therefore, appropriate softness can be presented by combining with a flexible resin.

3 SOFTNESS ADJUSTMENT

3.1 *Compression of single cell*

Figure 4 shows a single cell geometry filled urethane resin. When the displacement δx will add to the cell, the volume becomes as follows:

$$V = 2xh(l + \sqrt{l^2 - x^2}) \tag{4}$$

where $x = x_0 - \delta x/2$ is the size of cell and x_0 is the initial value. l is the length of the edge and

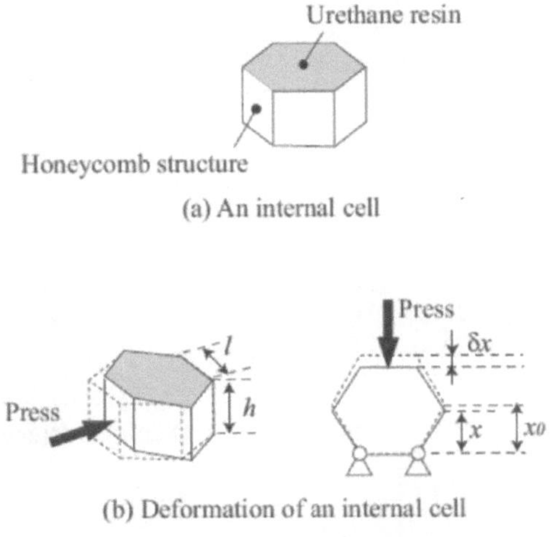

Figure 4. Model of an internal cell and planer models.

h is the thickness. Let K be the volume elasticity modulus as

$$K = -V\left(\frac{\partial p}{\partial V}\right)_T \tag{5}$$

From the Eqs. (4) and (5), we can see that K can be controlled by the displacement δx.

3.2 *Compression of cell matrix*

Figure 5 shows a model of the cell matrix set Δ with m cells. Here we simplify the tessellation problem. Figure 5(a) shows a standard row where cells are connected to construct a line. Then the cell matrix will be constructed to make a rectangle like shape (the n_x column and the n_y row). S_i is the subset of the set Δ and it can be inductively defined as follows.

1. $S_1 = \{a_1\}$, $\{a_1\}$ is the arbitrary member of the set Δ
2. $S_2 = \{a_1, a_2\}$, $\{a_1\}$ *and* $\{a_2\}$ share one edge
3. If $2 \le i$ and $S_i = \{a_1, a_2, \ldots, a_i\}$ can be defined, $S_{i+1} = \{a_1, a_2, \ldots, a_i, a_{i+1}\}$ $1 \le h \in Z < k \in Z \le i$,
   i. $\alpha_\eta, \alpha_\kappa \in \Sigma_t$
   ii. $\{\alpha_\eta, \alpha_{\eta+1}\}$, $\{\alpha_\kappa, \alpha_{\kappa+1}\}$ σηαρε ονε εδγε

Let $\acute{N}_i$ be the set of node points j at most near the point i other than i. Then $\acute{N}_i$ becomes adjacent points of i and the number of $\acute{N}_i$ becomes 2 or 3. Let we call the vertex of S_n satisfies $\acute{N}_i = 3$ as the good-vertex, and let the detail of such the good-vertex be λ_{S_n}. Then we call the edge $\varepsilon_{i,j}$ of the S_n consists of the good-vertex i as good-edge, and let the detail of such the good-edge be Γ_{S_n}. Here we call such a set S_n as magic-triangle, and then, the number of edges consists of vertex of S_n satisfies $\acute{N}_i = 3$ becomes 4 m.

Here we assume that the edges as the rigid linkage structures and the nodes as the rotary joints. When the displacement δx will add to the cell matrix, the volume becomes as follows:

$$V_n = 2mxh(l + \sqrt{l^2 - x^2}) \tag{6}$$

From the Eqs. (5) and (6), we can see that the stiffness of the cell matrix can be controlled by the displacement δx.

3.3 *Stiffness adjustment*

This section explains the stiffness adjustment strategy by compression. Figure 6 shows a stiffness model of an internal cell as simple-supported beam. Here, x is the effective length. The beam is assumed to be massless and its deflection δz is small enough so that the stiffness is linear in z. An actuator compresses the internal cell and x is changed.

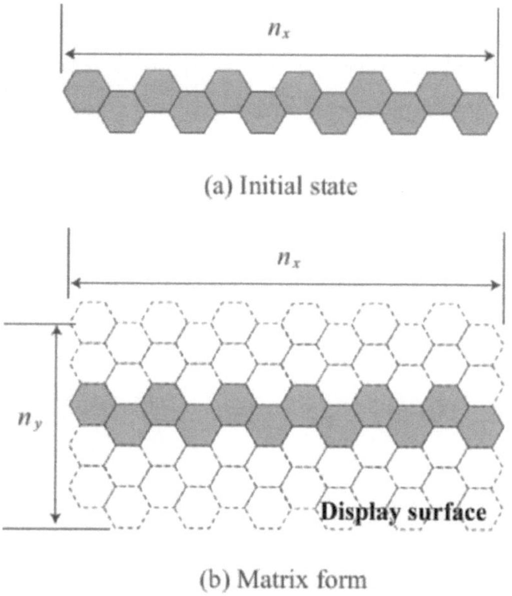

(a) Initial state

(b) Matrix form

Figure 5. Model of the cell matrix.

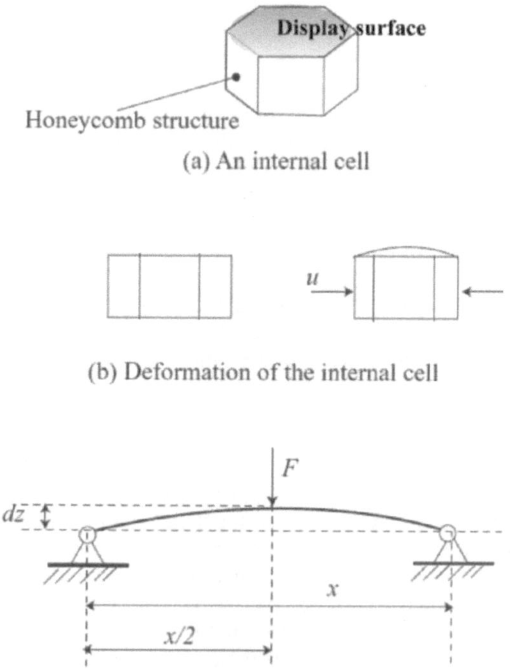

(a) An internal cell

(b) Deformation of the internal cell

(c) Stiffness model

Figure 6. A 2DOF stiffness adjustment model.

Here, we assume that reaction force F as

$$F \simeq \frac{48EI}{x^3} z. \tag{7}$$

where E, and I are the Young's modulus, and the geometrical moment of inertia, respectively. All of the parameters are constant. Here, we assume that the compression distance of the effective length as δx, we obtain

$$F \simeq \frac{48EI}{(x - \delta x)^3} z. \tag{8}$$

In the Eq. (8), we can control the stiffness

$$k = \frac{48EI}{(x - \delta x)^3} \tag{9}$$

from the compression distance δx.

Actually, the relationship between the deflection z and the reaction force F is nonlinear. Moreover, the constant part of Eq. (9) is a hexagonal cylinder an actual cell shape, and depends on the bulk-modulus of the resin when the compression distance is large and changes. These characteristic is experimentally shown in the following section.

4 EXPERIMENTAL RESULTS

In the proposal system, the stiffness is adjusted by expanding and contracting the resin. Experiments were conducted to verify the effects of the principal.

4.1 *Experimental setups*

Figure 7 shows an experimental system and Figure 8 shows a display material. The display is composed of the urethane resin and the honeycomb (aluminum core, cell size 9.0 [mm]). The index of softness is assumed to be a stiffness of the normal direction. We control the distance δx of the material and try to investigate how the stiffness along z axis will be changed. The force sensor probe is pushed into the surface of the display ($-z$ direction in the photo), and the stiffness is obtained from the measured force F and the pushed distance along z axis.

4.2 *Characteristic of the display material*

Figure 9 shows relation between distance along z axis and reaction force F at different three cells. These are measurements in three places where the position is different. Moreover, the stiffness value

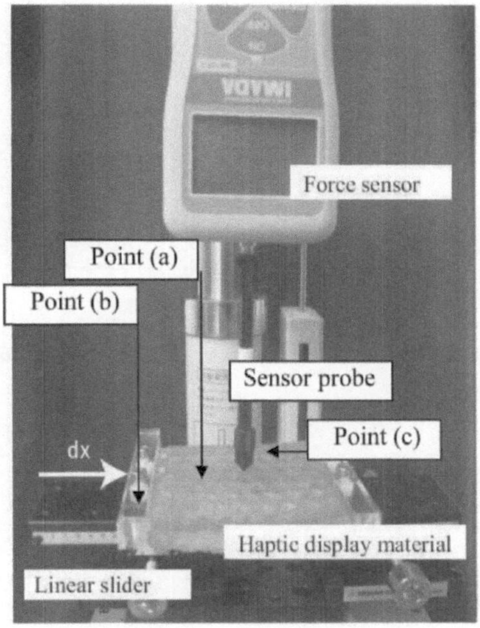

Figure 7. An experimental system.

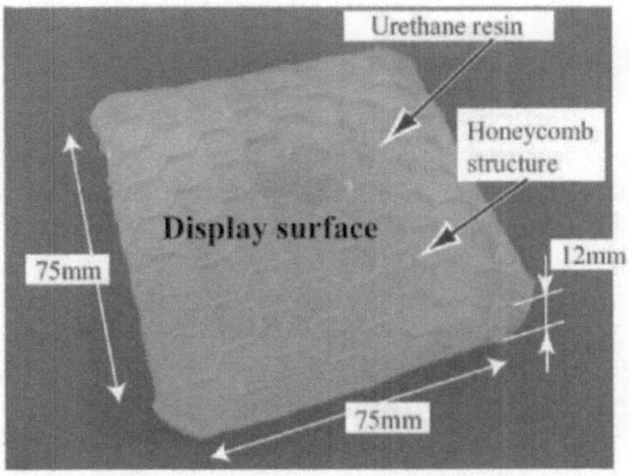

Figure 8. A display material.

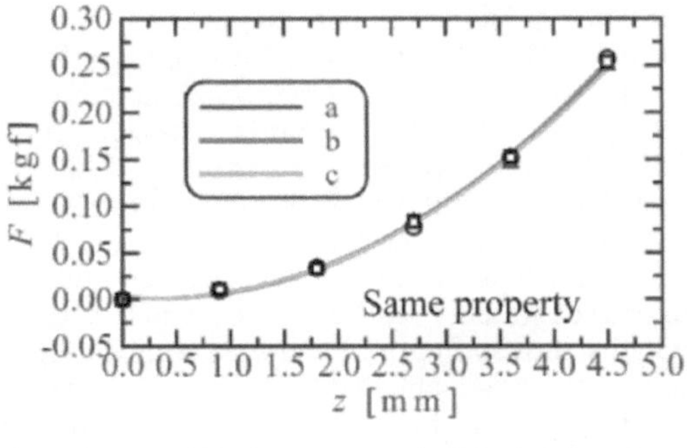

Figure 9. Relation between length along z axis and F at different three cells.

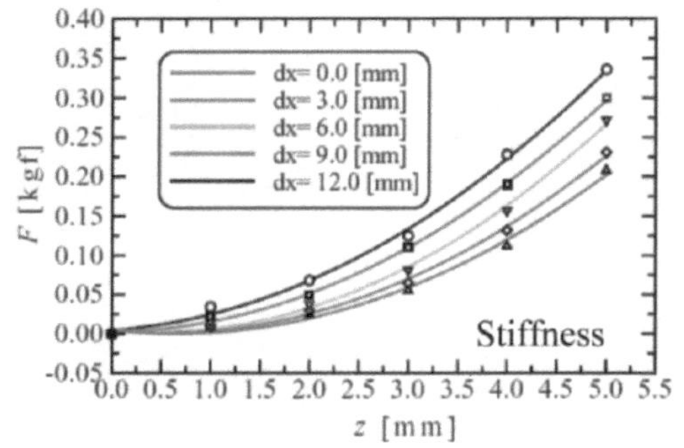

Figure 10. Relation between length along z axis and F.

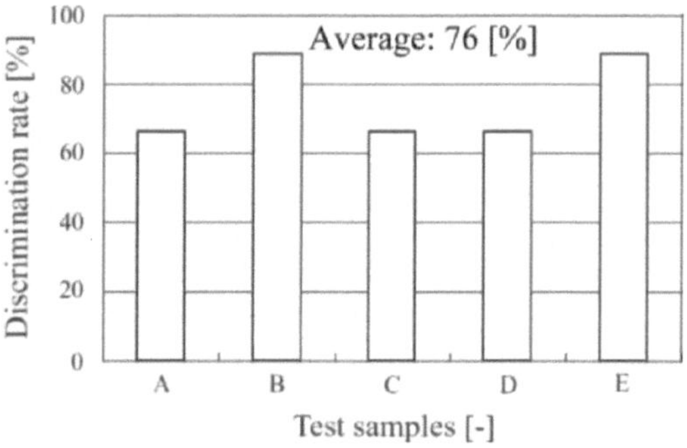

Figure 11. Result of the stiffness detection.

possesses the nonlinearity, and is guessed to originate in the equal volume transformation of the resin from the shape. Figure 10 shows the stiffness change when assuming $0.0 \leq \delta x \leq 12.0$ [mm]. Here, it can be confirmed that the stiffness value has changed by compressing x axially. Moreover, it can be confirmed that the deeper amount z of pushing is, the bigger the stiffness change is.

4.3 *Verification of the softness control*

The discrimination test has been conducted to confirm whether the sense of touch changed by the proposed principle. In the examination, subjects identified the difference of softness by a different compression distance. Compression distance δx was settled as $\delta x = 0.0, 6.0, 12.0$ [mm] and we call them as Soft, Medium, and Hard. Two were chosen at random from among three compression distances. Testees pressed the center part of the cell in the depth of about 2.0 [mm] with the forefinger, and the difference of softness was identified. At all the examinations, testee's finger was covered with the cloth, and gave aural a white noise. Figure 11 shows the identification rate by five testees (A-E). The average of the identification rate was 76 [%]. It is thought that the error factor originates in the difference of

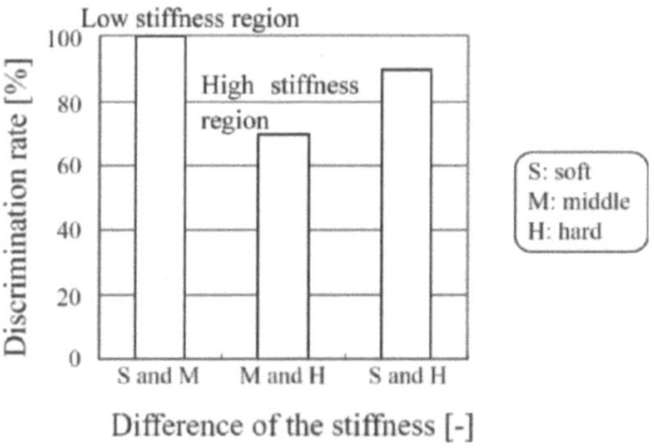

Figure 12. Result of the stiffness detection of each sample.

man's perception characteristic. Figure 12 shows the discrimination rate of the detection of the three stiffness levels (Soft, Middle, Hard). We can see that the discrimination rate becomes difference with the stiffness level.

5 CONCLUSIONS

This paper, proposed a principle that used a core material and a flexible resin of the mesh as a presentation display of softness. In the proposed procedure, the adjustment of softness is handily achieved by a little driving force. In a basic examination, it was confirmed that the surface trait of the display changed by compression. The improvement of the device that considers a just noticeable difference who corresponds to the flexibility change etc. local, adjusts of the frictional property on the surface, and is human etc. will be advanced in the future.

ACKNOWLEDGMENTS

This research was partially supported by the Ministry of Education, Science, Sports and Culture, Grant-in-Aid, The projects are supported by the staff of Nagoya Inst. Tech., Tokyo Metropolitan Univ., Ritsumeikan Univ. and Georgia Tech. Atlanta.

REFERENCES

Bianchi, M., Scilingo, E.P., Serio, A. & Bicchi, A. 2009, A new softness display based on bi-elastic fabric, *Proc. Worldhaptics*: 382–383.

Bicchi, A., De Rossi, D.E. & Scilingo, E.P. 2000, The role of the contact area spread rate in haptic discrimination of softness, *IEEE Trans. on Robotics and Automation* 16(5): 305–310.

Colgate, J. 1996. Nonholonomic Haptic Display, *Proc. ICRA:* 539–544.
Fujita, K. & Ohmori, H. 2001, A new softness display interface by dynamic fingertip contact area control, *Proc. WMSCI*: 78–82.
Fujita, K., Ohmori, H. & Katagiri, H. 2000, Development of Softness Display Device Based on Fingertip Contact Area Control, *Proc. the Virtual Reality Society of Japan 5th Annual Conference*: 251–254 (in Japanese).
Hoshino, H. & Tachi, S. 1999. A Method to Represent an Arbitrary Surface in Encounter Type Shape Representation System, *Trans. the Virtual Reality Society of Japan* 4(2): 445–454.
Inoue, K., Uesugi, R., Arai, T. & Mae, Y. 2003, Development of Haptic Device Using Flexible Sheet', *Journal of Robotics and Mechatronics* 15(2): 121–127.
Iwata, H., Yano, H. & Kawamura, R. 2002, Array force display for hardness distribution, *Proc. HAPTICS.*
Iwata, H., Yano, H., Nakaizumi, F. & Kawamura, R. 2001, Project FEELEX: Adding Haptic Surface to Graphics, *Proc. SIGGRAPH.*
Kawai, M. & Yoshikawa, T. 2002, Haptic Display of Movable Virtual Object with Interface Device Capable of Continuous-Time Impedance Display by Analog Circuit, *Proc. ICRA:* 229–234.
Kawamura, R., Yano, H. & Iwata, H. 2000, Development of surface type haptic interface for presentation of rigidity distribution, *Proc. the VirtualReality Society of Japan 5th Annual Conference*: 51–54.
Kawamura, S., Yamamoto, T., Ishida, D., Nakayama, Y., Tabata, O. & Sugiyama, S. 2002, Development of Passive Elements with Variable Mechanical Impedance for Wearable Robots, *Proc. ICRA:* 248–253.
Mitsuda, T., Kuge, S., Wakabayashi, M. & Kawamura, S. 2002, Haptic display implemented by controllable passive elements, *Proc. ICRA:* 4223–4228.
Papadopoulos, E., Vlachos K. & Mitropoulos, D. 2002. Design of a 5-DOF Haptic Simulator for Urological Operation, *Proc. ICRA:* 2079–2084.
Saga, S. & Deguchi, K. 2008, Dilatant fluid based tactile display (2nd report), *Proc. of the 2009 JSME Conference on Robotics and Mechatronics*: 953–954.
Sakaguchi, M., Fukusumi K. & Furusyo, J. 2000. Development and basic experiments of passive force display using ER brakes, *Proc. Japan-USA Symposium on Flexible Automation.*
Srinivasan, M.A. & Lamotte, R.H. 1995, Tactual Discrimination of Softness, *J Neurophysiol.* 73(1): 88–101.
Tachi, S., Maeda, T., Hirata, R. & Hoshino, H. 1994. A construction Method of Virtual Haptic Space, *Proc. ICAT*: 131–138.
Wagner, C.R., Lederman, S.J. & Howe, R.D. 2003, Design And Performance of a Tactile Shape Display Using RC Servomotors, *Haptics-e* 3: 2004.

E-Commerce, E-Business and E-Service – Lee (Ed)
© 2014 Taylor & Francis Group, London, ISBN 978-1-138-02646-9

Design and implementation of a micro-inverter for grid-connected applications

Ming-Tsung Tsai, Ching-Lung Chu & Jhe-Yu Lin
Department of Electrical Engineering, Southern Taiwan University, Tainan, Taiwan

ABSTRACT: A full bridge high frequency isolated inverter is proposed for renewable distributed power supply applications. The researched system is consisted of a full-bridge high-frequency DC/DC converter with the proposed symmetric phase-shift modulation algorithm to achieve the ZVS switching function, and a line frequency unfolding bridge. It replaces the traditional two stages of independently control algorithms with a one-stage control to obtain high conversion efficiency. Finally, a prototype of 350 W system through the implementation of the TMS 320F2812 digital signal processor is settled to verify the proposed idea.

Keywords: grid-connected; high frequency isolated; symmetrical phase-shift modulation

1 INTRODUCTION

Recently, renewable energy feeding to the distributed power systems has been increased and more visibility, while the world's power demand is increasing, especially for the small scale residential applications with a range below 1 kW [1]. Therefore, the cost of the grid-connected inverter is becoming more visible in the future. There are many existing power inverter topologies for interfacing renewable energy to the used terminal. Grid-connected is mainly used in the aero where the utility can be served. Inverters connected to the grid involve the major function to inject a sinusoidal current into the grid.

Development of grid-connected inverter system is divided into two categories, including centralized inverter type and micro-inverter type [2–7]. The former framework is usually to adopt a stabilize DC bus design, and uses a large capacity of electrolyte capacitor to obtain a stable DC voltage, its advantages is more flexible design, while the latter uses the pulsating DC bus design and a small volume electrolyte capacitor is adopt. Thus, each of renewable energy modules can have a better running performance. However, each team of module requires a special invertor to transfer the energy to the electricity.

A single-stage high-frequency inverter topology for decentralized PV systems has been presented in this paper for small scale residential applications. The researched system is consisted of a full-bridge high-frequency DC/DC converter with the proposed symmetric phase-shift modulation algorithm to achieve the ZVS switching function, and an unfolding DC/AC inverter. With the proposed control algorithm, it meets the requirement of a high efficiency conversion.

2 STRUCTURE

Grid-connected inverter system is roughly divided into isolated and non-isolated two major categories. This study was to investigate the design of high-frequency isolated structure. For such a architecture, it is basically divided into two control designs depending on the available large electrolyte capacitor or not, and can be described as follows.

2.1 *Comparison of two-stage and single-stage control based grid-connected inverter*

Two-Stage control based inverter system is basically consisted of a high-frequency DC/DC stage whose output connected to a stiff dc bus voltage which is with large electrolyte capacitors. Then the second DC/AC stage operated in sinusoidal pulse-width modulation switching transfers the energy to the utility. It can be shown as Figure 1(a). In contrast to this, the single-stage control based system is basically consisted of a high-frequency DC/DC stage whose output connected to a pulsating dc bus voltage which is with no electrolyte capacitors. Then an unfolding full-bridge inverter switched in 60 Hz transfers the energy to the utility. It can be shown as Figure 1(b).

2.2 *The researched architecture based on the single-stage control system*

The considered single-stage control based grid-connected inverter system in this paper is shown as Figure 2. This architecture has the characteristic of with no DC electrolyte capacitor, therefore, a pulsating waveform presented in its output side. With the proposed PWM switching scheme, it can

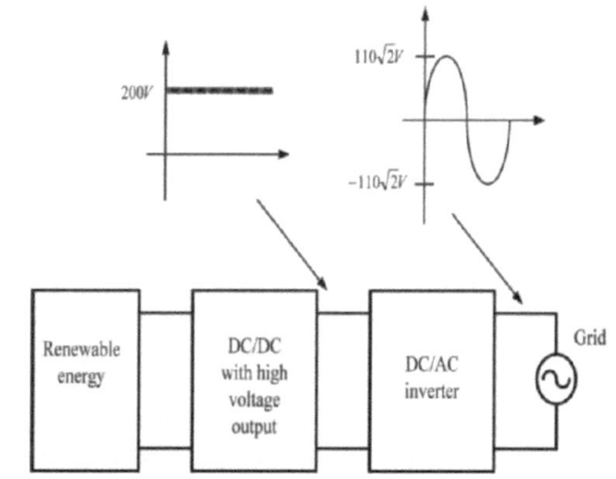

Figure 1(a). Two-stage control based PV system.

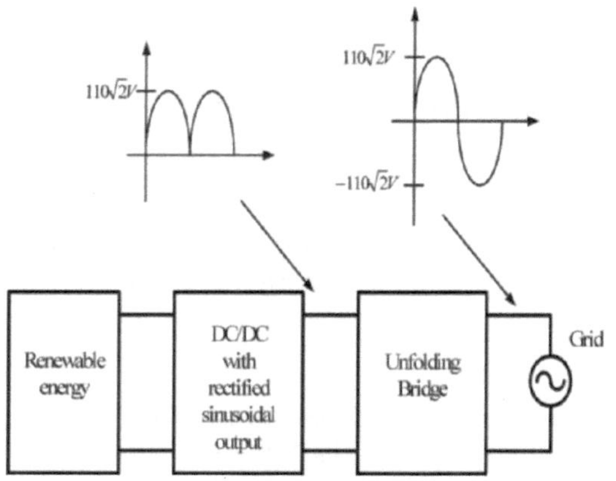

Figure 1(b). Single-stage control based PV system.

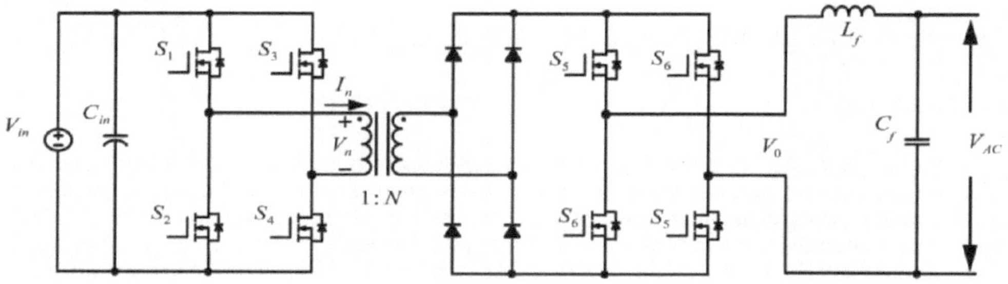

Figure 2. The single-stage control based grid-connected inverter system architecture in this paper.

feed an unit power factor current into the utility, and the control responsibility is in charge at the former stage (S_1–S_4), while the latter stage (S_5–S_6) is responsible only to low frequency (eg 60 Hz) switching to lower the switching losses. Under this situation, feeding a non-unit power current into the mains will cause a distorted current waveform, so cannot be fitted in this structure.

3 SYMMETRIC PHASE-SHIFT CONTROL ALGORITHM

This paper proposes a symmetric phase-shift control to fulfill the DC/AC function. The control algorithm is shown as follows:

In Figure 2, the modulation function achieved by S_1–S_4 before high frequency transformer, denoted as $H_F(t)$ can be described in (1), where α denoted as the desired phase shift angle:

$$H_F(t)=\begin{cases} +V_{in} & \alpha/2 \le t \le T_s/2-\alpha/2 \\ -V_{in} & T_s/2+\alpha/2 \le t \le T-\alpha/2 \\ 0 & elsewhere \end{cases} \quad (1)$$

Thus, the primary voltage V_n and the control command V_C can be shown as follows:

$$V_n = V_{in}\left(1-\frac{2\alpha}{T_s}\right) \qquad \begin{matrix} \alpha=0 & V_n=V_{in} \\ \alpha=T_s/2 & V_n=0 \\ \alpha=T_s & V_n=-V_{in} \end{matrix} \quad (2)$$

$$V_C=\left(1-\frac{2\alpha}{T_s}\right) \quad (3)$$

To obtain a sinusoidal output, the command $V_C(t)$ should be a $\sin\omega t$ waveform, that is:

$$V_C(t)=V_C \sin\omega \quad (4)$$

Therefore, the output after S_5–S_6 can be expressed as follows:

$$V_o = NV_{in}\frac{V_C}{V_{tri}}\sin\,\omega t \quad (5)$$

3.1 *Operation principle*

Figure 3 shows the conducting status in the proposed control algorithm.

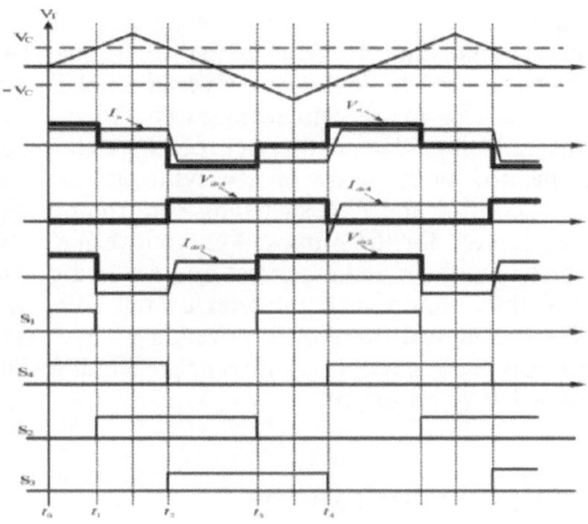

Figure 3. The conducting status due to the proposed control algorithm.

Interval ($t_0 \le t < t_1$)

In this status, S_1 and S_4, are ON, and S_2, S_3 are OFF. The transformer primary voltage is equal to the DC input voltage, and input current flows through the transformer primary side and the switches to form a current loop, making the energy from the input source through the transformer to the secondary side, and then through the S_5 or S_6 to the load dependent on the positive or negative cycle.

Interval ($t_1 \le t < t_2$)

In this status, S_2 and S_4, are ON, and S_1, S_3 are OFF. The energy flow through S_2 and the transformer primary side to form a flywheel current loop, and the transformer primary voltage V_n is in short status.

Interval ($t_2 \le t < t_3$)

In this status, S_2 and S_3, are ON, and S_1, S_4 are OFF. The transformer primary voltage is equal to the negative DC input voltage, and input current flows through the transformer primary side and the switches to form a current loop, making the energy from the input source through the transformer to the secondary side, and then through the S_5 or S_6 to the load dependent on the positive or negative cycle.

Interval ($t_3 \le t < t_4$)

In this status, S_1 and S_3, are ON, and S_2, S_4 are OFF. The energy flow through S_1 and the transformer primary side to form a flywheel current loop, and the transformer primary voltage V_n is in short status.

3.2 *Implementation*

The proposed single-stage control is based on the DSP TMS320 F2812 processor as shown as Figure 4(a) and (b) shows the current control block diagram.

4 EXPERIMENTAL RESULTS

Figure 5(a) shows the experimental results, including the driving signals of the switching devices, S_1, S_4, and the corresponding voltage waveform and current waveform of S_4. It shows the switches can achieve ZVS function. Figure 5(b) shows the primary input voltage and current of the high frequency transformer. It shows no bias current existed there. Figure 6 shows the inverter output current corresponding to the utility source voltage. It shows the proposed micro inverter can achieve the in-phase current function due to the current control loop, and with low harmonics. The overall system efficiency at 350 W is about 90%.

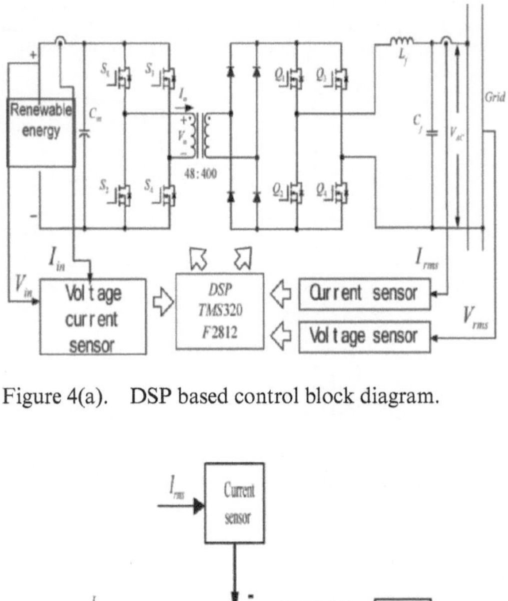

Figure 4(a). DSP based control block diagram.

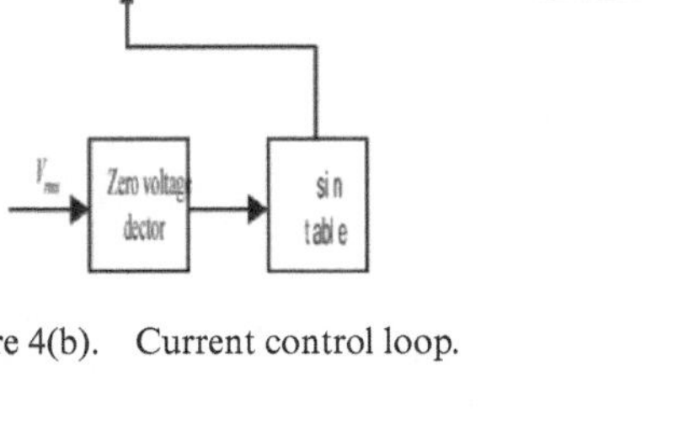

Figure 4(b). Current control loop.

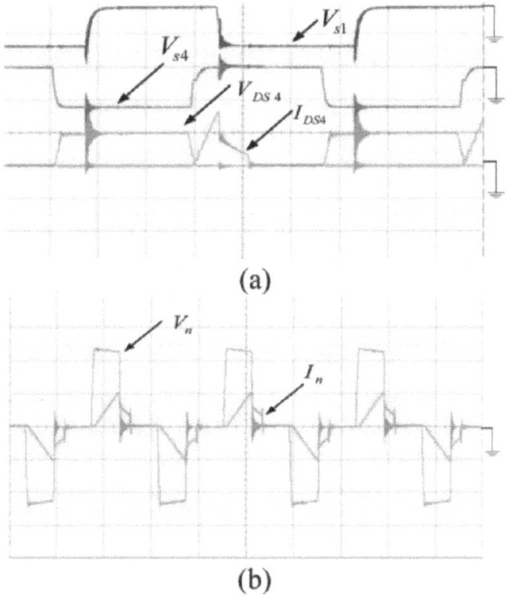

Figure 5. The experimental results of DC/DC stage, (a) The driving signals V_{S1}, V_{S4} and the corresponding voltage V_{DS4} and current I_{DS4} of S_4, (b) The primary input voltage and current of the high frequency transformer.

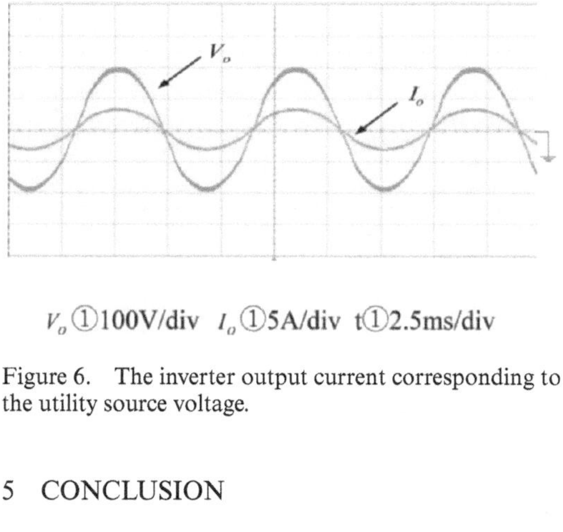

Figure 6. The inverter output current corresponding to the utility source voltage.

5 CONCLUSION

This paper discusses the steady-state behavior of the single-stage control based grid-connected inverter when controlled via a symmetrical phase shift modulation. The single-stage control based algorithm to replace the traditional two stage control to the micro inverter applications can reduces the system complexity and increase the reliability due to the lack of the electrolytic capacitor. With the use of new symmetrical phase shift control, the ZVS switching performance can be achieved for the proposed micro inverter, so as to reduce the switching stress and switching loss, and thus improves the inverter overall efficiency. The theoretical framework is validated by means of computer simulations and experimental results on a 350 W prototype.

ACKNOWLEDGEMENTS

This work has been supported by National Science Council, Republic of China under research project NSC100-2221-E-218-002.

REFERENCES

[1] Perry Tsao,"Solar Magic in Microinverter Application," National Semiconductor Application Note 2116 February 28, 2011.
[2] Gu Jun-yi, Wu Hong-fei, Chen Guo-cheng, and XING Yan, "Research On Photovoltaic Grid–Connect Inter Base On Soft–Switching Interleaved Flyback Converter", *IEEE Conference on Industrial Electronics and Applicationsis,* pp. 1209–1214, 2010.
[3] D. Chen and J. Liu, "The Uni-Polarity Phase-Shifted Controlled Voltage Mode AC–AC Converters with High Frequency AC Link", *IEEE Trans. on Power Electronics,* vol. 21, no. 4, July 2006, pp. 899–905.
[4] D. Chen and L. Li, "Novel static inverter with high frequency pulse dc link," *IEEE Trans. Power Electron.*, vol. 19, no. 4, pp. 971–978, July 2004.
[5] H. Choi, J.W. Kim, and H.B. Cho, "Novel zero-voltage and zero-current switching (ZVZCS) full-bridge PWM converter using coupled output inductor," in *Proc. IEEE APEC'01*, 2001, pp. 967–973.
[6] J. Zhou, M. Xu, J. Sun, and F. Lee, "A self-driven soft-switching voltage regulator for future microprocessors," *IEEE Trans. Power Electronics.*, vol. 20, no. 4, pp. 806–814, July 2005.
[7] Zhongming Ye, Praveen K. Jain, and Paresh C. Sen, "A Full-Bridge Resonant Inverter With Modified for High-Frequency AC Power Distribution Systems," *IEEE Transactions on Industrial Electronics*, vol. 54, no. 5, October 2007.

E-Commerce, E-Business and E-Service – Lee (Ed)
© 2014 Taylor & Francis Group, London, ISBN 978-1-138-02646-9

Characterization and properties of PLA/ENR/MWCNTs nanocomposite

Mou'ad. A. Tarawneh & Sahrim Hj. Ahamd
Materials Science Programme, School of Applied Science, Universiti Kebangsaan Malaysia, Bangi, Selangor, Malaysia

ABSTRACT: In this study the mechanical properties of TPNR (thermoplastic natural rubber) nanocomposites reinforced by MWCNTs (multi-walled carbon nanotubes) investigated. The TPNR nanocomposites were prepared from PLA (polylactic acid), ENR (epoxidised natural rubber) and LENR (Liquid epoxidised natural rubber) as a compatibilizer, respectively, with 2wt%–8wt% of MWCNTs. The Young's modulus and tensile strength increased by almost 79% and 84% respectively, at 2%wt MWCNTs. The elongation at break decreased with MWCNTs. The maximum impact strength was recorded at 4wt% of MWCNTs which was increased 209% as compared with a pristine TPNR sample. SEM micrographs confirm the effect of good dispersion of MWCNTs and their interfacial bonding in TPNR.

Keywords: epoxidised natural rubber; multi-walled carbon nanotubes (MWCNTs); thermoplastic natural rubber

1 INTRODUCTION

The nanoparticles have attracted much consideration in manufacturing polymeric nanocomposites using different nanoparticles as reinforcement. Experimental results on polymeric nanocomposites have opened a broad range of variation in their properties. The perfect properties of Carbon Nanotubes (CNTs) make them ideally suited as filler for reinforcing polymer composites and also best reinforcing filler for the manufacture of the next generation of filler-matrix composite materials. The efficiency of reinforcement depends on different factors such as the adhesion between the filler and the matrix, the filler aspect ratio, and the filler mechanical properties.[1]

Polylactic Acid (PLA) is gaining popularity due to its renewability and comparable properties with petroleum based polymers. Previously, such as resorbable sutures, Polylactic Acid (PLA) uses have been limited to specialty biomedical niches.[2] At present, a lot of development work on plastic materials has been focused on the application of biodegradable materials, especially PLA. It is now accepted in many application areas such as bags and sacks, food packaging, loose-fill packaging agricultural film and some specialty applications.[3] Epoxidized Natural Rubber (ENR) is a chemically modified form of the cis-1,4-polyisoprene rubber, whereby some of the unsaturation is converted into epoxide groups, which are randomly distributed along the polymer chain. It is known as a compatibilizer for incompatible blends as well as a processing aid. The incorporation of ENR-50 into the rubber blends has been found to improve processability, resilience, stiffness and also wet grip performance, good damping, oil resistance also shows excellent reduced air permeability[4] With an increasing polarity of ENR the compatibility of ENR with other polymers increases.[5] Both PLA and Epoxidized Natural Rubber (ENR) are derived from renewable sources and can be processed using most conventional polymer processing. In view of their complementary properties, blending PLA with ENR is a good choice to improve PLA properties, such as toughness and elongation at break without compromising its biodegradability.

The effect of MWCNTs on the mechanical properties of TPNR nanocomposites is reported in this paper to obtain the optimize value in tensile strength, tensile modulus, strain at break and impact strength by varying MWCNTs content.

2 MATERIALS AND METHODS

2.1 *Materials*

Polylactic acid of Natureworks Ingeo™ Biopolymer 3251D grades supplied by Unic Technology Ltd, China, is the thermoplastic resin used in this study. It has a density of 1.24 g/cm^3 and melt flow index of 30–40 g/10 min. *(190°C/2, 16 Kg)* and melting temperature between 160–170°C. Natural rubber of SMR L grade and Epoxidised

Natural Rubber (ENR50) were obtained from the Malaysian Rubber Board. Liquid Epoxidised Natural Rubber (LNR) was synthesized in our laboratory using the photochemical oxidation technique on natural rubber. The MWNTs were manufactured by (CVD) Chemical Vapor Deposition, and supplied from Chengdu Organic Chemicals Co., Ltd Chines Academy of Seiences. The specification of MWCNTs is as follow: purity >95%, length 10–30 μm, diameter 20–30 nm.

2.2 *Sample preparation*

TPNR nanocomposites were prepared by melt blending of PLA, ENR and LENR with MWCNTs in a ratio of 60 wt% PLA, 35 wt% ENR and 5 wt% LENR as a compatibilizer and 2, 4, 6 and 8 wt% MWCNTs. The Indirect Technique (IDT) was used to prepare nanocomposites, this involved mixing the MWCNTs with LENR separately, before it was melt blended with PLA and ENR in the internal mixer (Haake Rheomix 600P). The mixing temperature was 180° C, with 13 min mixing time and a rotor speed 100 rpm.

2.3 *Characterizations*

Tensile properties were measured using Testometric universal testing machine model M350-10CT with 5 kN load cell according to ASTM 412 standard procedure using a crosshead speed 50 mm min^{-1} and test specimens of 1 mm thickness. As a minimum, 5 samples were tested for each composition, and the average value was recorded. The sample dimension was 63 × 12 × 3 mm. The Izod impact test was approved out using Ray Ran Pendulum Impact System according to ASTM D 256-90b. The velocity and weight of the hammer were 3.45 m/s and 0.898 kg, respectively. The sample dimension is 63 × 12 × 3 mm. At least 5 samples were tested for each composition, and the average value was reported. SEM (scanning electron microscope, model Philips XL 30) used to study the morphology of TPNR nanocomposites.

3 RESULTS AND DISCUSSION

The tensile strength of TPNR and TPNR/MWCNTs with different percent is shown in Figure 1 (A). For TPNR/MWNTs nanocompsites, the optimum results is obtained at 2% wt which increased 84% respectively, compared with TPNR. The tensile strength completely increases as the percent of MWCNTs increases. An excellent interface between the matrix and MWCNTs is very important for a material to stand the load. Under load, the matrix distributes the force to the MWCNTs which take most of the applied load. When the percent of MWCNTs is higher than the optimum

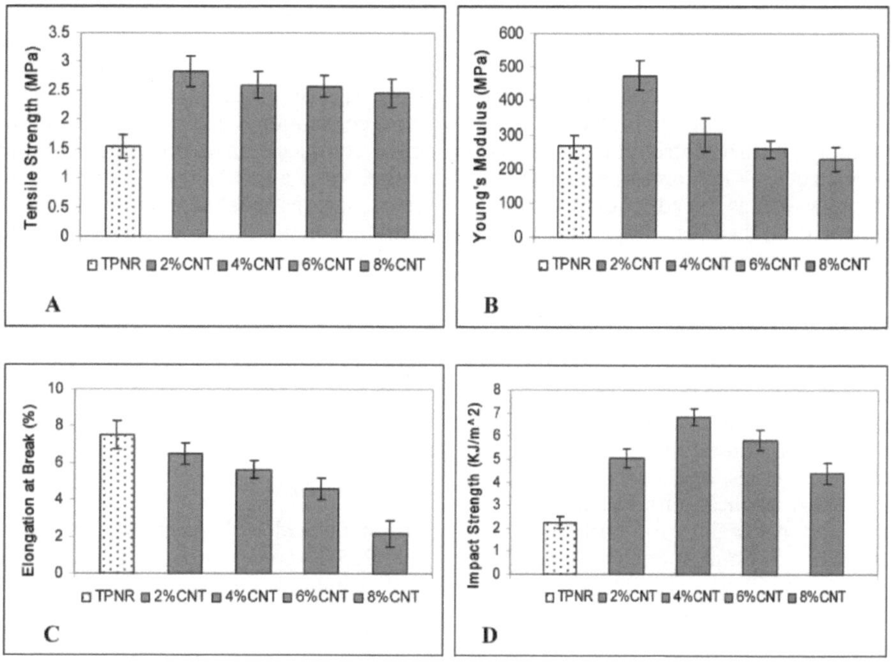

Figure 1. The mechanical properties of TPNR and TPNR/MWCNTs nanocomposites.

amount, the MWCNTs could not disperse well in the TPNR matrix and agglomerate to big cluster due to the surface energy of MWCNTs, and caused the reducing of the tensile strength.[6]

The Young's modulus of TPNR nanocomposites with various percentages of MWCNTs is depicted in Figure 1 (B). The Young's modulus increases with increasing the percentage of the MWCNTs as shown from the result. The maximum result is achieved at 2% wt, which increased about 79%. The improvement of modulus is due to high modulus of MWCNTs with good dispersion of MWCNTs inside TPNR. Well-dispersed nanotubes displaying perfect stress transfer.[7] Young's modulus at higher concentration such as 2 wt% decreased due to increase of localized clusters or aggregation of MWCNTs in TPNR, that can exposed in the SEM analysis. The elongation at break of TPNR with MWCNTs is indicated in Figure 1 (C), elongation at break decreased with MWCNTs. Its can be deduced that the reinforcing effect of MWCNTs is very clear. As the MWCNTs content in the TPNR increases, the stress level gradually increases but at the same time the strain of the nanocomposites decreased. This is because of MWCNTs into TPNR matrix which behaves as cross-linking points restrict the movement of polymer chains. It is an indicator of the material flexibility, which shows that the inclusion of MWCNTs makes the TPNR stronger but more brittle.

The impact strength of TPNR/MWCNTs nanocomposites are shown in Figure 1 (D). The impact strength of TPNR has improved with inclusion of MWCNTs. The results exhibited better impact strength of MWNT at 4 wt% and increased by almost 209%, due to the high dispersion of MWCNTs in the TPNR which generated a noteworthy toughening effect on the TPNR/MWCNTs nanocomposite. On the other hand, when the load is transferred to the physical network between the matrix and the filler, the debonding of the chain segments from the filler surface facilitates the relaxation of matrix entanglement arrangement, foremost to higher impact hardiness. However, due to their large aspect ratio and higher surface energy of MWCNTs, it will be difficult for them to disperse in the TPNR at high percentage; as a result the higher agglomeration of MWCNTs can cause the properties of the nanocomposites to get worse.[8]

SEM micrograph of TPNR is shown in Figure 2 (A). This figure shows the TPNR without filler inside it, and so the surface is smooth. The homogenous dispersion of MWCNTs in the

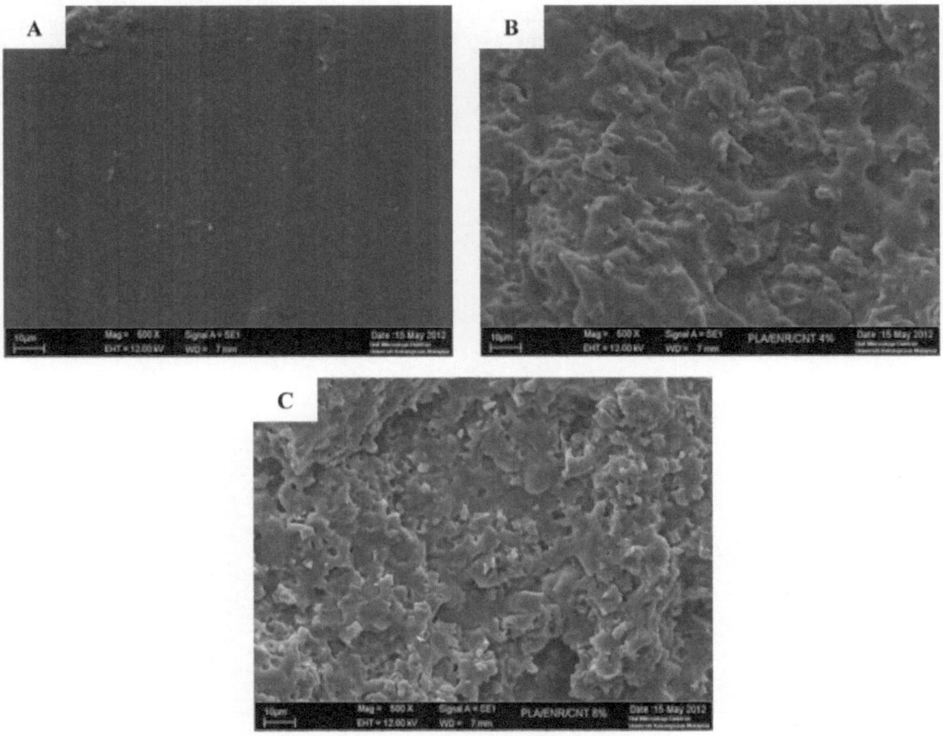

Figure 2. SEM micrograph of (A) TPNR (B) TPNR/4 wt% MWCNTs (C) TPNR/8 wt% MWCNTs.

composites is confirmed by scanning electron microscopy. Figure 2 (B) is shown the naocomposites with 4 wt% MWCNTs, which is well dispersed in the matrix, also showed the nanotubes in matrix were still coated with polymer also the bright spots inside the TPNR, suggestion a strong polymer nanotubes interfacial. Strong interfacial adhesion is essential for efficient stress transfer from the matrix to the nanotubes; this is supporting our observation of higher efficiency of carbon nanotubes assisted in enhancing the mechanical properties of TPNR. The poor dispersion of 8 wt% of MWCNTs in TPNR as depicted in Figure 2 (C). The Figure shows clearly a large number of unbroken carbon nanotubes, indicating a poor polymer/nanotube adhesion which attributed to reduce the properties of TPNR/MWCNTs nanocomposites.

4 CONCLUSION

MWCNTs reinforced PLA/ENR/LENR composites have been fabricated by using the melt blending method. The tensile modulus, tensile strength, also impact strength are improved significantly while with sacrificing high elongation at break by incorporating MWCNTs compared with TPNR. Homogeneous dispersion of MWCNTs in TPNR matrix and strong interfacial adhesion between the matrix and MWCNTs are proposed to be responsible for significant mechanical enhancement as confirmed by SEM images.

ACKNOWLEDGEMENTS

The authors would like to thank the Malaysian Government and National University of Malaysia (UKM) for financial support under Science Fund Grant UKM-OUP-2012-135.

REFERENCES

[1] Haggenmueller, R., Du, F., Fischer, JE. Winey, KI. Interfacial in situ polymerization of single wall carbon nanotube/nylon 6,6 nanocomposites. Polymer 2006:47: 2381–2388.
[2] Gripma, D.W., H. Altpeter, M.J. Bevis, and J. Feijen. Improvement of the mechanical properties of poly (D, L-lactide) by orientation. Polym. Int 2002:51: 845–851.
[3] Petchwattana, N., S. Covavisaruch, and N. Euapanthasate, 2012, Utilization of ultrafine acrylate rubber particles as a toughening agent for poly(lactic acid): Materials Science and Engineering: A 2012:532: 64–70.
[4] C.S.L. Baker, I.R. Gellingand R. Newell. Epoxidized Natural Rubber. Rubber Chem Technol 1985:58: 67–85.
[5] Ibrahim A. and M. Dahlan. Thermoplastic natural rubber blends. Prog. Polym. Sci 1998:23: 665–706.
[6] Sang Hyun Jin a, Young-Bin Park b, Kwan Han Yoon. Rheological and mechanical properties of surface modified multi-walled carbon nanotube-filled PET composite. J. Composites Science and Technology 2007:67:3434–3441.
[7] Potschke P., Fornes T.D., Paul D.R. Rheological behavior of multi-walled carbon nanotubes/polycarbonate composites. Polymer 2002;43:3247–3255.
[8] Jianfeng Shen, Weishi Huang, Liping Wu, Yizhe Hu, Mingxin Ye, The reinforcement role of different amino-functionalized multi-walled carbon nanotubes in epoxy nanocomposites, Composites Science and Technology 2007:67: 3041–3050.

Section 4: ICT for business and management

E-Commerce, E-Business and E-Service – Lee (Ed)
 ISBN 978-1-138-02646-9

Financial crisis management for contractors in construction market: Case study in China

Hongyu Chen
Institute of International Education, Wuhan University of Technology, Wuhan, China

Limao Zhang, Qian Liu & Yi Liu
School of Civil Engineering and Mechanics, Huazhong University of Science and Technology, Wuhan, China

ABSTRACT: Considering the financial data of listed companies in China has a characteristic of high correlations and dimensions, this paper combines principal component analysis and logistic analysis, and develops an early warning approach for financial crisis management for construction enterprises in China based upon 24 enterprise financial variables. Empirical analysis results show that this approach can realize the dimension reduction of multivariables with minimal loss of the variable information, and solve the problem that principal components of original data are irrelevant to each other. This model not only has a good identification of early warning ability to predict business failure, but also can provide effective decision support for identifying the financial advantage and weakness of construction enterprises and establishing future development strategies.

Keywords: construction enterprises; financial crisis; principal component analysis; logistic regression analysis

1 INTRODUCTION

As one of the most important industries in the national economy, construction industry has achieved rapid development in the recent 20 years.[1,2] Having characteristics of long cycle, high investment, high risk, high yield, construction industry plays an essential role in the economic and social development. In recent years, construction enterprises are faced with more and more complex environment, growing severe challenges and high-competitive construction market. Macro-control on real estate industry by the government and financial institutions is gradually strengthening. Asset-liability ratio in the construction business is generally more than 70% so that construction enterprises are under enormous financial risks and increasing likelihood of financial crises.[3]

Financial crisis, also called financial distress, reflects one kind of economic phenomena that an enterprise is unable to pay debts or fees due to financial management failures. Based on accounting information, early warning for financial crisis management implements real-time monitoring and predictive alerts to the financial crisis which the enterprise may or will be faced with by setting and observing the changes in the sensitivity of early warning indicators. It greatly helps business owners keep abreast of the problems in the financial management, effectively prevent and resolve financial crisis.[4,5] However, due to the fact that information in the financial statements seriously lags behind, and the financial statements are easily manipulated and reflect information in one-sidedness, there are some drawbacks to this approach. Univariate analysis is also called linear model, with individual financial ratio to evaluate financial condition. For example, in the univariate model which is proposed by an American scholar, William Beaver, in 1996, many ratios, such as cash flow and the ratio of the total amount of liabilities, show very good prediction abilities.[6] However, many different researchers have different opinions in choosing what kinds of financial indexes as evaluation indexes for early warning for financial crisis management so that their conclusions unlikely reach an agreement.

Multiple variables analysis is based on the opinion that these variables are related and could not be considered respectively so that we need to consider these variables integrally and choose core variables to analyze comprehensively. Ohlson firstly introduced logistic regression analysis into multivariate analysis for early warning of financial crisis. And he analyzed enterprises bankruptcy probability interval distribution the relationship between the two types of errors and dividing point with the prediction accuracy rate of 96.12%.[7] Logistic regression analysis methodology improves early warning for financial crisis a lot and it avoids many problems of traditional analysis methodology.[8,9] Combined with the current situation of the construction industry in China, this paper chooses 40 listed construction companies as samples and chooses 24 financial indexes which can reflect company financial condition comprehensively as analysis variables. Using logistic regression analysis methodology, it analyzes principal components from multiple financial variables and develops an early warning approach for financial crisis management for construction enterprises with a logistic regression model to analyze and evaluate financial abilities of construction enterprises.

2 METHODOLOGY

Regression analysis is a mathematical tool to study how the changes of a set of variables affect another variable. Logistic regression model, an effective methodology to solve multivariate 0–1 regression problem, is able to avoid assumptions that variables need to obey the multivariate normal distribution and bankrupt and non-bankrupt companies obey the same covariance matrix. It is necessary to judge an enterprise is whether in financial crisis in early warning for financial crisis management. So it can do the following definition: assume that p represents the probability of an event occurring in (financial crisis exists), then $p = P(Y = 1)$; and the probability of an event not occurring in (financial crisis not exists) is p, then $p = P(Y = 0) = 1 - p$. Introduce the logistic conversion of p, as formulae (1) and (2) show. $\theta(p)$ represents the logic conversion value of p, usually referred to logic value (logit (p)), and it is in symmetry using logit $(0.5) = 0$ as the center, changing from $-\infty$ to $+\infty$ at $p = 0$ or $p = 1$. So when $p \geq 0.5$, it can assume that listed companies are in financial crisis, and when $p < 0.5$, listed companies not in financial crisis.

$$\theta(p) = \text{logit}(p) = \ln[p/(1-p)] = \beta_0 + \beta_1 X_1 + \beta_2 X_2 + \beta_3 X_3 + \cdots + \beta_n X_n + \xi \tag{1}$$

$$p = P(Y=1) = \frac{\exp\begin{pmatrix} \beta_0 + \beta_1 X_1 + \beta_2 X_2 + \beta_3 X_3 \\ + \cdots + \beta_n X_n + \xi \end{pmatrix}}{1+\exp\begin{pmatrix} \beta_0 + \beta_1 X_1 + \beta_2 X_2 + \beta_3 X_3 \\ + \cdots + \beta_n X_n + \xi \end{pmatrix}} \tag{2}$$

In the case of the model containing a number of explanatory variables, the goal of logistic regression analysis is to get a "saving model" equation which can predict business conditions. The model equation needs to meet the following conditions: (1) it contains less explanatory variables as possible; (2) it can achieve the best measure results; (3) it needs to consider original data information as much as possible; (4) it is persuasive in economic sense and so on. Typically, logistic regression analysis uses stepwise selection method to estimate the "saving model" equation. However, the methodology is totally based on a statistical approach without considering economic relationship between variables.

In order to estimate the equations linear and solve original data loss problem, this paper combines logistic regression analysis with principal components analysis so that this approach can realize the dimension reduction of multivariables with minimal loss of the variable information and solve the problem that principal components of original data are irrelevant to each other.[10] At the same time, its judgment accuracy is relatively high. In this paper, to do the research of early warning for financial crisis management in construction enterprises, logistic regression analysis model combined with principal components contains the following five steps:

- Select a number of construction enterprises in normal financial condition and in financial crisis as total samples, and choose appropriate financial ratios according to the principles of comprehensiveness, effectiveness, operability and some financial indicators that other scholars used.
- Screen out selected financial ratios through significant differences analysis, and obtain analysis variables which can effectively reflect financial abilities of construction enterprises.
- Extract the study samples' principal components using principal components analysis methodology, and combine with analysis variables' degree of importance and contribution and then choose those irrelevant principal components whose cumulative variance contribution rates are at 70%–85% as explanatory variables when do logistic regression analysis to obtain comprehensive index factors that can reflect early warning model for financial crisis management in construction enterprises.
- Develop an early warning approach for financial crisis management for construction enterprises based on principal component analysis and logistic analysis, and predict the sample group, check its fitting results.
- Combined with the results of the test sample data, test the prediction accuracy of early warning model for financial crisis management in construction enterprises.

3 DATA

Considering the specific circumstances of construction enterprises in China and the feasibility of collecting sample data, this paper selects listed construction companies as samples. In order to facilitate seek threshold point to detect the validity of the model, this paper select as more construction enterprises which are in financial distress ST as samples as possible. This paper will deal with ST construction enterprises as financial crisis enterprises. In order to obtain a sufficient number of samples, it selects some construction and development enterprises based on diversified businesses. To November 2013, it selects a sample of 40 listed companies to do construction enterprises financial analysis and evaluation, as shown in Table 1, which includes 32 enterprises which are in normal financial condition and 8 in financial crisis (ST segment and *ST segment).

When selecting relevant financial ratios as variables, indicators should be selected to reflect financial capacity. Modern financial management theory is generally believed that the financial condition of

Table 1. Samples of listed construction enterprises.

Order	Code	Name	Order	Code	Name
1	SZ000002	Vanke A	21	SH600732	Shanghai Xin Mei
2	SH600383	Golden Group	22	SH600052	Zhejiang Guangsha
3	SH600823	Shimao shares	23	SH600048	Poly Real Estate
4	SH600246	Vantone	24	SZ002305	Southern home
5	SZ000926	Fuxing shares	25	SZ000961	South Building
6	SH600649	City Investment Holdings	26	SZ000567	Hyde shares
7	SH600185	Gree Real Estate	27	SZ000024	China Merchants Property
8	SH600266	Beijing Urban Construction	28	SH600533	Qixia construction
9	SZ002208	Hefei Urban Construction	29	SH600683	Metro Land
10	SH600173	Wolong Real Estate	30	SZ000043	CATIC Real Estate
11	SH600675	China Enterprises	31	SZ000671	Sunshine City
12	SZ002244	Riverside Group	32	SH600113	Zhejiang Dongri
13	SZ000046	Oceanwide	33	SH600634	ST Chenghai
14	SH600376	First open shares	34	SH600555	* ST Kowloon
15	SH600240	Chinese real estate industry	35	SZ000505	* ST Pearl River
16	SH600743	Huayuan Real Estate	36	SZ000056	* ST BCCI
17	SZ000031	COFCO Property	37	SH600358	* ST Joint
18	SZ000838	A country estate	38	SH600603	* ST SocieteGenerale
19	SZ000736	Zhongfang Real Estate	39	SH600234	* ST Dragon
20	SH600638	NHP	40	SZ000594	* ST Kwok-hang

Table 2. Construction enterprises financial capacity analysis and evaluation system.

Profitability	V1 Sales margin	**Operating capabilities**	V11 Accounts receivable turnover ratio
	V2 Sales net interest rates		V12 Inventory turnover
	V3 Return on equity (weighted)		V13 Current assets turnover
	V4 Return on equity (weighted-deduction)		V14 Total assets turnover
	V5 Return on total asserts		V15 Long-term debt to equity ratio
Growth	V6 Revenue growth	**Solvency**	V16 Quick ratio
	V7 Operating profit growth		V17 Current ratio
	V8 Net profit growth		V18 The gearing ratio
	V9 Expansion rate of fixed asset investment		V19 Multiples of interest earned
	V10 Total asset growth		V20 Cash ratio
Cash flow	V21 Sales cash ratio		
	V22 Operating cash flow growth		
	V23 Main business cash ratio		
	V24 Operating activities earnings quality		

an enterprise depends on the company's profitability, growth, operating capabilities, solvency and cash flow and so on. Therefore, on the premise of using theories of researchers at home and abroad for reference, combined with basic realities of domestic listed construction enterprises and basic characteristics of enterprises which are in financial crisis, this paper selects 24 indicators from five dimensions, namely, profitability, growth, operating capabilities, solvency and cash flow, to constitute a financial capacity analysis and evaluation system, as shown in Table 2.

4 RESULTS AND ANALYSIS

4.1 *Difference analysis of financial variables*

There was a significant difference between the normal financial enterprises and financial crisis enterprises, which is a basic condition to build the financial variables of early warning model. This paper uses the SPSS statistical analysis software for T-test, and then based on the probability of significant bilateral T to determine whether the indicators are totally different between the ST and non-ST enterprises. Assuming that there is difference indeed, T-test using a 90% confidence level, so it's considered as a significant difference between the sample data when the probability of significant bilateral T is less than 0.1.T-test was made on 24 financial variables of 40 financial enterprises above, and found that 10 financial variables including V2, V7, V8, V9, V10, V12, V16, V21, V23, V24, whose probability of significant bilateral T is more than 0.1, as a result these indicators should be removed because there is no significant difference.

4.2 *Principal component extraction*

To determine whether the financial indicators is suitable for principal component analysis, KMO and Bartlett test are used, when KMO is greater than 0.9 is generally considered the best effect, while above 0.7 is the better one, 0.6 is ordinary, but 0.5 is not appropriate for principal component analysis. KMO and Bartlett test results of paper selected samples are shown in Table 3, KMO statistic is 0.780, Bartlett test is rejected, which shows that the 14 remaining financial variables after removing by difference analysis are not independent. Moreover, the probability of P close to 0, indicating that the correlation matrix is quite different from the unit matrix. Therefore, the sample data researched on this paper is suitable for principal component analysis.

According to principal components scree plot (Fig. 1), the eigenvalues of the former 6 principal components are all greater than 1 with obvious changes, the eigenvalues after the 7th principal component is very small with slow changes, that is, the former 6 principal components should be extracted as they make a largest contribution to explanatory variables. According to the results of the total variance explained (Table 4) shows that the financial variables after rotating to explain the overall financial capacity of the contribution rate is listed as 15.384%, 13.935%, 11.951%, 11.251%, 10.867%, 7.797%, up to 71.184% cumulatively, which representing the vast majority of information in construction enterprises financial crisis risks.

When building factor loading matrix, it should be rotated so that the structure is compendious. With the help of variance to the power of 8 largest rotation method, each square of factor loadings is polarized by columns to 0 or 1 to get off the noise effect. Factor loading matrix of rotated principal component (see Table 5), we can get the mathematical expressions of the 6 selected principal components factor (F1–F6), such as formula (3)–(8).

Table 3. Financial variables KMO & Bartlett test results.

KMO & Bartlett Test	
Enough sampled Kaiser-Meyer-Olkin measurement	0.780
Bartlett's test of Sphericity	
The Approximate Computation of Chi-Square	176.716
df	91
Sig.	0.000

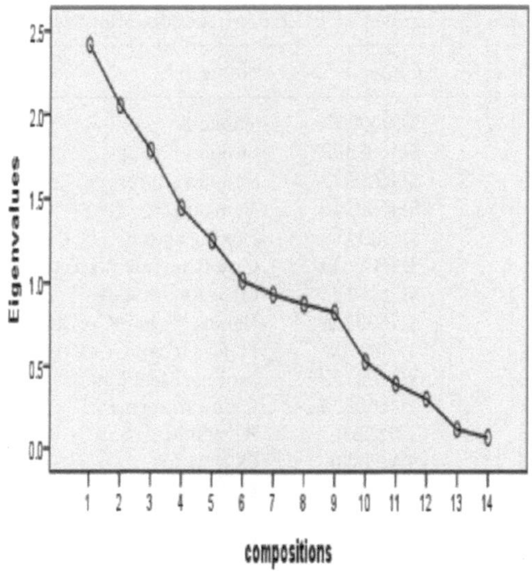

Figure 1. Scree plot.

Table 4. Factor analyze total variance explanation.

	Initial eigenvalues			Extracted sum-of-squares loading					
Components	Total	Variance (%)	Accumulation (%)	Total	Variance (%)	Accumulation (%)	Total	Variance (%)	Accumulation (%)
1	2.416	17.256	17.256	2.416	17.256	17.256	2.154	15.384	15.384
2	2.055	14.681	31.938	2.055	14.681	31.938	1.951	13.935	29.319
3	1.789	12.782	44.720	1.789	12.782	44.720	1.673	11.951	41.269
4	1.445	10.320	55.040	1.445	10.320	55.040	1.575	11.251	52.520
5	1.249	8.922	63.961	1.249	8.922	63.961	1.521	10.867	63.387
6	1.011	7.222	71.184	1.011	7.222	71.184	1.092	7.797	71.184
7	0.930	6.641	77.825						
8	0.872	6.230	84.055						
...	...	...	...						
13	0.114	0.816	99.535						
14	0.065	0.465	100.000						

Table 5. Factor loading matrix of principal components.

Financial variables	Principal components F1	F2	F3	F4	F5	F6
V1	0.108	−0.044	−0.043	−0.136	0.371	−0.214
V3	0.415	−0.101	0.034	0.026	0.075	0.044
V4	0.391	−0.008	−0.238	−0.101	−0.095	−0.017
V5	0.200	0.081	0.275	0.068	−0.219	−0.001
V6	0.040	−0.142	−0.058	0.016	−0.523	−0.102
V11	0.011	−0.042	−0.034	−0.067	0.028	0.868
V13	−0.046	0.501	−0.033	−0.032	0.183	−0.082
V14	−0.015	0.459	0.078	0.002	0.008	0.039
V15	−0.048	0.053	0.516	−0.013	−0.029	−0.141
V17	0.023	−0.059	−0.139	−0.568	0.014	0.137
V18	−0.173	−0.111	0.023	0.238	0.190	0.052
V19	0.240	0.034	0.135	0.163	0.332	−0.007
V20	0.035	−0.112	−0.258	0.449	−0.093	0.073
V22	−0.014	−0.155	0.322	0.059	0.152	0.289

$$F1 = 0.108V1 + 0.415V3 + 0.391V4 + 0.200V5 + 0.040V6 + 0.011V11 - 0.046V13 - 0.015V14 - 0.048V15 + 0.023V17 - 0.173V19 + 0.240V19 + 0.035V20 - 0.014V22 \quad (3)$$

$$F2 = -0.044V1 - 0.101V3 - 0.008V4 + 0.081V5 - 0.142V6 - 0.042V11 - 0.501V13 + 0.459V14 + 0.053V15 - 0.059V17 - 0.111V18 + 0.034V19 - 0.112V20 - 0.155V22 \quad (4)$$

$$F3 = -0.043V1 + 0.034V3 - 0.238V4 + 0.275V5 - 0.058V6 - 0.034V11 - 0.033V13 + 0.078V14 + 0.516V15 - 0.139V17 + 0.023V18 - 0.135V19 - 0.258V20 + 0.322V22 \quad (5)$$

$$F4 = -0.136V1 + 0.026V3 - 0.101V4 + 0.068V5 + 0.016V6 - 0.067V11 - 0.032V13 + 0.002V14 - 0.013V15 - 0.568V17 + 0.238V19 + 0.163V19 + 0.449V20 + 0.059V22 \quad (6)$$

$$F5 = 0.371V1 + 0.075V3 - 0.095V4 - 0.219V5 - 0.523V6 + 0.028V11 + 0.183V13 + 0.008V14 - 0.029V15 + 0.014V17 + 0.190V19 + 0.332V19 - 0.093V20 + 0.152V22 \quad (7)$$

$$F6 = -0.214V1 + 0.044V3 - 0.017V4 - 0.001V5 - 0.102V6 + 0.868V11 - 0.082V13 + 0.039V14 - 0.141V15 + 0.137V17 + 0.052V19 - 0.007V19 + 0.073V20 + 0.289V22 \quad (8)$$

4.3 *The principle components-logistic regression method*

Take the 6 principal influence factor extracted from principal components as the explanatory variables for logistic regression to build a logistic regression model based on principal components. Statistics obtained after iteration of the last logistic regression model are shown in Table 6. Ultimately, we got the mathematical model of early warning approach for financial crisis management for construction enterprises shown as formula (9) below.

$$p = \frac{\exp(-2.236 - 1.099F1 - 0.221F2 - 1.497F3 + 1.908F4 + 0.840F5 - 0.320F6)}{1 + \exp(-2.236 - 1.099F1 - 0.221F2 - 1.497F3 + 1.908F4 + 0.840F5 - 0.320F6)} \quad (9)$$

As can be seen from the formula (9), the principal components F1, F2, F3, F6 have a negative correlation with enterprises financial crisis probability of p, indicating that with the higher ROE (Rate of Return on Common Stockholders' Equity), the higher Asset Turnover, the more Cash Flow, the higher Assets Utilization, and the higher Receivables Turnover Ratio, the less possibility to meet financial crisis. While, the principal components F4, F5 have a positive correlation with p, indicating that the higher Debt Asset ratio, the lower Inventory Turnover plus the more Earnings Management Behavior, then it's more likely to have a financial crisis.

4.4 *The test of warning model ability*

In order to test the validity of the model obtained in this paper, reserved testing samples are introduced (20 construction enterprises included, 13 of which with a good financial capacity and 7 are suffering from a financial crisis) to test the model. First of all, the financial data for each sample are normalized, followed by taking the normalized data we

Table 6. Statistics of the logistic regression model equation.

Principal components	B	S.E.	Wals	df	Sig.	Exp (B)
F1	−1.099	0.562	3.821	1	0.051	0.333
F2	−0.221	0.451	0.241	1	0.623	0.801
F3	−1.497	0.884	2.868	1	0.090	0.224
F4	1.908	1.528	1.558	1	0.212	6.740
F5	0.840	0.610	1.900	1	0.168	2.317
F6	−0.320	1.074	0.089	1	0.765	0.726
Constant	−2.236	0.756	8.755	1	0.003	0.107

Table 7. Test judgment results.

	Test		
	ST group		
Prediction	non-ST	ST	Accuracy rate
ST Group			92.3
non-ST	12	1	
ST	1	6	85.7
Total percentage			90.0

got above into the principal component equation F1–F6 (see formula (3)–(8)), and then according to the model (see formula (9)) to calculate the regressed value of p, which, a financial crisis is identified if $p \geq 0.5$, $p < 0.5$ otherwise. Test results are shown in Table 7. As we can see, 12 in 13 judged as good compare with 6 in 7 judged as crisis, a final accuracy rate of 90% tells this model can be applied in practice with a good effect on judgment to achieve early warning approach for financial crisis management for construction enterprises.

5 CONCLUSIONS

1. Considering the financial data of listed companies in China has a characteristic of high correlations and dimensions, this paper combines principal component analysis and logistic analysis in developing an early warning approach for financial crisis management for construction enterprises in China. Empirical analysis results show that this approach can realize the dimension reduction of multivariables with minimal loss of the variable information, and solve the problem that principal components of original data are irrelevant to each other to get a relatively high financial crisis prediction to determine the correct rate.
2. By difference test, we eliminate 10 non-significant variables from 24 enterprise financial variables which symbolizing the financial condition of listed companies in the construction industry, and get 14 variables to determine the ST and non-ST companies, then we used logistic regression analysis which improved from principal component analysis to get 6 principal influence factor on the financial capacity of construction industry. This model has a good identification of early warning ability to predict business failure.
3. Empirical analysis of financial data with 40 listed companies in construction industry, showed that the main component analysis combined with logistic regression analysis method has a high accuracy rate of financial crisis early

warning, which can provide effective decision support for construction enterprises to establish future development strategies.

REFERENCES

[1] Zeng S.X., Tam V.W.Y., Tam C.M. Towards occupational health and safety systems in the construction industry of China [J]. Safety science, 2008, 46(8): 1155–1168.

[2] Wei X., Lin Z. Present status and development of construction industry in China [J]. Journal of Harbin Institute of Technology, 2004, 1: 34–45.

[3] Liu J., Li B., Lin B., et al. Key issues and challenges of risk management and insurance in China's construction industry: an empirical study [J]. Industrial Management & Data Systems, 2007, 107(3): 382–396.

[4] Shinong W., Xianyi L. A Study of Models for Predicting Financial Distress in China's Listed Companies [J]. Economic Research Journal, 2001, 6: 46–55.

[5] Sun J., Hui X.F. Financial distress prediction based on similarity weighted voting CBR[M]//Advanced data mining and applications. Springer Berlin Heidelberg, 2006: 947–958.

[6] Beaver W.H. Financial ratios as predictors of failure [J]. Journal of accounting research. 1966: 71–111.

[7] Ohlson J.A. Financial ratios and the probabilistic prediction of bankruptcy [J]. Journal of accounting research. 1980, 18(1): 109–131.

[8] Chen M.Y. Predicting corporate financial distress based on integration of decision tree classification and logistic regression [J]. Expert Systems with Applications, 2011, 38(9): 11261–11272.

[9] Ashin Chen. Financial distress prediction using logistic regression with corporate governance measure [J]. Management Science and Statistical Decision, 2007, 4(1): 8–17.

[10] Shih V., Zhang Q., Liu M. Comparing the performance of Chinese banks: a principal component approach [J]. China Economic Review, 2007, 18(1): 15–34.

E-Commerce, E-Business and E-Service – Lee (Ed)
© 2014 Taylor & Francis Group, London, ISBN 978-1-138-02646-9

The determinants of local fiscal distress in Taiwan: A Poisson regression analysis

Y.H. Ho & C.J. Huang
Department of Public Finance, Feng Chia University, Taichung, Taiwan

ABSTRACT: This study uses Poisson regression to explore the determinants of fiscal distress on the local governments in Taiwan. Results suggest that political budget cycles, economic growth, increase in central government expenditures, and legislative checks and balances, all have insignificant effects on the local governments' fiscal health over the 1984 to 2014 period. More importantly, this study finds that an increase in central government revenues significantly decreases local governments' fiscal health. This finding is consistent with that of Chang & Ho (2002), whom suggested that a unidirectional causality runs from government revenues to government expenditures, which further supports the tax-and-spend hypothesis for Taiwan. The major conclusion drawn from this study is that in order to improve the fiscal distress of local governments, Taiwan's central government should elect a fiscal policy that emphasizes spending cuts instead of heavier taxes and other means of raising revenue.

1 INTRODUCTION

The fiscal health of local governments is important as it is an indication of the ability of local governments to meet its financial and service obligations. Fiscal health may not be the ultimate measure of success for local governments, but without a healthy financial condition, the level and quality of public services will suffer. Additionally, there are many disparate factors affecting the fiscal health of local governments. Local governments generally benefit from a booming national economy. More specifically, in addition to the general economic performance, demographic factors, nature disasters, central and local elections factors, local administrative efficiency, as well as the effective local legislative checks and balances, all affect the fiscal health of a local government.

There are some commonly used terms found in the fiscal health literature that describe the fiscal health of local government. Fiscal stress, fiscal strain, fiscal crisis, fiscal health, fiscal distress, financial condition, and financial position, are all used and have different meanings attached to them. Ladd & Yinger (1989) define fiscal health as the underlying or structural ability to deliver public services to its residents, independent of the budgetary decision made by city officers. They further explain that local health is the difference between revenue-raising capacity and expenditure need, expressed as a revenue-raising capacity and expenditure need, expressed as a percentage of capacity. Clark & Appleton (1989) define fiscal strain as an institutional lack of adaption to a changing environment. Meanwhile, Berry's (1994) definition of fiscal health describes the extent to which a government's financial resources exceed its spending obligations.

According to Lin & Raman (1998), local government's fiscal health is related to its financial condition, which they describe as having to do primarily with fiscal effort for the relative level of taxation and spending. They point out that a government could be in good financial position but be in poor financial condition. In their study, a poor financial condition meant that local governments are less likely to sustain the current level of service at acceptable levels of taxation. Groves et al. (2003) state that a local government's financial condition is the result of a number of very diverse factors and can be measured by looking at cash solvency, budgetary solvency, long-run solvency and service level solvency. Kloha et al. (2005) define fiscal distress in terms of whether a local government is sufficiently meeting the needs of its community. Badu & Li (1994) define fiscal stress as the imbalance between the revenue raising capacity and expenditure needs of a local government. Badu and Li's definition is very much associated with the concept of tax effort, which shows the ratio of actual yield to that obtained by the standard tax system.

The variety of definitions around fiscal health has inevitably led to the development of a diverse amount of mechanisms for measuring or predicting local government's fiscal health. Brown (1993) provides a concise and easy-to-use 10-point test of financial condition, where he suggests that 10 ratio measures be computed, equally weighted, and aggregated to provide an overall picture of a government's financial condition. Kleine et al. (2003) discuss some of the limitations of Brown's test, and developed a more simplified 10-point scales of fiscal distress index based on nine variables to assess the performance of local government. Kloha et al. (2005) applied the model to predict local fiscal stress in a sample of Michigan local governments. They found that the 10 point scale of fiscal distress appears to perform considerably better than Brown's 10 point test and provides function as an early warning system of fiscal difficulties.

Hendrick (2004) presents a theoretic framework for assessing local government fiscal health based on dimensions of properties of government's environment, balance of fiscal structures with environment, properties of the government's fiscal structure, and population and institutional factors. Hendrick points out that the dimensions are related but often in indirect or nonlinear ways, and hence must be measured separately rather than combined into a comprehensive indicator of fiscal health. Wang et al. (2007) define financial condition as the level of financial solvency. They develop and test a measure of financial condition for state governments. The measure includes the dimensions of cash solvency, budget solvency, long term solvency, and service levels solvency, and eleven indictors. Their findings show that the measure they used is relatively reliable and valid.

Political budget cycle was first demonstrated by Rogoff & Sibert (1988), who proposed a model of adverse selection that emphasizes the idea of competency (ability to handle the economy) coupled with asymmetric information. In a political budget cycle model, opportunistic policy-makers, regardless of their ideology, try to use expansionary fiscal policies prior to elections in order to please voters, maximize their popularity, and increase their chances of re-election (see, e.g. Schuknecht 1996, Hallerberg & von Hagen 1999, Brender 2003, Alesina & Tabellini 2005, Shi & Svensson 2006).

Early empirical studies on this topic find evidence in favor of election-driven fiscal policy manipulations (see, e.g. Alesina & Roubini 1992, Alesina et al. 1999, Gonzàlez 2002). Nevertheless, some recent studies in the empirical literature show that the magnitude of political budget cycles in developing countries is significant and higher than that in developed countries (see, e.g. Schuknecht 2000, Alesina et al. 1999, Drazen 2001, Persson & Tabellini 2002, Shi & Svensson 2006, Diallo 2009). Brender & Drazen (2010) suggest that the results of these studies are driven by the first few elections in countries that are "new democracies", where fiscal manipulation may have worked because inexperienced voters in established democracies lack information.

From a theoretical point of view, Political budget cycles arise in equilibrium when rational voters are imperfectly informed about an incumbent's competency and the incumbent enjoys discretionary power over the budget. Without discretionary power, asymmetric information alone is not sufficient for Political budget cycles. Previous empirical studies on Political budget cycles were conducted under the assumption that the executive has full discretion over fiscal policy, implicitly neglecting the role of the legislature in the budgetary process. As a result, the degree of discretion of the executive has been overlooked in the empirical literature on Political budget cycles, perhaps because theoretical papers on opportunistic cycles usually model fiscal policy in terms of a single policy-maker. Persson & Tabellini (2003, 2004) find that checks and balances in the formulation and implementation of the budget are epiphenomena, but have real effects on budgetary outcomes (see, e.g. von Hagen & Harden, 1995, Hallerberg & von Hagen 1999, Alesina et al. 1999). Henisz (2004) concludes that checks and balances that limit the discretion of policy-makers reduce the volatility of government expenditures and revenue. Fabrizio & Mody's (2006) finding is that budget institutions' mechanisms and rules of the budget process that create checks and balances have significant value in curbing fiscal pressures. Streb et al. (2009a) argue that when there is separation of powers, appropriate checks and balances may work as a commitment device that eliminates electoral cycles in fiscal policy, making all players better off. Streb et al. (2009b) find that stronger effective checks and balances explain why Political budget cycles are weaker in developed and established democracies.

As mentioned above, scholars have provided several techniques to measure a local government's fiscal health. It is difficult to argue that one particular indicator (or one set of indicators) is the best sign of fiscal health without knowing the intended purpose, the target audience, the practical constraints on the analysis, and the availability of account information. This study is designed to add to the existing body of work, by exploring factors affecting local government fiscal health. Firstly, we compute the fiscal health indicators for 21 local governments in Taiwan over the 1984 to 2010 period, based on existing work of Kleine et al. (2003) and Kloha et al. (2005). Secondly, we use a Poission regression analysis to examine the effect of economic variables, political budget cycles, and legislative checks and balances on the extent of the local government's fiscal distress. This paper is organized as follows. Section two describes the computation of fiscal health measures, section three gives the explanatory variables used, section four describes empirical methodology and findings, and section five presents the conclusions.

2 FISCAL HEALTH INDICATOR

In this study, Kleine et al.'s (2003) 10-point scale of fiscal distress are applied to a sample of 21 local governments in Taiwan over the 1984 to 2010 period. The 10-point scale of fiscal distress is a composite of nine budgetary and socioeconomic indicators capable of measuring the different elements that make up the local government's fiscal condition in Taiwan. Following Kleine et al.'s (2003) works, a performance standard of each indicator is established and then used for the grading of local government's fiscal condition. If the local government meets the indicator standard, score a 0 in that indicator, otherwise, it was assigned a 1.

The nine indicators are equally weighted, and aggregate to provide an overall picture of a government's fiscal health (hereafter, FH), resulting in a total points ranging from 0 to 10. Local government with the FH score of 10 indicates severe fiscal distress and a score of 0 or 1 indicates little or no distress. The FH score is then used as a dependent variable in the Poisson regression analysis. The definition and performance standard of the nine indicators are described as follows.

2.1 *Population growth*

It is clear that an increase in population will bring in more taxpayers and will result in a better fiscal condition of a government. If a local government has a negative growth rate in population scored a 1; otherwise scored a 0. Annual data on population growth rate are taken from Year Book of Population Statistics of Taiwan.

2.2 *Actual tax collection growth*

Local government's actual tax collection generally benefit from a booming national economy. If the government experienced negative growth in the actual tax collection, it scored a 1; Governments that experienced a decline below the threshold in the growth of actual tax collection received a 1 and others were marked 0. The annual data on actual tax collection are taken from Year Book of Financial Statistics of Taiwan.

2.3 *Decreases in actual tax collection growth*

Fiscal distress is likely to occur if local government experienced a severe decrease in the growth of actual tax collection. The value of a one-half standard deviation below sample median of the growth rate was chosen as the threshold of severe decrease in the growth of actual tax collection. For this indicator, governments that measured below the threshold value scored a 1; otherwise marked 0.

2.4 *Government expenditures as a percentage of actual tax collection*

Ratios of annual expenditures to the actual tax collection indicates the scale of a tax base needed for supporting a government functioning. A higher ratio represents a higher possibility for fiscal distress. To assess the possibility of fiscal distress, value of one-half standard deviation above the sample median of these ratios was used to serve as the threshold for evaluating the tax scale for operating a government. For this indicator, local governments with ratios above the threshold value received a 1, indicates that these government scale are fairly large relative to the tax base that supports them, hence have a higher possibility to experience fiscal distress; governments below the threshold value received a 0. Annual data on annual government expenditure and actual tax collection are taken from Annual Report of Municipal, County and City Final Accounts of Taiwan.

2.5 *Fiscal deficit*

If the local government does not have budget deficit, or if the deficit relative to revenue is less than 1%, its indicator score is 0; governments with deficit greater than 1% of revenue, indicating local governments has a nontrivial deficit, the indicator scored 1. Annual data on revenues and expenditures are taken from Annual Report of Municipal, County and City Final Accounts of Taiwan.

2.6 *Prior fiscal deficit*

A single year deficit is considered only a minor sign of fiscal distress, however as the deficit accumulated over time, it should be more concerned as a serious sign of fiscal distress. This indicator captures this type of concern by measuring whether the local government had a deficit during the past two years. The sixth indicator uses the same data as those used by the fifth indicator; the only change is that a different assessment standard is used. Follows the previous definition of deficit, if the local government had deficits for both two previous years, it would receive a score of 2 for this indicator. If local government had one deficit during the past two years, it would score a 1 on this indicator; otherwise, scored a 0.

2.7 *Tax revenues a percentage of government expenditures*

Tax revenues have been an important financial source of government expenditures in Taiwan. Increase in tax revenues will bring in a sounder finance of government, and it is a sign of fiscal distress for governments if the tax revenues are less than the annual expenditures. Governments typically find it beneficial to keep their ratio of tax revenues to annual expenditures from declining too much, as this will inhibit the ability of local governments to meet their financial and service obligations.

The variable constructed for this indicator is the tax revenues as a percentage of government annual expenditures. There is some debate as to how much this ratio should be. This paper uses a one-half standard deviation above the sample median to serve as the threshold of the standard. Local governments with ratios below the standard, received a score of 1, indicating governments are fairly lack of their own finance, hence have a higher possibility to experience fiscal distress; governments with ratios above the standard received a 0. The annual data on tax revenues and government expenditures are taken from Annual Report of Municipal, County and City Final Accounts of Taiwan.

2.8 *Revenues from aid and assistance as a percentage of government expenditures*

Revenues of aid and assistance are one of the sources for local government in Taiwan to meet their financial and service obligations. If the high proportions of local expenditures are coming from revenues of aid and assistance, indicates that local governments are over rely on the intergovernmental grants to finance their expenditures, hence are easily to experience fiscal distress.

The variable constructed for this indicator is the revenues of aid and assistance as a proportion of annual government expenditures. A one-half standard deviation above the sample median is used to serve as the threshold of the standard. Local governments with ratios above the standard received a score of governments with ratios above the standard received a score of 1; governments with ratios below the standard received a 0. The annual data on revenues of aid and assistance and government expenditures are taken from Annual Report of Municipal, County and City Final Accounts of Taiwan.

2.9 *Revenues from loans as a percentage of actual tax collection*

Large loans relative to government's actual tax collections signified a clear fiscal distress for the local government. This indicator is constructed by taking revenues from loans and dividing it by the actual tax collection. A one-half standard deviation above the sample median is used to serve as the threshold of the standard. Local governments with ratios above the standard received a score of 1 on this indicator; indicates that governments are highly possible to experience fiscal distress; governments with ratios below the standard received a score of 0. The annual data on revenues from loans and actual tax collections are taken from Annual Report of Municipal, County and City Final Accounts of Taiwan.

The nine indicators are equally weighted, and aggregate to provide an overall picture of a government's fiscal health. We compute each local government's annual fiscal health score by pooling the annual scores for all governments across all years and computing the standard deviation of the pooled indicators from the sample median. The median for our research sample is 2, with a standard deviation of 2.1723. We elected to use the median instead of the mean value for this study because the distribution of health indicators takes on a skewed shape.

3 EXPLANATORY VARIABLES

The independent variables include macroeconomic growth rate (EGR), central government revenues (TREV) and expenditures (TEXP), dummy for political budget cycles (PBC) and effective legislative checks and balances (CHECKS). The data of EGR is taken from Statistics Bureau of Directorate-General of Budget, Accounting and Statistics of Taiwan. Data of TREV and TEXP are taken from Year Book of Financial Statistics of Taiwan. Following Streb et al. (2009), PBC takes value 1 in central election year, −1 in the following year, and 0 otherwise. The election data is obtained from the Central Election Commission of Taiwan.

A measure of effective checks and balances of the Legislative Institutions in Taiwan is constructed based on the work of Streb et al. (2009). We also use the Henisz (2005) Political Constraints Index (POLCON) to measure the veto player variable (VETO), and the International Country Risk Guide (ICRG) Law and Order index (LOI) to measure compliance with the law for Taiwan. The combination of the legislative veto player with the law dummy (DLAW) for compliance with the law is used as a proxy for the effective checks and balances on the executive budgetary process in Taiwan.

4 EMPIRICAL METHODOLOGY AND RESULTS

This study leverages both cross-sectional and time-series characteristics of fiscal health indicators in order to identify the occurrence of fiscal distress in Taiwan's local governments over the 1984 to 2010 period. Additionally, an explanatory analysis is conducted to examine the effect of political budget cycles, legislative checks and balances, and macroeconomic variables, on the extent of the local government's fiscal distress from a Poisson regression context.

To identify the local governments with fiscal distress for each year, three evaluation criteria are used in this study respectively. Criterion C_1 is set as the sample median, C_2 is set as a half standard error above median, and C_3 is set as one standard error above the median. Local governments with a health score above these criteria are assumed to have fiscal distress in that year. We do this by summing the number of fiscal distress of all localities in each year and repeating this for every year. The resulting variable, N_t, represents the number of having fiscal distress of sampled localities in a given year.

We base our inference on a regression of the dependent variables N_t on the set of political and economic variables X_t, discussed in the previous section. Because the dependent variable is integer-valued, least square analysis would result in inefficient coefficients estimates, inconsistent standard errors, and thus invalid hypothesis. One appropriate alternative is the Poisson regression model, which is an extension of the simple Poisson probability model. The specification of this model is

$$\Pr(N_t = n_t) = \frac{e^{-\lambda(X_t)}(\lambda(X_t))^{n_t}}{n_t!}, \quad n_t = 0,1,\dots, t = 1,2,\dots,T \quad (1)$$

where $\lambda(\cdot)$ is finite and positive and X_t = set of explanatory variables. We wish to estimate the expected counts of fiscal distress conditional on the explanatory, i.e. $E(N_t \mid X_t) = \lambda(\cdot)$. Assumes that $\lambda(\cdot)$ is log-linearly depend on the model's explanatory variables. The estimated equation can be expressed as

$$\ln E(N_t|X_t) = \beta X_t \quad (2)$$

where β = vector of regression coefficients.

Table 1 reports the results for C_1 through C_3 evaluation criteria. Results of this study suggest that political budget cycles, macroeconomic growth and increase in central government expenditures, all have positive effects on the elimination of fiscal distress for local government, even though most of the effects are statistically insignificant. While legislative checks and balances have ambiguous effects on the elimination of fiscal distress among three criteria, they are statistically insignificant.

Table 1. Poisson results based on various time series of fiscal distress counts using different criteria.

Explanatory variable	Dependent variable: N_t		
	C_1	C_2	C_3
Intercept	1.1123***	0.2197	−1.8160***
	(0.2680)	(0.3066)	(0.6358)
PBC	−0.3046	−0.3307	−0.9526*
	(0.2382)	(0.3066)	(0.5553)
CHECKS	0.0184	−0.1465	0.1524
	(0.5066)	(0.5684)	(0.7641)
PBC_ CHECKS	0.4830	0.5931	1.5978*
	(0.3964)	(0.4944)	(0.8458)
EGR	−1.38E-09	−4.44E-08	−1.04E-07
	(9.41E-08)	(1.05E-07)	(1.31E-07)
TREV	1.69E-06**	2.22E-06**	3.56E-06***
	(7.43E-07)	(8.86E-07)	(1.33E-06)
TEXP	−1.03E-06*	−1.02E-06	−1.48E-06
	(5.32E-07)	(6.64E-07)	(1.07E-06)
R^2	0.7614	0.7631	0.8203
Adj. R^2	0.6899	0.6920	0.7664
Log-likelihood	−66.6948	−66.9284	−51.2804

Notes: *, **, and *** denote significance at the 10%, 5%, and 1% levels. Numbers in parentheses are standard error.

More importantly, this study finds that an increase in central government revenues significantly increase the number of local fiscal distress, and thus decreases local governments' fiscal health. This finding is consistent with that of Chang and Ho (2002), whom suggested that there an unidirectional causality runs from government revenues to government expenditures, which further supports the tax-and-spend hypothesis for Taiwan. The major conclusion that we draw from this study is that in order to improve the fiscal distress of local governments, Taiwan's central government should elect a fiscal policy that emphasizes spending cuts over heavier taxes and other means of raising revenue.

5 CONCLUSIONS

This study leverages both cross-sectional and time-series characteristics of fiscal health indicators in order to identify the occurrence of fiscal distress in Taiwan's local governments over the 1984 to 2010 period. To support this study, we conduct an explanatory analysis to examine the effect of political budget cycles, legislative checks and balances, and macroeconomic variables, on the extent of local government's fiscal distress.

Fiscal health indicators for 21 local governments in Taiwan were derived based on studies from Kleine et al. (2003) and Kloha et al. (2005). To identify the local governments with fiscal distress for each year, we used three evaluation criteria in this study. Local governments with a health score exceeding these criteria are assumed to have fiscal distress in that year. The sum of the number of local governments experiencing fiscal distress across localities was calculated for each year within the study period.

We base our inference on a regression of the fiscal distress on the set of political and economic variables. Given that the dependent variable is integer-valued, and the fact that an ordinary least squares analysis would result in inefficient estimates, we use Poisson regression model in our study. Results of this study suggest that political budget cycles, macroeconomic growth and increase in central government expenditures, all have positive but insignificant effect on improving local governments' fiscal health, while legislative checks and balances has a negative but insignificant effect on local governments' fiscal health.

More importantly, this study finds that an increase in central government revenues significantly decreases local governments' fiscal health. This finding is consistent with the results of a prior study conducted by Chang and Ho (2002), which suggested that a unidirectional causality runs from government revenues to government expenditures, which further supports the tax-and-spend hypothesis for Taiwan. The major conclusion that we draw from this study is that in order to improve the fiscal distress of local governments, Taiwan's central government should elect a fiscal policy that emphasizes spending cuts instead of heavier taxes and other means of raising revenue.

REFERENCES

Alesina, A. & Roubini, N. 1992. Political cycles in OECD economies. *Review of Economic Studies* 59: 663–688.

Alesina, A. & Tabellini, G. 2005. Why do politicians delegate? *NBER Working Paper*. No. 11531.

Alesina, R. Hausmann, H. Hommes, R. & Stein, E. 1999. Budget institutions and fiscal performance in Latin America. *Journal of Development Economics* 59: 253–273.

Badu, Y. & Li, S. 1994. Fiscal stress in local government: a case study of the Tri-Cities in the commonwealth of Virginia. *The Review of Black Political Economy* 22(3): 5–17.

Brender, A. 2003. The effect of fiscal performance on local government election results in Israel: 1989–1998. *Journal of Public Economics* 87: 2187–2205.

Brender, A. & Drazen, A. 2005. Political budget cycles in new versus established democracies. *Journal of Monetary Economics* 52: 1271–1295.

Brown, K. 1993. The 10-point test of financial condition: toward an easy-to use assessment tool for smaller cities. *Government Finance Review* 9(6): 21–26.

Chang, T. & Ho, Y. H. 2002. Tax or spend, what causes what: Taiwan's experience. *International Journal of Business and Economics* 1(2): 157–165.

Clark, T. & Appleton, L. 1989. Coping in American cities. In S. Clarke (ed.), Urban Innovation and Autonomy: 31–68. Newbury Park, CA: Sage Publications.

Diallo, O. 2009. Tortuous road toward countercyclical fiscal policy: lessons from democratized sub-Saharan Africa. *Journal of Policy Modeling* 31:36–50.

Drazen, A. 2001. The Ppolitical business cycle after 25 years. In: B. Bernanke & K. Rogoff (eds.), *NBER Macroeconomics Annual*: 75–117. MIT Press.

Drazen, A. & Eslava, M. 2010. Electoral manipulation via voter-friendly spending: theory and evidence. *Journal of Development Economics* 92(1): 39–52.

Fabrizio, S. & Mody, A. 2006. Can budget institutions counteract political indiscipline? *Economic Policy* 21: 689–739.

Gonzàlez, M.d.l.A. 2002. Do Changes in democracy affect the political budget cycle? evidence from Mexico. *Review of Development Economics* 6(2): 204–224.

Groves, M. Godsey, W. & Shulman, M. 2003. Evaluating financial condition: A handbook for local government (3rd edn). *working paper* no.1531, 2003.

Hakkio, S. & Rush, M. 1991. Is the budget deficit "too large"? *Economic Inquiry* 29: 429–445.

Hallerberg, M. & von Hagen, J. 1999. Electoral institutions, cabinet negotiations, and budget deficits in the European Union. In: Poterba, J. & von Hagen J. (eds.), *Fiscal Institutions and Fiscal Performance*, University of Chicago Press, Chicago.

Hendrick, R. 2004. Assessing and measuring the fiscal health of local governments: focus on Chicago suburban municipalities. *Urban Affairs Review* 40(1): 78–114.

Henisz, J. 2004. Political institutions and policy volatility. *Economics and Politics* 16: 1–27.

Henisz, J. 2005. *POLCON 2005 Codebook*, Manuscript, University of Pennsylvania.

Kleine, R. Kloha, P. & Weissert, C. 2003. Monitoring local government fiscal health: Michigan's new 10-point scale of fiscal distress. *Government Fiscal Review* 19(3): 169–190.

Kloha, P. Carlo, S. & Kleine, R. 2005. Developing and testing a composite model to predict local fiscal distress. *Public Administration Review* 65(3): 313–323.

Ladd, F. & Yinger, J. 1989. *America's ailing cities: fiscal health and the design of urban policy*. Baltimore, MD: John Hopkins University Press Wang, X. Dennis, L. & Tu, Y.S. 2007. Measuring financial condition: a study of U.S. States. *Public Budgeting and Finance*, 27(2): 1–21.

Lin, W. & Raman, K. 1998. The housing value-relevance of governmental accounting information. *Journal of Accounting and Public Policy* 17: 91–118.

Persson, T. & Tabellini, G. 2002. Do electoral cycles differ across political systems. mimeo, IIES, Stockholm University.

Persson, T. & Tabellini, G. 2003. *The Economic Effect of Constitutions: What Do the Data Say*, MIT Press, Cambridge, MA.

Persson, T. & Tabellini, G. 2004. Constitutional rules and fiscal policy outcomes. *American Economic Review* 94: 25–45.

Rogoff, K. & Sibert, A. 1988. Elections and macroeconomic policy cycles. *Review of Economic Studies*, 55: 1–16.

Schuknecht, L., (1996), "Political Business Cycles and Fiscal Policies in Developing Countries," *Kyklos*, 49, 155–170.

Schuknecht, L. 2000. Fiscal policy cycles and public expenditure in developing countries. *Public Choice* 102: 115–130.

Shi, M. & Svensson, J. 2006. Political budget cycles: do they differ across countries and why? *Journal of Public Economics* 90: 1367–1389.

Streb, M. Lema, D. & Torrens, G. 2009. Checks and balances on political budget cycles: cross-country evidence. *Kyklos* 62: 426–447.

Vergne, C. 2009. Democracy, elections and allocation of public expenditures in developing countries. *European Journal of Political Economy* 25: 63–77.

von Hagen, J. & Harden, I.J. 1995. Budget processes and commitment to fiscal discipline. *European Economic Review* 39: 771–779.

Wang, X. Dennis, L. & Tu, Y.S. 2007. Measuring financial condition: a study of U.S. States. *Public Budgeting and Finance* 27(2): 1–21.

E-Commerce, E-Business and E-Service – Lee (Ed)
 ISBN 978-1-138-02646-9

A construction method of disclosure index of listed company in China

Shaoshuang Xu
College of Economics and Management, West Anhui University, Lu'an, China

ABSTRACT: Theoretically, there is no consensus on how to evaluate disclosure quality of listed company. However, in practice, the study of each institute has its own characteristics. Disclosure index is the main method in theoretical and practical circle to evaluate disclosure of listed company. At present, most of the researchers on disclosure index show great differences, especially in component of disclosure index. This situation may hinder the academic and practical development in the field. This paper proposes a construction method of disclosure index of listed company, based on measurement theory and comprehensive evaluation theory. Theoretically, the researchers in the field may communicate with each other based on the construction method of disclosure index of listed company.

1 INTRODUCTION

The resource allocation efficiency of capital market depends on the pricing efficiency of security, which depends primarily on the extent of information asymmetry between security buyer and seller. A capital market is essentially an information market. Disclosure can reduce information asymmetry among capital market traders, so it is a major deciding factor that influences the function of capital market. The supervisory board of China's capital market attaches great importance to the disclosure of listed company just at the moment of the establishment of the capital market. On the basis of the mature capital markets' experiences, under the control of The Securities Law of The People's Republic of China and The Company Law of The People's Republic of China, the supervisory board of China's capital market has established a system of disclosure rule that includes the criterion of the disclosure content and format of listed company, the compilation criterion of the disclosure of listed company, the answers of frequently asked questions and the analysis of the case of disclosure of listed company. The system of disclosure rule is already in line with international standards. However, the Chinese market is still an undeveloped market in a transitional economy and the disclosure of listed company still faces many problems. The disclosure regulation of the Chinese capital market has a long way to go. Evaluation of disclosure quality of listed company, a "moderate" external mechanism that can improve disclosure quality of listed company, is given to increase importance by theoretical circles and regulatory authorities. However, the theoretical research and practice of evaluation of disclosure quality of listed company has fallen far behind. Chinese scholars began to pay attention to this field since the end of 1990s and few scholars have produced positive results in the basic theory and approach of this field. In practice, the evaluation of disclosure quality is executed only by the Shenzhen Stock Exchange, who uses four grades of Excellent, good, qualified and unqualified to distinguish the disclosure quality of listed company. Accordingly, the Shanghai Stock Exchange used three grades of Excellent, qualified and unqualified, which ended in 2001.

Table 1 shows that the number of unqualified listed companies reduces, "excellent" and "good" grade increases obviously during the period of 2001–2012. This situation indicates rating behavior of the Shenzhen Stock Exchange promotes

Table 1. The rating results of disclosure in main board market by the Shenzhen stock exchange.

Year	Excellent	Good	Qualified	Unqualified
2001	5.84%	39.11%	48.44%	6.61%
2002	7.84%	46.86%	38.83%	6.48%
2003	8.07%	52.56%	34.25%	5.12%
2004	5.98%	60.36%	29.28%	4.38%
2005	8.08%	57.17%	27.68%	7.07%
2006	8.37%	50.61%	34.69%	6.33%
2007	8.61%	47.54%	39.34%	4.51%
2008	9.47%	53.09%	33.95%	3.50%
2009	10.52%	65.77%	21.24%	2.47%
2010	11.16%	66.94%	19.21%	2.69%
2011	11.57%	66.94%	18.18%	3.31%
2012	11.64%	68.40%	17.26%	2.70%

Data source: The Shenzhen stock exchange.

improvement of the disclosure quality of listed companies which feel the pressure from the exchange, the media and investors. But we find also that more and more listed companies cluster together in "good" grade. Therefore, discrimination of rating behavior of the Shenzhen Stock Exchange is not enough. This difficult may be solved by disclosure index which is quantified.

Theoretically, there is no consensus on how to evaluate information disclosure quality. However, in practice, the study of each institute has its own characteristics. Disclosure index is the main method in theoretical and practical circle to evaluate information disclosure. But we find disclosure index has the problems in ignoring definition of connotation of disclosure, big difference in component of disclosure index, no consensus on the set of weight of indicators and so on [1].

In this paper, based on measurement theory and comprehensive evaluation theory, we propose a construction method of disclosure index of listed company to solve the problems mentioned above.

The paper is organized as follows. In the next section, we propose the definition of connotation of disclosure quality that we research in this paper, In Section 3, the construction method of disclosure index is presented. Section 4 presents a example of disclosure index of listed company in China. Finally, we conclude our paper in section 5.

2 THE DEFINITION OF CONNOTATION OF DISCLOSURE QUALITY

As is well known, Disclosure quality is difficult to define due to its subjective nature. There is no consensus on definition of disclosure quality Just as summarized by Laidroo (2008) [2], for example, Singhvi and Desai (1971) claim that disclosure quality refers to completeness, accuracy, and reliability [3]. Brown and Hillgeist (2006) suggest that it reflects the overall informativeness of firm's disclosures and depends on the amount, timeliness, and precision of disclosed information. Diamond and Verrechia (1991) believe that disclosure quality is measured by the Bayesian investors' beliefs about security value after receiving disclosure. The definition of disclosure quality is based on the element of "news" (unexpectedness) contained in it (Laidroo 2008). Similarly, timeliness, reliability and relevance are suggested by Yuejun Tang et al. (2008) in China [4].

In fact, Definition of disclosure captures the release of relevant information either in the form of press releases, public announcements, or financial reports. the disclosure quality is of special interest to investors and capital market regulators, because the expectation is that the higher is the quality of disclosure, the lower are the information asymmetries and the smaller are the agency conflicts between investors and managers. Therefore, this paper believe that disclosure quality can be measured by the extent of satisfication of the information users, which may be agreed by the majority of people. In a word, in order to reduce differences, the definition of disclosure quality should be started in here firstly.

3 THE CONSTRUCTION METHOD OF DISCLOSURE INDEX

Disclosure is obligations of the listed company and an important basis for investors to make investment decision. Investors and security regulatory authorities expect that disclosure quality of listed company is high. However, we are confronted with difficulty in evaluation of disclosure quality of listed company. A actual question is that the connotation of high disclosure quality is not same for different people. Many investors are accustomed to classify the disclosure phenomenon in capital market for decision-making. They are mostly confined to the needs of their interests. What is the high quality disclosure is often difficult to define. The main reason for this phenomenon lies in that the "disclosure quality" does not exist in the real world just like a stone. As a term, "disclosure quality" is abstract. Only when we use the term "disclosure quality" to express our views about the behavior of the disclosure of listed company, "disclosure quality" is concrete. In fact, the high quality disclosure is a concept in our brains. So the term "disclosure quality" almost has no single, clear significance from person to person. Just mentioned above in this paper, disclosure quality can be measured by the extent of satisfication of the information users. But how can we assign a number to satisfication of the information users? The psychophysicist Stevens (1946) formulated a concept of measurement, according to which measurement is defined simply as the assignment of numbers to things and properties according to a certain rule [5]. So assignment a number to satisfication of the information users is considered to be solved by measurement theory. Therefore, in order to increase the accuracy and objectivity of our understanding of the disclosure quality of listed company, we need to measure the disclosure of listed company based on measurement theory.

People measuring an abstract thing may start a measurement procedure with conceptualization based on measurement theory, secondly, present a clear definition of the connotation of the abstract thing, thirdly, operationalization procedure is executed to obtain a set of indicators, collect data of the set of indicators finally. Perhaps some people

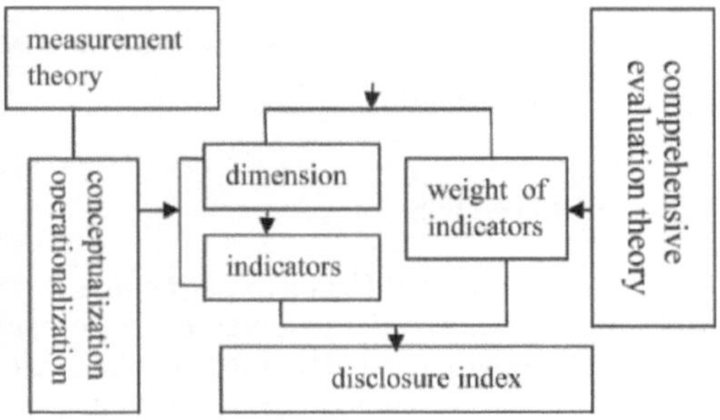

Figure 1. The construction method of disclosure index.

do not agree with the conceptualization and operationalization procedure that we use, but they can still understand these specific indicators through these procedures. That will help to communicate with each other in the field clearly.

A construction method of disclosure index of listed company has two key factors: indicators and weight of indicators. Weight of indicators is researched by comprehensive evaluation theory.

A brief description of the construction method of disclosure index in this paper is provided in Figure 1.

4 A EXAMPLE IN CHINA'S CAPITAL MARKET

To demonstrate the construction method of disclosure index in this paper, we conduct the procedure above. Firstly, conceptualization step of disclosure quality is done. We review the related literatures to get measurement objection. We find that many researchers agree that disclosure quality is composed mainly of disclosure behavior quality and information quality. Therefore, we design a questionnaire which contains 14 questions based on the Likert-Scale five-point measure questionnaire to consult with nine specialists, who come from Tianjin University of Finance and Economics, Anhui University of Finance and Economics, Shandong University of Finance and Economics, Tianjin Capital Environmental Protection Group Company Limited, Industrial Securities Co., Ltd., Bohai Securities Co., Ltd and Qingdao bureau of China Securities Rugulatory Commission. Table 2 is provided below.

M_j represents average, $M_j = (1/m_j)\sum_{i=1}^{m_j} C_{ij}$, m_j represents the number of experts who mearsure indicator j, C_{ij} represents the number of rank of indicator j that is ranked by i expert. K_j represents ratio of level max, $K_j = (m'_j / m_j)$, m'_j represents the number of experts who give the highest rank to indicator j, m_j represents the number of experts who rank indicator j. σ_j represents the variance of indicator j, V_j represents relative variance, $V_j = \sigma_j / M_j$. Sum of rank is represented by S_j, $S_j = \sum_{i=1}^{m_j} R_{ij}$, Rij represents the rank of indicator j. A order of rank of all of indicators that is ranked by i expert is ranged, the number of the order is R_{ij}.

Table 2. Results of the questionnaire.

Dimension	Average	Variance	Ratio of level max	Relative variance	Sum of rank
Legality	4.78	0.44	89%	0.14	47.5
Fairness	4.89	0.11	89%	0.07	49.5
Timeliness	4.89	0.11	89%	0.07	49
Accessibility	4.33	0.50	44%	0.16	33
Accuracy	4.67	0.25	67%	0.11	41
Integrity	4.67	0.25	67%	0.11	41
Format	3.67	0.75	11%	0.24	17
Usefulness of information	4.78	0.19	78%	0.09	45.5

The accommodation coefficient of the investigation is 0.25, $\chi^2 = 15.49 > \chi^2_{0.05}(7) = 14.067$, p = 0.03. These figures indicate that the results of investigation can be used.

In general, the indicator who meets two conditions below can be retained.

- ratio of level max is greater than 50% or average greater than four
- relative variance is smaller than 0.30.

Therefore, Accessibility and Format should be discarded from the dimensions of discloure quality.

Secondly, operationalization step is done. We review the related literatures and consult the specialists of disclosure of listed company. We design a questionnaire based on the Likert-Scale five-point measure questionnaire to conduct Delphi method similarly. The complete process is ignored because this paper discusses of the construction method of disclosure index mainly. We recommend the Delphi method in screening indicators process. The reliability of the conceptualization and operationalization procedures depends upon the appropriateness of the specialists when we conduct the Delphi method.

Some specialists advise the five principles should be met when we screen indicators. Firstly, the principle of purposiveness indicates that the indicators should reflect the measurement objective of the dimension. Secondly, the principle of comprehensiveness indicates that the indicators should reflect all aspects of the dimension. Thirdly, the independence principle indicates that the indicators should not overlap each other in measurement. Fourthly, the feasibility principle indicates that the figures of the indicators should be gained accurately. Fifthly, the dynamic principle indicates that the indicators should be established dynamically.

We hand out the questionnaire again and again to help the specialists to reach a consensus about the indicators. At last, we obtain a set of indicators that meet the screening conditions of the Delphi method. Some indicators are provided below as examples.

Fairness of disclosure consists of three indicators.

Have the listed company and the disclosure obligors disclosed non-public information to the appointed people in advance?

Have the listed company disclosed non-public information in the internal publications or internal network in advance?

Does the listed company keep the important information that should be disclosed in secrecy when its share price fluctuates dramatically?

Timeliness of disclosure consists of two indicators.

The reporting lag time of the listed company.

Has the listed company delayed disclosure and broken the disclosure rules of China's capital market?

Accuracy of disclosure consists of three indicators.

Have the terms of the finance report been interpreted necessarily?

Do informal information clash with the formal information about the listed company?

Are the formula and data of the finance report correct?

Based on comprehensive evaluation theory, we may get the weights of indicators by the Analytic Hierarchy Process (AHP) and Delphi method and so on.

At last, we obtain a set of indicators and corresponding weights. Disclosure quality of a listed company can be measured using the Equation 1. See for example Equation 1 below:

$y = \sum_{j=1}^{m} w_j x_j, w_j$, x_j represents indicator and w_j represents corresponding weight.

5 CONCLUSIONS

Disclosure quality is difficult to define due to its subjective nature. In order to increase the accuracy and objectivity of our understanding of the disclosure quality of listed company, we need to measure the disclosure of listed company based on measurement theory. A conceptualization procedure is done to get the dimensions of discloure quality, and then a operationalization procedure is conducted to obtained the indicators. Based on comprehensive evaluation theory, for example AHP, we may get the weights of indicators. Perhaps some people do not agree with the conceptualization and operationalization procedures that we use, but they can still understand these specific indicators through these procedures. That will help to communicate with each other in the field clearly.

ACKNOWLEDGEMENTS

The research work was supported by the Education Department of Anhui Province under Grant No. SK 2013 A 194.

REFERENCES

[1] Shaoshuang Xu. 2011. A Study on the Direct Evaluation Theory Model of Information Disclosure Quality of Listed Company. *West Forum* 21 (5): 56–63.
[2] Laidroo L. 2008. Measuring Public Announcements' Disclosure Quality on Tallinn, Riga and Vilnius Stock Exchanges. Working Papers., 181 (School of Economics and Business Administration, Tallinn University of Technology).
[3] Singhvi S.S, Desai H.B. 1971. An Empirical Analysis of the Quality of Corporate Financial Disclosure. *Accounting Review* 46 (1):129–138.
[4] Yuejun Tang et al. 2008. Counterbalance of Large Shareholders, Governance Strategy and Information Disclosure: Evidence from Chinese Listed Companies in 2003, *China Economic Quarterly* 7(2):647–664.
[5] Stevens. 1946. On the Theory of Scales of Measurement. Science 103(2684): 677–680.

E-Commerce, E-Business and E-Service – Lee (Ed)
 ISBN 978-1-138-02646-9

Analysis on the survey method of mystery shopping in hospitality management

Lifang Xu
School of Hotel Management, University of Jinan, Jinan, China

Shijie He
Buri Al Arab Hotel Dubai, The United Arab Emirates

ABSTRACT: Since the 1940s, the survey method of mystery shopping has been widely used in all different kinds of industries. Recently hotel mystery shopping has become a useful method to help the market research companies to evaluate the service of the hotel industry. However, there are some problems in mystery shopping, and the maturity and validity of this kind of method still need to be improved. This thesis illustrates the importance of the mystery shopping in hotel management, and also gives us the ways to solve these problems while the survey method of mystery shopping is used, helping understand the meaning of the combination in practice of mystery shopping in hotel management.

Keywords: mystery shopping; mystery shopper; hotel management

1 INTRODUCTION

Truly great customer service is the key to success in any business. Hotel mystery shoppers, secret shopping and customer experience management services are the way to measure the customer experience, guest satisfaction and increase occupancy rates and your hotels profit margin and guest loyalty. SSI's founder, Mike Albert, was quoted recently stating: "Great service is only a great service when it has been ordered, bought and consumed. Loyal customers are built from great service." Most services are generally delivered by people to other people and companies must strive to deliver top quality services to all of the customers all the time. A customer's opinion of your hotel is always based on their experience of the product and service they receive.

Mystery shopping is a well known marketing technique used by companies and marketing analysts to measure quality of service, and gather specific information about products and services. Mystery shopping is a well established methodology which was introduced in the early 1940s primarily by the management of banks and retail chain stores to assess the performance of their employees (Zikmund, Babin, Carr, & Griffin, 2009). Recently hotel mystery shopping has become a useful method to help the market research companies to evaluate the hotel industries. A mystery shopper is essentially someone who poses as a real customer for a hotel and secretly evaluates things about the business such as customer service, cleanliness of store environment, employee interactions, for example, usually mystery shoppers are hired by marketing companies as contractors who do the secret shop in their own time in their own city and state. The marketing companies they work for are hired by the hotels that are being "shopped" as a means to try and get an accurate portrayal of how their employees are behaving and how the hotel is running during a general every day.

Hotel mystery shoppers work with hotels for the purpose of assessing customer service, product quality, guest satisfaction, guest loyalty, profit margin and other specific details requested by clients. They also evaluate the front desk staff, restaurants, catering services, bars, reservations for clients to enable the client to gain advantage over competitors.

A number of studies have been written on mystery shopping applying in bank management, retail chain stores, etc. Tarantola, Vicard and Ntzoufras (2012) analyzed data from mystery shopping surveys via Bayesian Networks in order to examine and evaluate the quality of service offered by the loan departments of Greek Banks. Finn and Kayande used a generalizability theory approach to assess the psychometric quality of mystery shopping data. Ford, Latham and Lennox showed in their research that how information from mystery shoppers can be used by managers to overcome limitations of traditional performance appraisals. But the application of mystery shopping in hospitality industry hasn't been studied before. The purpose of writing this paper is to help readers recognize the role of mystery shoppers in hotel management. To begin with, I give relevant information about mystery shopping in hotel followed by problems or challenges existing in this industry, and then the countermeasures to solve the problems will be given. In the following part, three mystery shopper audits and some information about the mystery shopper audits used in Burj Al Arab Hotel will be introduced. The final part is the conclusion.

2 BASIC INFORMATION AND DISCUSSION ON MYSTERY SHOPPERS IN HOSPITALITY INDUSTRY

2.1 *Mystery shopping in hospitality industry*

Mystery shopping or a mystery consumer is a tool used by Mystery Shopping Providers and market research companies to measure quality of service and gather specific information about products and services. Mystery shopping has developed into a well known marketing technique used by companies and marketing analysts. Mystery Shopping should be performed by a person, so called Mystery Shopper, who is unknown to companies to be evaluated.

Hotel mystery shopping involves the use of a professional evaluator, the mystery shoppers who are unknown to the hotel to be evaluated, visits a hotel and offers feedback relating to the customer service, employee attitudes, room cleanliness, clean rating and availability of needed items. This may include an ample supply of towels, shampoo, soap, tissues, and so on. At the end of the stay, the hotel mystery shopping evaluator completes a report and submits to the company who hired them. Mystery shoppers are usually hired by large restaurants, hotels, movie theaters and financial institutions to evaluate the level of customer service and to motivate the business.

Some hotel owners hire marketing research companies to evaluate the quality of service in their hotels; these companies use mystery shoppers to get the information anonymously. They assign a mystery shopper to make a particular purchase in a hotel, for example, and then report on the experience. Theories in human resource management stress the necessity of providing employees feedback on their performance so they can set specific high goals. These theories also explain the importance of feedback in enabling employees to see the relationship between what they are doing and the outcome they can expect. According to Ford, Latham and Lennox (2011), employee performance feedback that is credible in the eyes of employees, the frequency for providing feedback that will result in an improvement in their job performance.

Many professionals in the field consider mystery shopping a part-time activity. And, they add, opportunities generally are posted online by marketing research or merchandising companies. Once the evaluation is completed, hotel mystery shoppers generate a narrative report based on the data gathered during mystery shopping. Reports generated by the shoppers help their clients to improve the training programs and recruitment process, service quality etc. Hotel mystery shoppers are paid for submitting the reports on time.

Many hotel mystery shoppers are not even required to travel to other cities, but may evaluate local accommodations. This allows for flexible hours, along with extra income. The majority of hotel mystery shopping professionals work part-time and still have many hours left in the day for other activities, such as a full time job.

2.2 *Features of a mystery shopper*

The goal of a mystery shopper is to test all departments within the hotel: front office, human resources, marketing, engineering, the room, housekeeping, laundry, all F&B outlets, spa, and so on.

During his stay, the mystery shopper will observe and notice every single detail. He will write everything what he sees, hears, talks. He will record your names, what you say, and take pictures from the suite or restaurant to collect information.

The mystery shopper is usually alone, sometimes two, maybe in two different suites. In a resort this could be a family.

The mystery shopper will stay two to three days.

The mystery shopper always asks a lot of questions, about the menu, the hotel, the city, and he always looks at you in the eyes.

Anyway, for these reasons, the best way to succeed is to treat all the guests as they would all be mystery shopper, and follow all the standards all the time.

2.3 *The importance of mystery shopping in hotel management*

For the most part, hotel owners are appreciative for the hotel mystery shopping industry. This method allows company leaders to know which areas need to improve without having to do the inspection themselves. By posing as regular hotel guests, hotel mystery shopping professionals are able to receive the same treatment as all other guests without otherwise being recognized. Customer service is a very important part of the hotel industry. As such, hotel owners want to ensure that their employees are treating guests with great consideration. Hotel mystery shopper is a direct and reliable method of assessing customer experience and satisfaction. No other method can replace simple, effective feedback from an anonymous hotel shopper more commonly known as a secret shopper. By visiting the hotels anonymously, posing as a member of the public, they complete a detailed critique of their visit, giving a complete evaluation of the employees' service and customers' experience. Different hotel shoppers make several visits over a period of weeks or months depending upon what type of service or hotel services they are evaluating.

2.4 *The problems of mystery shoppers*

2.4.1 *Hospitality industry mystery shoppers without professional knowledge or training*

Nowadays, there are hundreds of companies providing services related to mystery shopping surveys (see for example http://www.mysteryshop.org/). Some mystery shopper companies are not regular. They always hire the mystery shoppers without any training or experience. This kind of shoppers has no professional hospitality industry knowledge. They do not know which part of the services need pay more attention to observe and how to inspect the hotels. They can not provide professional reports to the managers.

2.4.2 *Mystery shoppers are too subjective*

Mystery shoppers are not as the same as the normal customers because they know that they are shopping with a critical mindset. They will be much more critical than the common customers, and they have to be very careful in framing their study or they are more likely to see a negative part in the results. However, if the mystery shoppers have good relationship with the servicers, they will ignore the mistakes.

What is more, mystery shoppers can lie. This is one of those things that are very difficult for clients to verify. Mystery shoppers can go into a hotel very easily, just get the external situation, and then leave without any interacting with the staffs. After the simple observe, they will find out their forms as if they had observed the hotel clearly. These things really happen in the hotel industry, because the mystery shoppers do not want to spend much time on inspecting a hotel.

2.4.3 *The managers can not use the reports effectively*

Many managers do not really know how to use the data of the report and they use the mystery shoppers' reports as leverage. After they read the reports, they will directly punish the employees who make mistakes immediately without any further researches. It will cause some wrong and unfair decisions.

2.4.4 *Hotel staffs are usually warned about shoppers ahead of time*

There is an ethical question as to whether or not remind the staffs before the mystery shoppers come. Ethically speaking, the managers should not remind the employees. For example, most mystery shopping studies give managers and district manages a general period in which a shopper might come, such as "the first two days of the month."

Here is the problem: if the managers announce the information, the mystery shopper will be difficult to observe the real situation. If district managers announces to the staffs that a mystery shopper will come, and then they would try their best to improve the normal working conditions for getting a perfect score. This may bring an artificial improvement in service just until the mystery shopping period is over. This will invalidates the data and make it more likely that the busy people who always doing their jobs correctly, but they just be caught in something stupid mistake, while the others who are good at disguising themselves, but not doing their jobs correctly are not to be found.

2.5 *Countermeasures to solve the problems*

2.5.1 *Hotels should carefully select the companies*

As more and more mystery shopper companies in the real world or in the Internet, the regulation of them can not be guaranteed. In order to make the report useful and valuable, hotels should carefully select the mystery shopper companies which could provide variety of training to mystery shoppers, and trained and excellent mystery shoppers to hotels.

2.5.2 *Develop research objectives of hotels*

There is no doubt that mystery shoppers are different from normal customers because they are shopping with a critical mindset. They are especially sensitive about the surrounding environment, cleanliness of the rooms, availability of facilities in hotels, freshness of the food, and so on. So the method to solve it is to develop research objectives. The advantage of creating this method is that it gives you a very nice outline for your survey instrument later on. Research objectives could help to narrow down what hotel managers really want to learn from the research. The way to develop these is to state general research objectives and determining which questions need to ask to meet them. Here are some examples listed as below which is used in LQA (Leading Quality Assurance) standard:

Q1.1: Are the customers being greeted within 30 seconds when they arrived?
Q1.2: Are the guest relations executive picking up the phone within 3 rings or 10 seconds?
Q1.3: If any staff is on the phone when the customer walks in, does the staff interrupt the call to briefly acknowledge the customer?
Q1.4: Did the colleague greet the guest with a smile prior to the guest greeting the colleague?
Q1.5: Did the colleague enthusiastically welcome/greet the guest?

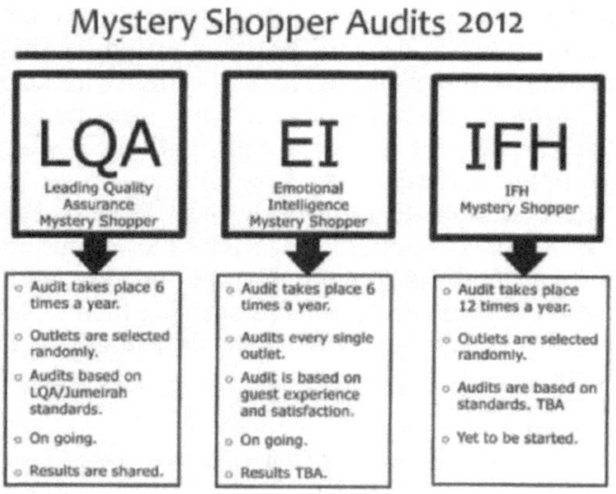

Figure 1. Three main mystery shopper audits.

Q1.6: Did the colleague seat the guest within one minute of their arrival?
Q1.7: Did the colleague use the guest's name during the interaction?
Q1.8: Was the guest escorted to the table as opposed to being directed?
Q1.9: Did the colleague offer chair assistance?
Q2.0: Did the colleague unfold the napkins for the guest?

There are different standards in different apartment, so there are a lot of different questions based on different situations.

2.5.3 *Hotels repeat regularly*

Emergency accidents are very ordinary in hotels. There is no doubt that things can not be totally controlled by people. So there are many problems everyday and no one knows what will happen in a day. As mystery shopper will come to do the research without advance notification, some accidents are inevitable. It can not reflect the real situation of the hotels. The method to this problem is to repeat the mystery shoppers once every quarter, or, at a minimum, once every year using the same process. Keep in mind, your repetitions are at regular intervals, but do not do them too frequently.

3 MYSTERY SHOPPING EXAMPLES

3.1 *Three main mystery shopper audits (Fig. 1)*

3.2 *Findings: summary report (Table 1)*

Table 1. Findings: summary report.

Reception check in summary 20 February 2006–21h38
The chauffeur called the hotel as they approached the hotel to announce the arrival of the agent. When the car came to a halt in front of the hotel at 21h38, a porter offered car door assistance and upon exiting the car, the agent was greeted by the Operations Manager and also by the Reception Manager (Mohammed). The Reception Manager escorted the guest into the hotel, and accompanied the agent to the room. Where a prepares registration card was ready for the agent to sign. However, whist all of the information was correct as provided during the reservation, the telephone number was not detailed and the guest was asked to provide this information. The employee then confirmed the number of nights that the agent was staying and also handed the guest a copy of a written confirmation and stated that the hotel was not able forward the confirmation to the AOL e-mail address that was given during the reservation (the home address was correct and the confirmation could have been sent by regular mail). The Reception Manager also asked foe the agent's passport and a credit card in order to obtain an authoritarian. On the way to the suite. Mohammed pointed out the restaurants and bars, and also explained the opening times. Moreover, he informed the agent that the Seafood restaurant was one of the best in the world and highly recommended to make a reservation. Furthermore, it was explained to the agent where the concierge desk was and how to gain access to the private beach as well as the Spa area. Before leaving, Mohammed introduced the private butler to the agent and wished him a pleasant stay.
The Reception Manager was very pleasant and professional in manner. Moreover, he was well presented and smiled when interacting with the agent. Of particular note was the use of the guest's name throughout the interaction. Overall, he gave a very good first impression of Burj Al Arab.

4 CONCLUSION

Service performance and its impact on the customer experience are critical to hotel management. The hotels should pay attention to not only the quality of the service provided but also to its effectiveness. One method for service evaluation, which is very popular in recent years, is mystery shopping. All in all, mystery shoppers play an important role in hotel management by evaluating the performance of hotel employees. This allows the hotel leaders to know which areas need improvements. However, as the article points out there are several problems mystery shopping may bring to hotel management. So the hotels should carefully select the companies, develop research objectives of hotels and repeat mystery shopping regularly, to make the survey method of mystery shoppers a helpful tool in hospitality management. With all of these problems solved, four-fold benefits can be achieved by using mystery shopping survey method as stated by Ford, Latham and Lennox (2011). First, this system offers an assessment of employee performance on predefine and measurable performance factors. Second, it is administered at seemingly random intervals so that employees don't wait until the annual appraisal time nears to demonstrate their performance. Third, the feedback is based on objective as opposed to subjective criteria so that the employee being evaluated can get relatively neutral third-party feedback on the quality of job performance. This minimizes the problem of "ad hominen" attributions towards a supervisor. Fourth, a Feedback system provides a basis for coaching by a manager, who can give the quiz score back along with an invitation to discuss the performance.

REFERENCES

Allen H. Munro. 2009. "Structures for Strategy: The Human Resources Management Plan". New York: Morgan James Pub.

Anna G., Rachel M., Joanna C. Loudon & Aileen M. 2010. "Community pharmacists providing emergency contraception give little advice about future contraceptive use: a mystery shopper study". *Contraception, Vol. 82, No. 6*, pp. 538–542.

Boothroyd, Jennifer. 2006. "Benefits of Mystery Shopping for Secret Shoppers". Issue 2. Lerner Publications, pp. 17.

Cathy Stucker. 2004. "Mystery Shopper's Manual". New York: Special Interests Pub.

Claudia Tarantola, Paola Vicard & Ioannis Ntzoufras. 2012. "Monitoring and improving Greek banking services using Bayesian Networks: An analysis of mystery shopping data". *Expert Systems with Applications, Vol. 39, No. 11*, pp. 10103–10111.

Denney G. Rutherford & Michael J. O'Fallon. 2006. "Hotel Management and Operations". California: John Wiley & Sons.

Jim Dees. 2008. "Mystery Shoppers: Dispelling the Mystery". Columbus: Mcgraw-Hill.

Robert C. Ford, Gary P. Latham, Gwen Lennox. "Mystery shoppers: A new tool for coaching employee performance improvement". *Organizational Dynamics, Vol. 40, No. 3*, pp. 157–164, 2011.

Shirley S. Wang. 2006. "Health Care Taps 'Mystery Shoppers'". Wall Street Journal, 08, pp. 13–14.

E-Commerce, E-Business and E-Service – Lee (Ed)
© 2014 Taylor & Francis Group, London, ISBN 978-1-138-02646-9

Study of the market position and regulation of Electronic Property Rights Trading Markets

Ming Chen
School of Business, Soochow University, Suzhou, China

ABSTRACT: China has established many Electronic Property Rights Trading Markets (EPRTMs) of agricultural products, raw materials, precious metals, artworks etc. in recent years. Their market position and economic benefits can be analyzed by specialization model of the New Classical Microeconomic framework. Because of the legal regulation failed to timely follow-up, they appear a variety of problems in operation process. It is more than two years since the State Council of China issued the decision on the clean-up of EPRTMs. Some deviations and resurgence are appeared in some places. Especially, it is needs correctly grasp in the aspects of cleaning scope, cleaning procedures etc. To this end, we have drawn some conclusions about the market position and legal regulation of EPRTMs, and then provide some advanced suggestions.

1 INTRODUCTION

In recent years, all kinds of Electronic Property Rights Trading Markets (EPRTMs) have sprung up in China. An estimated, in addition to the stock exchange, commodity futures exchange and financial futures exchange formally approved by China central government, there are more than 300 EPRTMs all over China. These EPRTMs is a mixed bag. Only a small portion of them managed properly. Most have not been effective management, appeared disorder, or even fraud phenomenon, seriously disrupted the normal economic order, and caused irrational price change radically.

There are a lot of studies on EPRTMs till now. Recently, E. Gil & A. Sanz (2013) [1] have done a research on the legal issues of EPRTMs evolution system; R. Postrel (2013) [2] has done a research on the electronic property rights transaction method and system based on the point redemption scheme; Lin Juan (2013) [3] has Pay attention to the so-called "2.5th" art trading markets; Wu Yueren (2013) [4] has studied on the struggling art share trading markets; Jiang Zhefeng (2013) [5] put forward The question of the supervision of cultural property rights trading markets. Especially, since the release of "The Decision on the Clean-up and Rectify All Types of Electronic Exchanges to Guard Against the Financial Risk" [6] by China State Council in November 24. 2011, China has formal been taking rectification work to all types of EPRTMs. But, over the past two years, the rectification work has suffered stubborn resistance, to little effect. Many of the original markets have made only superficial changes, stage a comeback [7]. We believe that effective legal regulations should be taken on these EPRTMs urgently. This paper intends to make a research on the question of market position and regulation of EPRTMs.

2 THE CLASSIFICATION AND CURRENT SITUATION OF EPRTMS

China's EPRTMs have been developing rapidly. From the trading varieties, these EPRTMs can mainly divided into the following categories:

2.1 *Agricultural products electronic trading markets*

Many of these electronic trading markets exist in the large trading center of agricultural products. The trading varieties include garlic, ginger, vegetables, livestock and etc. Such as Shandong Shouguang Vegetable Exchange and Tianjin Guosheng Agricultural Products Exchange. In recent years, many agricultural products electronic trading markets are sprouting everywhere, The operation of a lot of them are not standard, make rise to the a large number of speculation phenomenon of agricultural products prices.

2.2 *Bulk raw materials electronic trading markets*

The trading varieties of bulk raw materials electronic trading markets include petrochemical, coal, nonferrous metal etc. Such as Tianjin Bohai Commodity Exchange, Hebei Xinjiang Stainless Steel Exchange, Shanxi Coal Electronic Trading

Market, Tianjin Tianbao Bulk Coal Trading Market, Tianjin Rare Metal Exchange, and Zhejiang Zhoushan Commodity Exchange. Some of these raw material exchanges are of large scale, standardized operation, and have good reputation, Some others set up for a short time, blind mount a horse, and thus occurred various problems.

2.3 *Precious metals electronic trading markets*

Precious metals electronic trading markets are mainly engaged in the electronic trading of gold, silver and other precious metals. Such as Hunan Weicai Precious Metals Exchange etc. This kind of markets is often used in margin trading. As the international prices of gold and silver vary wildly, the traders can go broke in no time, lose everything.

2.4 *Artworks electronic trading markets*

Tianjin Artworks Exchange, Shandong Taishan Artworks Exchange are the famous artworks electronic trading markets in China. Because of the unique charm of artworks, the product is very difficult to copy, the value of it is difficult to estimate, its market liquidity is low [8]. Its price has large room to rise or decline. Therefore, the speculation of artworks trading is often very strong, their prices change radically, thus have caught many investors.

2.5 *Other types of electronic trading markets*

Other types of EPRTMs include the electronic trade of carbon emissions, solid waste, drugs, financial assets and etc [9]. In general, all types of electronic trading markets have been gradually appearing. It seems that as long as there is a price of things, it can be used to trade in electronic trading markets.

The deep reason of the rapid development of all types of EPRTMs is the clash over the financial resources of each region. Currently, here are dozens of city of China proposed to set up national or regional financial center. Financial center must be gathered to financial institutions and market place. But, because of Chinese financial market is not developed, the standard stock exchanges approved by the State are only in Shanghai and Shenzhen, commodity futures exchanges only in Shanghai, Zhengzhou, and Dalian, financial futures exchange only in Shanghai. While other parts of China, especially some big cities such as Tianjin etc., also have strong strength. In order to create a national or regional financial center, have embarked on a way of self built EPRTMs. Thus, other cities try to catch up, also built many EPRTMs to attract business resources.

3 MODEL ANALYSIS OF THE MARKET POSITION OF EPRTMS

From the point of view of the market position of EPRTMs, the EPRTMs have played an active role to the organization and achieving market integration of transactions of agricultural products, bulk raw materials, precious metals, art products etc., which were original dispersion and bilateral. Its market position in modern economic system can be analyzed by specialization model of the New Classical Microeconomic framework:

As for the property rights trading markets which specializing in trade integration, its product is "Trade Integration" itself. "Trade Integration" is needed by each market participant and therefore has high preference, and it can also be considered as a highly specialized product. We can prove by modeling that "Trade Integration" itself is a trade product (i.e. purchase and sale of "Trade Integration") can get high per capita income:

For convenience, assume that there are only two products in the market: x, y; and x is a general product, y is the product of "Trade Integration".

In the new classical microeconomic framework [16], Yang expressed the utilities of (x/y) and (y/x) as:

$$\ln u(x/y) = \alpha(2\ln x + \ln r - \ln M_{xy}) + \gamma \ln z + \alpha \ln K \quad (1)$$

$$\ln u(y/x) = \alpha(2\ln x + \ln r + \ln M_{xy}) + \gamma \ln z + \alpha \ln K \quad (2)$$

where $x = [2\alpha/(3\alpha+\gamma)]^{\alpha}/2$, $z = [\gamma/(3\alpha+\gamma)]^{\alpha}$, $r = [\alpha/(3\alpha+\gamma)]^{\alpha}$.

The utility equalization condition is:

$$E = \ln u(y/x) - \ln u(x/y) = 2\alpha \ln M_{xy} = 0 \quad (3)$$

Assume u^* is the real income, M^*_{xy} is the relative number of companies which choose different configurations at the equilibrium. From (1), (2), (3), it's can be obtained:

$$dM^*_{xy}/d\alpha\big|_{\alpha=\gamma} = -(\partial E/\partial \alpha)/(\partial E/\partial M_{xy})\big|_{\alpha=\gamma} = 0 \quad (4)$$

where $\partial E/\partial \alpha\big|_{\alpha=\gamma} = 2\ln M_{xy} = 0$, If $\alpha = \gamma$, thus $M_{xy} = 1$, thus:

$$\partial \ln u^*/\partial \alpha\big|_{\alpha=\gamma} > 0 \quad (5)$$

Therefore:

$$d\ln u^*/\partial\alpha\big|_{\alpha=\gamma} = \partial\ln u^*/\partial\alpha\big|_{\alpha=\gamma} + (\partial\ln u^*/\partial M_{xy})(dM^*_{xy}/d\alpha)\big|_{\alpha=\gamma} > 0 \quad (6)$$

Formula (6) shows using product with larger preference parameter as equilibrium can obtain high per capita income.

Assume that $\alpha = \beta = \gamma$, $a = b = h$. If:

$$x = [2a/(3a+c)]^a/2,\ z = [c/(3a+c)]^a,$$
$$M_{xy} = [2a/(3a+c)]^a \quad (7)$$

replace x, z, M_{xy} by $[2a/(3a+c)]^a/2$, $[c/(3a+c)]^a$, $[2a/(3a+c)]^a$ respectively, then:

$$\partial\ln u^*/\partial a\big|_{a=c} > 0$$
$$dM^*_{xy}/da\big|_{a=c} = -(\partial E/\partial a)/(\partial E/\partial M_{xy})\big|_{a=c} = 0 \quad (8)$$

where $\partial E/\partial a\big|_{a=c} = 0$, thus:

$$d\ln u^*/\partial\alpha\big|_{a=c} = \partial\ln u^*/\partial\alpha\big|_{a=c} + (\partial\ln u^*/\partial M_{xy})(dM^*_{xy}/d\alpha)\big|_{a=c} > 0 \quad (9)$$

Formula (9) shows using specialized product as equilibrium can obtain higher per capita income level.

Since the establishment of the above two conclusions, and "Trade Integration" is a specialized product with high preference parameter. The EPRTM is the provider of the "Trade Integration" product, and thus creates a high level of per capita income. Namely: Changing from the bilateral trading market of such as agricultural products, raw materials, precious metals, artworks, etc. to central trading market. As shown in Figure 1.

To sum up, because the electronic technology improves the transaction efficiency, it makes the product of "Trade Integration" from provide for oneself to specialized production. When provide for oneself, there is no professional platforms. Due to the division, there appear the EPRTMs. The rapid progress of information technology and the popularity of information infrastructure since the 1990's have created conditions for high efficiency of electronic trading, and make the emergence of EPRTM possible. At the same time, the rapid progress of economy in China has a huge demand for electronic transactions of property rights. So, it has inevitability to develop EPRTM. The market effect of EPRTM should be affirmed.

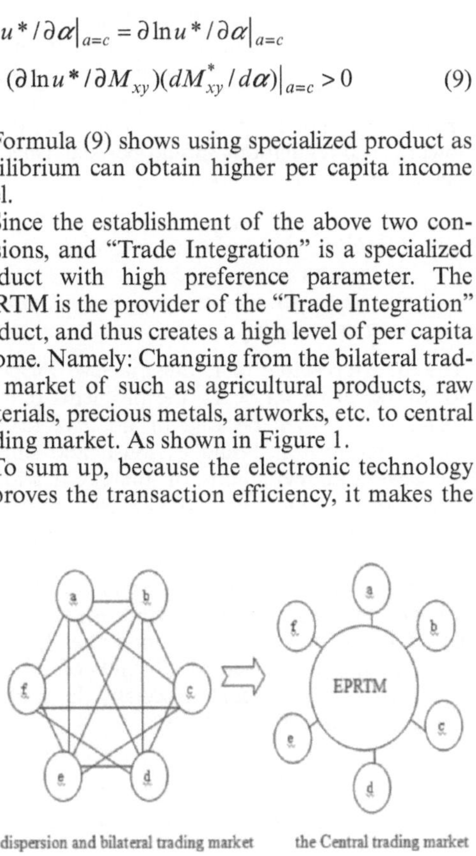

Figure 1. From dispersion and bilateral trading market to central trading market: EPRTM.

4 THE EXISTING PROBLEMS OF EPRTMS

Because of the legal regulation failed to timely follow-up, these EPRTMs have occurred various problems in their operation processes. Mainly in the following aspects.

4.1 *Commodity trading becomes game for profit, deviate from the intrinsic value of goods*

According to the law of value, commodity price should fluctuate around their fair value. But in EPRTMs, due to the change of investors trading purposes, the purpose of trading is not to have the goods itself, but to buy low and sell high, get the difference. Meanwhile, because a lot of money are gathered in the EPRTMs, these funds can continue to buy/sell, continue to push up/push down the price in a period of time. Thus, artificially causes "rational bubble". In this condition, rational investors can only follow the trend. Otherwise, if adhere to the value judgment, they will lose largely. At the same time, the "rational bubble" develops to a certain extent, it will cause the "non rational bubble", attract a large number of outside investors blindly follow the trend, buy in at a high price. As soon as wind sways grass, all are into panic, at same time to sell. The price suddenly turns for the worse. Therefore, in EPRTMs, commodity prices tend to be substantially deviate from its fundamental value. For example, the first batch of artworks (paintings) traded in Tianjin Artworks Exchange, the price was fired up to 10 times the original when trading in a not long time. Then, its price continued to fall, no one shows any interest in them. This pattern of trade, it's more "pass the parcel" game for profit than normal commodity trading business. It not only caused huge losses for many investors, but also has undermined the social atmosphere.

4.2 *Lack of legal norms, there are serious price manipulation and tunneling behavior*

If we say the "pass the parcel" game mentioned above is still in the legal range, then, the severe price speculation and price manipulation in many EPRTMs is beyond the scope of law. Some people take advantage of the loopholes in the law, using their influence to manipulate prices and take advantage. Some electronic exchanges are not provide fair, transparent trading mechanism to each investor, but try various devices, through a variety of insider methods, do manipulation in the name of trading. Some electronic exchanges artificially inflated the initial purchase price, conveying interests for some people, and damage the interests of investors. Some electronic trading market are not strict on the trading goods, mix the spurious with the genuine, mislead investors, and so on.

4.3 *Contrary to the law, implementing "gambling", defrauding investors*

A small number of EPRTMs actually have been in serious illegal situation. Their shareholders directly involved in the trading, gambling against investors. In these electronic markets, investors trading activity has been completed on the surface, but did not actually true. When the price goes to the direction negative to investors, investors are forced to close out. The loss is of their own, and the proceeds are obtained by these electronic exchanges. While the price goes to the favorable direction of investors, by all means such as network blocking, slow speed etc. these electronic exchanges stop investors unwind their trading, thus, render the profit of common investors can not be achieved, until the profit become loss.

4.4 *The illegal crimes of EPRTM management personnel*

The most serious is the illegal crimes of EPRTM management personnel. Such as misappropriating client funds, and absconded with money. It is beyond the scope of law, and is a serious criminal offense. In this case, common investors tend to lose everything, cause great property damage and very bad social influence.

5 CONCLUSIONS: THE LEGAL REGULATION OF EPRTMS

On the one hand, the various types of EPRTM established in recent years have promoted the market transactions, provided convenience for the vast number of consumers and users, also contributed to the local social and economic development. But on the other hand, there also exist a variety of problems of them mentioned above. Some of the problems are quite serious, to which much attention should be paid.

It is more than two years since the State Council of China issued the "Decision" in Nov. 2011. All around has been carrying out the clean-up to all types of EPRTM. It should be said has made no small achievement. However, we should also see, some deviations and resurgence are appeared in some places [11]. Tendency of expansion appears in some other places. These situation and tendency are to be corrected. Especially, it is needs correctly grasp in the aspects of cleaning scope, cleaning procedures etc. So, it is necessary to strengthen the construction of law regulation of EPRTMs,

The author believes that, cleaning up the EPRTMs, is not to fully close them. After all, many of them have been in the performance of their certain, indispensable social function. They indeed provide certain help for the investors to make investment in commodities. We should put emphasis on the relevant laws and regulations to establish, perfect EPRTMs, make them rapidly normalized under legal regulation. Especially, here must be laws to follow in the prevention of financial risks. Thus, it helps to guard against financial risks, establish EPRTMs of financial security, investors trusting, and good reputation. To this end, the following legal regulations should be established and perfect as soon as possible.

5.1 *Establish and improve the laws and regulations of the registration of establishment, alternation or cancellation as soon as possible*

The first step is that there must be laws for people to follow. In recent years, the problems of many EPRTMs, is often because of no laws to abide by. Therefore, we should establish and improve the laws and regulations of the registration of establishment, alternation or cancellation as soon as possible. Especially in the establishment of EPRTM, should establish a higher barrier to them. Laws and regulations should specify the funds and other conditions of establishment of each type of EPRTM. To prevent from fountainhead, consideration should be given to all kinds of problems appeared in recent years, especially the problems related to financial security.

5.2 *Formulate the national and regulatory trading rules of all types of EPRTM as soon as possible*

At the same time, we also should formulate the national and regulatory trading rules of all types of EPRTM as soon as possible. Trading rules should adopt standardized contract transaction modes such as market maker, bidding, electronic matchmaking etc., or prohibiting anonymous transactions, control the number of equity holders. All types of EPRTM shall comply with once opening. Thus, it can avoid the possibility of various transaction traps.

5.3 *Establish the financial safety regulations, implement the daily reporting system, monitor real-time trading data*

To guard against systemic, regional financial risks, to ensure social stability, is an important target of China State Council to clean up and rectify all types of EPRTM. So, we should establish the financial safety regulations, monitor real-time data of various kinds of EPRTM daily operation. It can be considered to implement the daily reporting system for the EPRTM. So, once the regulatory authorities found financial risk situation of the EPRTM, can timely discover, timely prevention, timely intervention, and solve the problem in the bud.

5.4 *Establish the grading system of EPRTMs*

Meanwhile, it can be considered to establish the grading system of EPRTMs, implement the hierarchical management of EPRTMs as soon as possible. Based on the unified standards, in accordance with different specific quantitative standard, EPRTMs can be divided into different grades. Each electronic property rights trading market may apply different level according to its characteristics and conditions. The regulatory authorities verify and approve the applicant which satisfies the requirements, strictly controlling the EPRTMs engaged in financial products transactions. Meanwhile, the regulatory authorities should put forward different requirements of initiate the establishment, membership recruitment, trading rules, daily operation, safety assurance, liquidation accounting etc. to the EPRTM of different level, also conducive to investors to choose their suitable EPRTMs according to their different levels, thus reduce the investment risk.

5.5 *Perfect information disclosure mechanism of EPRTMs*

The information disclosure mechanism is the key to even bargain of all trading markets, also is the key to long-term development of EPRTMs. Therefore, we should perfect information disclosure mechanism of all EPRTMs. Put forward clear requirements of contents, aging, mode and responsibility of information disclosure for EPRTMs. On the one hand, it can guarantee the interests of investors, on the other hand, it also can greatly reduce the possibility of "financial bubble", avoid the "pass the parcel" game, thereby effectively defuse financial risks.

5.6 *Amend, perfect relevant laws, rules and regulations, strengthen the fight against all kinds of illegal and criminal related to electronic trading*

The reason why misappropriating client funds often appeared in the EPRTMs is because the relevant laws and regulations are not sound enough. Especially, it is lack of pertinence in the law enforcement process, leading to its crackdown is not enough, resulting in its low illegal cost. In future, we should timely amend, perfect relevant laws, rules and regulations, strengthen the fight against the illegal crime of EPRTM management personnel. So they can not escape the punishment by law and suit the punishment to the offence. Thus, we can avoid the occurrence of such affairs, safeguard the legitimate rights and interests of investors, and achieve a good development of the electronic property rights trading market.

ACKNOWLEDGEMENT

Thanks for the Philosophy and Social Sciences Foundation of Jiangsu, China (04EYB022).

REFERENCES

[1] Gil E, Sanz A G C. Legal Issues Concerning P2P Exchange of Educational Materials and Their Impact on E-Learning Multi-Agent Systems [J]. International Journal of Interactive Multimedia and Artificial Intelligence, 2013, 2 (Regular Issue).

[2] Postrel R. Method and system for electronic exchange of reward points: U.S. Patent Application 13/855,239 [P]. 2013–4-2.

[3] Lin Juan. The Art Fair: "2.5th" market gorgeous breakthrough [J]. Art Market. 2013 (2).

[4] Wu Yueren. Art share trading in an awkward situation [J]. Shanghai Economy. 2013 (4).

[5] Jiang Zhefeng. How to supervise Cultural property market [J]. Culture Monthly. 2013 (3).

[6] Ma Jinyu. The curtain of All kinds of exchange clean up and rectify [N]. Shanghai Securities News. 2011.11.25.

[7] Qu Jian. "Trading platform" Nirvana rebirth [J]. The New Production. 2013 (4).

[8] Ge Dan., The Clean-up of electronic property rights trading market [EB]. People's Daily Online. http://www.people.com.cn. 2011.11.26.

[9] The State Council issued rectification opinions in electronic property rights trading market [EB]. Chinese Broadcasting Network. http://www.cnr.cn. 2011.11.20.
[10] Yang Xiaokai, Huang Youguang. Specialization and Economic Organization [M]. Economic Science Press. 2000. pp 149.
[11] Worrying about the enlargement of electronic property rights trading market clean-up and rectify [EB]. Daily Economic News. http://gov.finance.sina.com.cn. 2012.1.14.

E-Commerce, E-Business and E-Service – Lee (Ed)

Virtual organization manufacturing resource planning: Flexible multi-objective mathematical programming

Yao Hsu
Department of Business and Entrepreneurial Management, Kainan University, Taoyuan, Taiwan

Jih-Jeng Huang
Department of Computer Science and Information Management, Soochow University, Taipei, Taiwan

Jhao-Syuan Chen
Ph.D. Program in Business, Feng Chia University, Taichung, Taiwan

ABSTRACT: Optimizing production resource planning significantly affects the profits and competitive advantages of factories by increasing the size and complexity of organizations. Mathematical programming models are mainly used to optimize the allocation of traditional production planning. However, using traditional mathematical programming models to optimize certain types of organizations is difficult because of rigid parameters and new organizational types, such as virtual organizations. Studies have proposed the De Novo programming model and the changeable parameter model, which allow parameter modification. This study proposes a multi-objective flexible mathematical planning model, and combines the De Novo programming model and changeable parameter model to assist virtual organizations with the optimization planning of production resources for factories. The empirical results show that the proposed models produce better resource-planning results than do previous models. This study also provides decision makers with a more flexible operational model.

Keywords: production resource planning; De Novo programming; variable parameters; virtual organizations

1 INTRODUCTION

Scientific and technological progress and the popularization of the Internet have made virtual organizations an important industry. Therefore, companies distribute their internal non-core business to external cooperation partners (including third-party and satellite manufacturers). The virtualization of traditional office systems and enterprises provides organizations with the freedom to expand beyond traditional business boundaries. Companies may even become virtual corporations. The current emphasis on supply chain enterprises has increased product outsourcing. In other words, outsourcing has become an important strategy for enterprise management (Baiman et al., 2001). McGoon (1995) predicted that as distributed enterprise networks mature, future organizational work patterns will ensure that workers from various areas will cooperate based on mutual trust and form virtual organizations.

Ching, et al. (1996) stated that virtual organizations can reduce operational risks and increase their organizational development. Burn et al. (2002) found that the characteristics of a successful virtual organization—including increased productivity, profit, revenue, and market share, as well as efficient resource planning—can attract partners. Saabeel et al. (2002) concluded that virtual organizations can quickly adjust to market changes to respond to market opportunities or demands, but cannot perfectly predict market trends. The characteristics of virtual organizations tend to reduce market complexity and uncertainty.

Studies have explored the effectiveness of planning and development to apply the concept of resource planning to virtual organizations. Zeleny (1990) proposed that the most common problem encountered by enterprises involved in resource production planning was a tradeoff between quality and cost. He proposed using De Novo programming to eliminate the tradeoff between quality and cost by using an optimized system design to explain De Novo programming, and eliminated its limitations and disadvantages to produce ideal solutions. Bernard et al. (2003) used traditional mathematical programming to solve the interaction and support problem between multiple manufacturing plants and production items. They used mathematical models to construct an efficient heuristic procedure, to solve the problem of the limited capacity of multiple plants and level zed. Yu, P.L. (2007) used variable parameters and multi-objective decision-making to "approach the multi-resource constraint level of a multi-criteria linear programming model." The multiple criteria and multiple constraint levels linear programming model is known as the MC^2 LP model. Long Yu Po used this model to examine operating parameters (including product unit profit and available resource level), capital investment, and time changes to design an effective quantitative model for analyzing and identifying the best decisions.

However, the model cannot provide effective solutions for a virtual resource-planning organization. De Novo programming parameters are limited to virtual organization member changes, increased economies of scale, and capital investment. A variable parameter model allows for parameter changes, but does not consider cost/price elements and how to configure resources. Therefore, this study uses virtual organizations to combine the programming model, empirical analysis, and traditional mathematical programming. The results of this study show that De Novo programming and the variable parameter model generate better profit and quality results than previous models. The adjustment coefficient or goal improvement allows the model to reflect more complete production systems and to reach production targets.

2 MATHEMATICAL PROGRAMMING AND RELATED EXTENSIONS

Traditional resource planning uses mathematical programming model processing to optimize resource allocation to various competing activities. For example, better product quality increases costs and reduces profit (see Fig. 1). Therefore, high quality and high profits cannot be simultaneously achieved. This is known as a tradeoff.

The traditional resource-planning multi-objective mathematical programming model is expressed as follows:

$$\begin{aligned} & \max \boldsymbol{Cx} \\ & s.t.\ \boldsymbol{Ax} \leq \boldsymbol{d} \\ & \boldsymbol{x} \geq \boldsymbol{0}, \end{aligned} \tag{1}$$

where $\boldsymbol{x} = (x_1, x_2, \dots, x_n)^T$ is a decision vector, C is an objective coefficient matrix, $A = (a_{ij})$ $(i = 1, \dots, m;$ $j = 1, \dots, n)$ is a resource consumption matrix, and $\boldsymbol{d} = (d_1 d_2 \dots d_m)^T$ is a resource availability vector.

Traditional mathematical programming models use given resource constraints to optimize resource allocation. Zeleny (1981) stated that traditional resource planning should be able to overcome resource constraints. He proposed De Novo programming to overcome the limitations of the traditional target system and the resource tradeoff relationship.

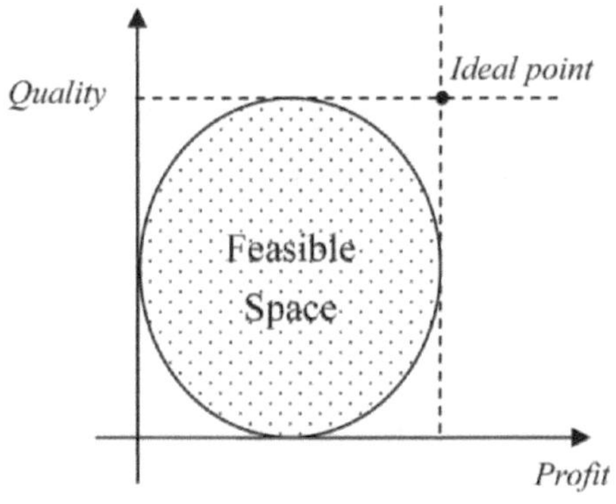

Figure 1. De Novo programming model.

The multi-target De Novo mathematical programming model is expressed as follows:

$$\begin{aligned} & \max \boldsymbol{Cx} \\ & s.t\ \boldsymbol{Ax} \leq \boldsymbol{b} \\ & \boldsymbol{p}'\boldsymbol{Ax} \leq \boldsymbol{B} \\ & \boldsymbol{x} \geq 0, \end{aligned} \tag{2}$$

where $\boldsymbol{C}$ is the target coefficient matrix, $\boldsymbol{p}$ is a factor price vector, and $\boldsymbol{B}$ is the total budget.

De Novo programming considers price factors in the traditional mathematical programming model by adjusting the level of available resources to avoid a tradeoff relationship between targets. Therefore, the difference between the traditional planning model and De Novo programming is that the traditional planning model allows only a fixed level of available resources. The De Novo programming model extends the traditional mathematical programming model, which restricts resource tradeoffs, but the target coefficient and resource consumption matrix coefficients are set at fixed parameters. However, a technological breakthrough or outsourcing may enable organizations to expand their capabilities, causing mathematical programming parameters to fluctuate. Thus, Yu, P.L. (2007) stated that the variable model parameters allow the model to react to changes in the external environment.

If the target mode coefficient vector (c) is a linear function of the input variable differences (i.e., decision makers invest more and increase profit per unit), this relationship is represented using the following formula:

$$c_j = f_{cj}(y) = c_{j,0} + c_{j,1}y, \quad j = 1, \dots, n \tag{3}$$

where $c_{j,0}$ is the original unit profit, $c_{j,1}$ is the unit investment to increase profit, and y is an investment to improve profit per investment unit. If the available resource level vector (d) from Eq. (1) is a linear function of the investment level and decision makers invest by purchasing more production resources to generate profit, this relationship is expressed as follows:

$$d_i = f_{di}(z) = d_{i,0} + d_{i,1}z, \quad i = 1, \dots, m \tag{4}$$

where $d_{i,0}$ is the original level of resources available, $d_{i,1}$ is a unit investment resource, and z represents additional resource investment units. By changing the definitions of c and d, Eq. (1) is rewritten as follows:

$$\begin{aligned} max\ \ & (c_{1,0} + c_{1,1}y)x_1 + (c_{2,0} + c_{2,1}y)x_2 \\ & +\dots + (c_{n,0} + c_{n,1}y)x_n \\ s.t.\ \ & a_{11}x_1 + a_{12}x_2 + \dots + a_{1n}x_n \le (d_{1,0} + d_{1,1}z) \\ & a_{21}x_1 + a_{22}x_2 + \dots + a_{2n}x_n \le (d_{2,0} + d_{2,1}z) \\ & a_{m1}x_1 + a_{m2}x_2 + \dots + a_{mn}x_n \le (d_{m,0} + d_{m,1}z) \\ & x_j \ge 0, j = 1, 2, \dots, n, \quad y, z \ge 0 \end{aligned} \tag{5}$$

The MC^2 LP planning mode can be solved by using the MC^2Simplex method to calculate the best potential solution set.

The variable parameter model formula is still subject to certain limitations, preventing it from being used for optimal virtual resource planning. The single-target resource-planning problem model cannot implement multi-objective resource planning and cannot consider tradeoffs between multiple targets. The model also does not include the unit price elements, and therefore, better solutions may exist. The flexibility of the model fixes the fixed resource consumption coefficient.

3 FLEXIBLE MULTI-OBJECTIVE MATHEMATICAL PROGRAMMING

Research has suggested that multi-target, multi-criteria, or multi-attribute choices affect decision making. This is because the model parameters of individual goal-planning problems are set before the optimal solution is determined (Zeleny, 1975). Zeleny stated that productive resources should not be individually or separately engaged because resources are not independent (Huang et al., 2005). The most important issue for multi-criteria De Novo programming is to determine the best path for optimal performance by designing a special resource budget-level optimization system. Zeleny used the basic methods proportion of the optimal path system design—or multi-standard performance related to a given budget level—to determine the best system design. De Novo programming uses various factors to exceed limits and achieve an ideal solution. Therefore, the difference between traditional mathematical programming and De Novo programming is that De Novo programming uses system redesign, reconfiguration, or restructuring to redefine its scope and form a new feasible solution (Zeleny, 1990).

During system optimization design, parameters can be changed (parameters include unit profit, available resources, production function, and input and output coefficients). Parameters can also indirectly change capital investment or time, but cannot affect cost/price elements or resource configuration issues. The original variable parameter space model considers only a single-objective, and practical optimization problems are often multi-target issues. Therefore, this study uses a flexible programming model to achieve the best design and integration.

This study combines the basic multi-objective mathematical programming, De Novo programming, and the variable parameter model to create an integrated mathematical programming model. This model is expressed using the following formula:

$$\begin{aligned} \max\ \ & f_1' = (c_{11} + \hat{c}_{11})x_1 + \dots + (c_{1n} + \hat{c}_{1n})x_n \\ \max\ \ & f_m' = (c_{m1} + \hat{c}_{m1})x_1 + \dots + (c_{mn} + \hat{c}_{mn})x_n \\ s.t. & \\ & p_1(a_{11}x_1 + \dots + a_{1n}x_n) + \dots + p_k(a_{k1}x_1 + \dots + a_{kn}x_n) \\ & \quad + (\hat{c}_{11}p_{11}^c + \dots + \hat{c}_{mn}p_{mn}^c) \le B + \hat{B} \\ & x_i \ge 0, i = 1, \dots, n, \end{aligned} \tag{6}$$

where P_k the kth unit prices are factors of production; P_{ij}^c the goal of the jth unit is the product improvement cost; $\hat{c}_{ij}$ is the number i of j products required to fulfill the first goal; B is the original total budget; and $\hat{B}$ is the new budget.

With this model, decision makers can consider parameter changes. The model includes factor prices and resource allocation. Results are produced by optimizing resource allocation according to the new planning system. The next section uses a numerical example for empirical analysis. The example uses a virtual production planning organization and then maximizes its profit, investment quality indicators, and outsourcing, and ensures that the target coefficient is below the total budget limit. The solution is used to solve the mathematical programming model.

4 NUMERICAL EXAMPLE

Assuming that a virtual organization produces two products (x_1 and x_2), its production and sale goals are maximizing profit and quality indicators. Producing 1 x_1 unit creates 12 units of profit and 7 units of quality. Producing one x_2 unit creates 12 units of profits and 10 units of quality. The company may use investments, outsourcing, or new technologies to improve the unit cost coefficient. The improved unit cost coefficient is \$1, and the total budget is \$2600 (see Table 1).

Producing the desired product requires resources including materials, manpower, machinery, and equipment. Products x_1 and x_2 used 5 and 6 \$10 units of material resources, respectively. They used 2.3 and 4 \$15 human resource units, respectively, and 4 and 2.5 \$12 machine resources, respectively. These resources can be used to purchase all b_1, b_2, and b_3 resources (see Table 2).

Therefore, with a budget of \$2600, this virtual company can achieve system optimization by producing and allocating resources and maximizing profit and quality indicators. Using the figures in Table 1, the empirical linear programming model is as follows:

$$\begin{aligned}
&max\ f_1 = (12+a_{11})x_1 + (10+a_{12})x_2 \\
&f_2 = (7+a_{21})x_1 + (12+a_{22})x_2 \\
&s.t. \\
&10(5x_1+6x_2)+15(2.3x_1+4x_2)+12(4x_1+2.5x_2) \\
&\quad +(a_{11}\times 8+a_{12}\times 5+a_{21}\times 4+a_{22}\times 7)\le 2600 \\
&x_1,x_2 \ge 0 \\
&a_{11},a_{12},a_{21},a_{22} \ge 0,
\end{aligned} \tag{7}$$

where f_1 represents profit targets and f_2 represents the quality targeted.

Table 1. Target coefficient, unit objectives to improve the cost and budget.

Units to improve cost	Objective	Products I coefficient	Products II coefficient	Budget
$a_{11}=8$, $a_{12}=5$	Profit	12	10	\$2600
$a_{21}=4$, $a_{22}=7$	Quality	7	12	

Table 2. Production and allocation of resources and the unit price.

Unit price	Resources	Product I	Product II	Resources available standard
\$10	Materials	5	6	b_1
\$15	Human	2.3	4	b_2
\$12	Machine	4	2.5	b_3

The empirical use of a compromised solution (Yu and Zeleny, 1982) in the mathematical programming model is solved using the Chebyshev distance definition, calculated using Lingo software optimization results, $f_1 = 235.472$ and $f_2 = 208$. The related resource distribution plan is as follows: production products are calculated as $x_1 = 9.52$ and $x_2 = 8.88$, and improved target coefficients are calculated as $a_{11} = 1.97$, $a_{12} = 1.55$, $a_{21} = 1.88$, and $a_{21} = 1.91$. The factors of production are then purchased, $b_1 = 100.88$, $b_1 = 57.42$, and $b_1 = 60.28$, spending the whole \$2600 budget.

If the compromise method is used and the same distance is defined for solving the De Novo model, $f_1 = 200.46$ and $f_2 = 177.06$. The De Novo model allocates resources as follows: purchasing costs are $b_1 = 101.40$, $b_2 = 58.73$, and $b_3 = 58.70$, and production costs are $x_1 = 8.58$ and $x_2 = 9.75$, spending the whole \$2600 budget. The proposed model generates better profit and quality goals than De Novo programming. The results show that by adjusting or improving the target coefficient, the production system is better designed, sets better goals, and reaches higher targets. Because the variable parameter model can only manage a single objective, it is not included in the empirical results of the variable parameter model. However, the empirical results show that the extended variable parameter model produces better solutions to the multi-objective problem.

5 CONCLUSIONS

Research has explored multistage De Novo programming to examine problems and solutions, and innovative extensions of fuzzy dynamic planning have been used to address changing algorithms to reflect more realistic problems. This study uses De Novo programming, traditional mathematical programming, and variable parameters to increase the feasibility of the model and to reduce its research restrictions. The elastic planning mode—the proposed extension of the traditional mathematical programming combined with De Novo programming and the variable parameter model—allows decision makers with various planning issues to optimize resource allocation results. The empirical results show that the flexibility of the model provides better performance than other models. The model is valid and can be used to solve practical availability problems.

REFERENCES

Baiman, S., Fischer, P.E. & Madhav, V. (2001). Design and Development. *Rajan management science*, 47(1), 173–188.

Bernard, A., Philippe, C. & Nicolas, M, Robustness. (2003). Evaluation of multisite distributed schedule with perturbed virtual jobshops. *Production planning and control*, 14, 55–67.

Burn, J., Marshall, P., & Barnett, M. (2002). *E-business strategies for virtual organizations,* Oxford: Butterworth-Heinemann.

Ching, C., Clyde, W.H. & Andrew, B.W. (1996). Toward IT support for coordination in network organizations. *Information & Management*, 30(4), 179–199.

ChiangLin, C.Y. & Yu, P.L. (2007). Generalized Theoretical Analysis on "Taking Loss at the Ordering Time and Making Profit at the Delivery Time"—*Programming Models with Changeable Parameters. Journal of Management*, 24(5), 569–584.

Huang, J.J., Tzeng, G.H. & Ong, C.S. (2005). Motivation and resource-allocation for strategic alliances through the de novo perspective. *Mathematical and Computer Modeling* 41, 71–72.

McGoon, C. (1995). 10 years from now. *Communication World*, 18–22.

Saabeel, W., Verduijn, T.M., Hagdorn, L. & Kumar, K. (2002). A model of virtual organization: a structure and process perspective. *Electronic Journal of Organizational Virtualness*, 4(1), 1–16.

Sasaki, M., Gen, M., & Yamashiro, M. (1995). A method for solving fuzzy de Novo programming problem by genetic algorithms. *Computers and Industrial Engineering* 29, 507–511.

Yu, P.L. & Zeleny, M. (1975). The set of all nondominated solutions in linear cases and multicriteria simplex method. Journal of Math. *Analysis and Applications*, 49(2), 430–468.

Yu, P.L. (1985). Multiple criteria decision making: concepts, techniques and extensions. Plenum, New York.

Zeleny, M. (1981). A case study in multiobjective design: de novo programming, In Multiple Criteria Analysis: Operational Methods, (Edited by P. Nijkamp and J. Spronk), *Cower Publishing Co., Hampshire,* 37–52.

Zeleny, M. (1986). Optimal system design with multiple criteria: De Novo programming approach. *Engineering costs and production economics*, 10(1), 89–94.

Zeleny, M. (1990). Optimal given system vs. designing optimal system: The de novo programming approach. *International Journal of General System,* 17(4), 295–307.

Zeleny, M. (1982). *Multiple criteria decision making.* McGraw-Hill, New York.

E-Commerce, E-Business and E-Service – Lee (Ed)
 ISBN 978-1-138-02646-9

Application of cluster analysis to western China population quality assessment

Xie Zhang
School of Public Finance and Management, Lanzhou University of Finance and Economics, Lanzhou, China

Zhendong Li
Jiuquan Vocational and Technical College, Jiuquan, China

ABSTRACT: In this paper, the author employed the clustering method to analyze great differences in the development of the population quality in western China, so as to provide strategic suggestions on improving it, which is based on the characteristics for western China population quality and the western China social and economic development; By considering the 12 western provinces (including autonomous regions and municipalities directly under the Central Government) as the fundamental research unit, the author built a suitable comprehensive assessment index system for the population of western China.

1 AN INTRODUCTION TO THE REGIONS RESEARCHED

Population quality is known as the ability and conditions that a population has in order to deal with and change the world, mainly including the qualities of ideology and morality, culture and science, as well as physical fitness, which consist an organic system, referred to as population quality system. It is necessary to improve population quality in ideology and morality, culture and science, as well as physical fitness in western China. Based on research into the coordination between the three factors and the systemic effect they have, the author has made a cluster analysis on the quality of the population in 12 western provinces, automomous regions and 1 municipality. In addition, this paper is aimed at laying a scientific ground for improving the quality of population in western China.

An introduction to the regions researched.

The regions of western China mentioned in this paper include 5 ethnic autonomous regions (Inner Mongolia, Guanxi, Tibet, Ningxia and Xinjiang), 1 municipality (Chongqing) and 6 provinces (Sichuan, Guizhou, Yunnan, Shaanxi, Gansu and Qinghai).

Western China has a large area of 6.867 million km^2, 71.5% of the national territory but it had a comparatively small population of 360.693 million in total in 2010, 27.0% of the national population. There are 50 ethnic groups living in western China, making up about 80% of the ethnic population.

Western China is enjoying a smaller economic aggregate. Its GDP was 8,140,850,000,000 RMB in 2010, 18.6% of the national total GDP and its export-oriented economy is less developed, only 4.3% of the total amount of the export and import trade.

Social services in western China are growing slowly. There were 564 higher education colleges in total in 2010, only 23.9% of the total number of 2358 nation wide. The student population in colleges and schools is 5.027 million, 22.5% of the national total. And the number of medical and health institutions is 288,631, only 30.6% of the total. The professionals in this industry are 1.468 million, 25.0% of the total.

People's incomes are increasing year by year. Disposable personal incomes in urban areas were 15,806 RMB yuan in 2010, with an increase of 9636.1 yuan, compared with 6169.9 Yuan in 2001, an average growth of 1070.7 yuan per year. Those in rural areas were only 4418 yuan with an increase of 2662.9 yuan, compared with 1755.1 yuan in 2001, an average growth of 295.9 yuan per year.

In brief, compared with the regions in eastern and central China, there are big gaps in the economy, and social services in western regions, even bigger in some industries.

2 THE RESEARCH METHOD

Cluster analysis or clustering is a statistical multi-analysis whose task is to group a set of objects in such a way that objects in the same group (called a cluster) are more similar (in one sense or another) to each other than to those in other groups (clusters). Those variables or samples which enjoy some sort of values can be classified by selecting modified distance data preprocessing and systemic clustering until the result achieves the desired properties.

Table 1. China west population quality evaluation index system.

Driving factors	Index
Physical fitness	Per capita GDP, urban disposable incomes, per-capita net income of rural residents, death rate, average life expectancies, infant mortality rate, percentage of disabled, licensed doctors, hospitals and clinics per 1000 population.
Science and culture	Average schooling years, illiteracy rate of 15 year olds and above, rural labor forces with college education and above, population with higher education, uneducated population, population with college education and above older than 6 years.
Ideology and morality	Birth rate, birth rate according to the policy, poverty rate, illiteracy rate among women 15 years old and above, maternal mortality rate.

Table 2. The choice of index of population quality evaluation in western China.

Codes of index	Data chosen	Codes of index	Data chosen
X_1	2007 per capita GDP	X_{11}	2000 average schooling years
X_2	2005 poverty rate	X_{12}	2007 illiteracy rate of 15 year olds and above
X_3	2007 urban disposable incomes	X_{13}	2007 rural labor forces with college education and above
X_4	2007 per-capita net income of rural residents	X_{14}	2007 population with higher education
X_5	2007 death rate	X_{15}	2007 uneducated population
X_6	2000 average life expectancy	X_{16}	2000 population with college education and above older than 6 years
X_7	2000 infant mortality rate	X_{17}	2007 birth rate
X_8	2006 percent of the disabled	X_{18}	2007 birth rate according to the policy
X_9	2007 licensed doctors per 1000 population	X_{19}	2007 illiteracy rate among women 15 years old and above
X_{10}	2007 percent of hospitals and clinics per 1000 population	X_{20}	2007 maternal mortality rate

Data processing.

Suppose that the number of samples is n, and p is the number of index reflecting each sample, thus the original data of the n samples form a matrix with n rows and p columns. Each sample can be seen as a point in a p dimensional space and the distance in the space can be used to measure the closeness between samples and so forth to make a cluster analysis. There are many definitions of distance in p dimensional space and the most used is the Euclidian distance, defined as the following:

To avoid the negative effect resulting from different dimensions, we can normalize them because there are often great differences between the values of index of samples due to the use of various dimensions. After working out the distance between points of samples, a distance matrix forms, which is a lower triangle matrix or upper triangle matrix with zero diagonals. Based on distance matrix, a cluster analysis can be made.

In this paper, two-step clustering analysis, also known as second order clustering, is used. First, finish the process of pre-clustering or quasi clustering, which is to create a cluster feature tree. Second, finish the process of hierarchically clustering the input data from the pre-clustering step.

2.1 *Creation of index*

Setting of cluster factors is to choose corresponding variables of clustering analysis according to research objectives. On the basis of population quality in western China, 20 evaluation factors are chosen as the variables of clustering analysis (Table 1).

2.2 *The data source*

The data of 20 categories of clustering variables have been obtained from *Handbook of Frequently-used Data for Population and Family Planning (1999–2011)*[Φ] and the data are according to the variables of evaluation factors in population quality of 12 provinces (or municipality,

[Φ]This handbook is edited by the National Family Planning Commission of China, Development and Information Department, published by China Population Press.

Table 3. The original data for population quality evaluation in western China.

Index	Chongqing	Sichuan	Guizhou	Yunnan	Tibet	Shaanxi	Gansu	Qinghai	Ningxia	Xinjiang	Inner Mongolia	Guangxi
X_1	14660	12893	6915	10540	12109	14609	10345	14257	14649	16999	25393	12555
X_2	1.6	2.0	9.0	7.0	7.2	7.2	8.1	11.5	3.0	5.1	6.6	3.2
X_3	12591	11098	10678	11496	11131	10763	10012	10276	10859	10313	12378	12200
X_4	3509	3547	2374	2634	2788	2645	2329	2684	3181	3183	3953	3224
X_5	6.30	6.29	6.60	6.22	5.10	6.16	6.65	6.13	5.04	5.01	5.73	5.99
X_6	71.73	71.20	65.96	65.49	64.37	70.07	67.47	66.03	70.17	67.41	69.87	71.29
X_7	21.75	21.57	66.05	70.32	43.01	33.04	52.98	50.55	27.33	40.10	32.06	31.10
X_8	6.05	7.57	6.40	6.46	7.00	6.69	7.20	5.54	6.83	5.31	6.39	7.23
X_9	1.2	1.3	1.0	1.3	1.4	1.6	1.3	1.7	1.8	2.1	2.0	1.2
X_{10}	2.19	2.30	1.83	2.53	2.39	2.90	2.51	2.92	2.95	4.27	2.73	1.94
X_{11}	7.3	7.1	6.1	6.3	3.4	7.7	6.5	6.1	7.0	7.7	7.8	7.6
X_{12}	8.00	10.62	16.59	16.13	36.77	8.89	19.33	18.40	13.80	4.29	8.23	5.82
X_{13}	0.8	0.7	0.7	0.4	0.1	1.2	1.3	0.4	0.8	0.8	0.9	1.0
X_{14}	3.77	4.10	3.22	4.02	1.21	7.73	3.83	7.07	7.43	8.97	7.46	4.01
X_{15}	7.33	9.94	14.14	14.29	34.44	8.38	17.14	16.69	12.06	4.59	7.98	5.60
X_{16}	3.0	2.7	2.2	2.2	1.5	4.5	2.9	3.6	4.1	5.6	4.1	2.6
X_{17}	10.10	9.21	13.28	13.08	16.40	10.21	13.14	14.93	14.80	16.79	10.21	14.19
X_{18}	75.66	87.91	90.04	96.43	99.95	98.03	88.95	94.31	91.54	99.84	97.43	80.94
X_{19}	11.65	15.24	24.91	22.74	44.77	12.64	26.60	25.91	19.83	5.43	11.65	9.20
X_{20}	50.2	48.6	66.7	50.4	265.4	36.6	62.6	78.7	47.4	73.7	35.4	26.0

Note: 2006's data. Source: *Handbook of Frequently-used Data for Population and Family Planning (1999–2011).*

autonomous regions) in western China. That is the original data for population quality evaluation in western China(Table 2).

Because the units of data are not unified, the original data have to be normalized (Table 3). With extremum method, the data has been standardized and after that made a cluster analysis.

3 A DESCRIPTION OF CLASSIFIED REGIONS AND SUGGESTED SOLUTION

Through the computer software, SPSS 17.0, the result of clustering analysis shows that the 12 provinces, autonomous regions and municipality can fall into three groups. Group 1 is Chongqing, Sichuan and Guangxi; group 2 is Shaanxi, Ningxia, Xinjiang and Inner Mongolia; group 3 is Yunnan, Guizhou, Tibet, Gansu and Qinghai.

Group 1 makes up 25% of western regions, enjoying the highest economic level and rapid development in education, science and technology, health care, cultural services and sports, so this group ranks No. 1 among the 3 groups. But this group needs further to develop its economy, to promote sustainable development in science and technology, education and culture, to strengthen ideology and morality education, focusing on civic virtue, personal conduct and awareness of law and to continue to reduce the gap with central and eastern regions. In addition, it should pay special attention to the advantages of geographical locations and resources, enlarging the effect on other areas and Northwest China.

Group 2 makes up 33% of western regions, enjoying more rapid economic development and a higher degree of education, science and technology, health care, cultural services and sports. But there is a distinct difference compared with central and eastern regions, so this group of 4 still needs to pay more attention to science and technology, education, culture, and morality promotion, to speed economy, to improve population quality, reducing the gap with southwest regions.

Group 3 includes 5 provinces and autonomous regions, which are Yunnan, Guizhou, Tibet, Gansu and Qinghai. This group makes up 42% of the 12, enjoying slower economic development and lower degree in education, science and technology, health care, cultural services and sports. Because population quality in Tibet, Gansu and Guizhou is lower in physical fitness and science and culture, it is necessary to change concepts, to adjust the economic structure, to better the infrastructure for science and technology, education and culture. Along with Yunnan and Qinghai, learning from the success experience of eastern areas and other western areas, Tibet, Gansu and Guizhou should promote social causes, the economy, science and technology, education and health care to develop in a healthy and stable way.

REFERENCES

Lv Hongjun, Zhang Chengzong, Lu Jiehua. A study on population in China ethnic groups [M]. China Social Science Press, 2010:85.

Liu Dawei, Li Qian. An application of cluster analysis to general evaluation of agricultural productivity [J]. Agriculture and Technology, 2006,2:36–38.

E-Commerce, E-Business and E-Service – Lee (Ed)
 ISBN 978-1-138-02646-9

Comparative evaluation of modulation strategies for a three-phase three-level unity power factor rectifier

Ming-Tsung Tsai, C.-L. Chu, Fang-Kai Guo & Bo-Jang Jiang
Department of Electrical Engineering, Southern Taiwan University, Tainan, Taiwan

ABSTRACT: This paper presents a comparative research for different pulse-width modulation schemes to realize the three-phase three-level boost rectifier with Power-Factor Correction (PFC) function. It uses three bidirectional switches connected to the common capacitive center point of the DC bus that allows the three-level operation and a low blocking-voltage stress on the power switches, and has the advantages of lower per-unit current ratings of the switching devices. Three modulation strategies are compared, and the validity of the proposed method is confirmed by simulation and experimental results through a 1 kW prototype.

Keywords: three-level; boost rectifier; voltage stress

1 INTRODUCTION

The diode rectifiers are used to interface with the three-phase utility voltage can result in various adverse problems as it injecting harmonics into the grid. Many topologies of switching converters are adopted to mitigate these problems [1,2]. Among these topologies, the three-level boost rectifier used three bidirectional switches connected in a star structure have the benefit, including a lower current harmonics than the two-level topologies, and a reduced switching stress, and these features make it pleasing for the use of three-phase boost type PFC structures [3–8].

This study is to design a desired three-phase three-level boost rectifier with such a high power factor and low current harmonics. Two dc capacitors combined with a three bidirectional switches with a star-connection provide the three-level function. It uses lower rating switching devices and let the mainly line current pass through the parallel three-phase diodes, resulting a lower switching devices cost. In addition, a lower input current distortion can also be obtained in a low output voltages (as low as 0.816 ratio of the source input peak voltage, rather than the conventional 1.414 ratio), however, this function is accomplished with the price of missing the bilateral power flow ability and increased devices numbers.

This paper makes a comparative research about three types of pulse-width modulation strategies, including the convention Sinusoidal Pulse-Width Modulation Strategy (SPWM), and two modified SPWM schemes. The operating modes of the researched topology are described in detailed in Section II. The proposed three-phase PFC controller is shown in Section III. Experimental verification of a 1 kW prototype is given in Section IV. Finally, a conclusion is given in Section VI.

2 CIRCUIT DESCRIPTION

2.1 *The three-phase three-level boost rectifier*

The main circuit configuration of the researched rectifier is shown in Figure 1. By controlling the three bidirectional power switches, the circuit can achieve the power factor correction function. In this study, the average current mode control is adopted to reduce the input current harmonics. There are two types of select methods to determine the third

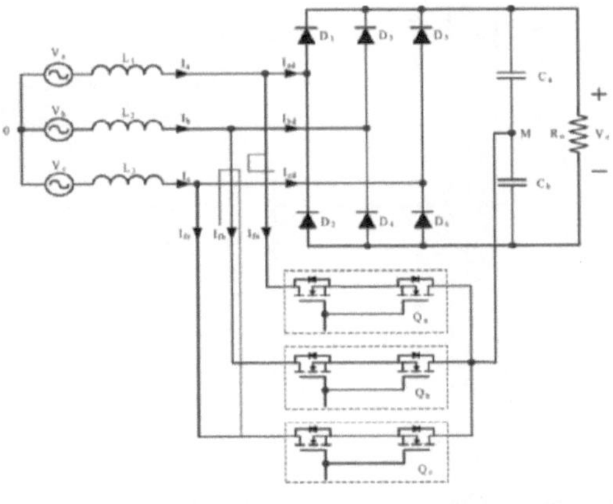

Figure 1. The three-phase three-level boost rectifier.

low switching loss element: one is to adopt the phase which has the maximum absolute value; the other method is to adopt the phase which has the minimum absolute value. The difference between these two methods will be shown as follows:

2.2 *Analysis of the two modified modulation strategies*

Figure 2 shows the line cycle can be divided into six sections. For the first method, it is to adopt the phase voltage which has the maximum absolute value to turn on, for an example, during (0~60) from Figure 2, the phase voltage V_b has the maximum absolute value, thus the bidirectional switch Q_b is turned on during this section, and the other two switches are controlled at PWM switching, resulting in the average inductor current in phase V_b will be regulated by the current controllers of the other two phases. The equivalent circuits for others using this method can also be found as the follows:

State 1 (0°~60°):

In this section, $V_a > 0$, $V_c > 0$ and $V_b < 0$, diodes of D_1, D_5 are forward bias when the corresponding switching device of Q_a and Q_c are turned off, and the phase voltage V_b has the maximum absolute value, thus the bidirectional switch Q_b is turned on during this section. The switching devices of Q_a and Q_c are pulse-width modulated by the current controllers of I_a and I_c, resulting in the source current I_b following by—(+). In this section, the source voltages of V_a and V_c provide energy to the capacitor C_a and load.

State 2 (60°~120°):

In this section, $V_a > 0$, $V_b < 0$ and $V_c < 0$, diodes of D_4, D_6 are forward bias when the corresponding switching device of Q_b and Q_c are turned off, the phase voltage V_a has the maximum absolute value, thus the bidirectional switch Q_a is turned on during this section. The switching devices of Q_b and Q_c are pulse-width modulated by the current controllers of I_b and I_c, resulting in the source current I_a following by—$(I_b + I_c)$. In this section, the source voltage V_a provides energy to the capacitor C_b and load.

State 3 (120°~180°):

In this section, $V_a > 0$, $V_b < 0$ and $V_c < 0$, diodes D_1, D_3 and D_6 are forward bias, and the phase voltage V_c has maximum absolute value, thus the bidirectional switch Q_c is turned on during this section. The switching devices of Q_a and Q_b are pulse-width modulated by the current controllers of I_a and I_b, resulting in the source current I_c following by—$(I_a + I_b)$. In this section, the source voltages of V_a and V_b provide energy to the load. The other states can be observed in similar manner.

For the second method, it is to adopt the phase which has the minimum absolute value to turn on, and the other two phases are controlled at PWM switching, will follow the above concept to adopt the corresponding switches during the related sections.

3 CONTROL CIRCUIT DESIGN

The current controller is consisted of an outer voltage loop and three independent inner current loops. It is achieved by sensing the three-phase input currents, and forces the average inductor current in each phase to follow the related sinusoidal phase voltage. It can be a conventional P or PI controller. The voltage loop control is done by feeding back DC bus voltage, and is also a conventional PI controller. Figure 3(a) shows the circuit diagram of considering system, and Figure 3(b) shows the selection logic.

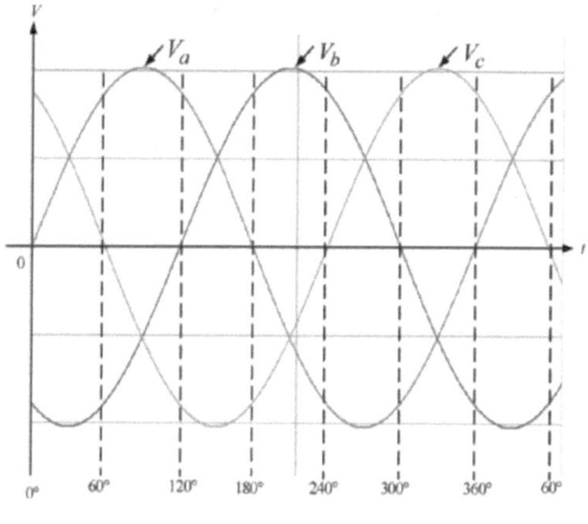

Figure 2. The waveforms of the idea three-phase source voltage.

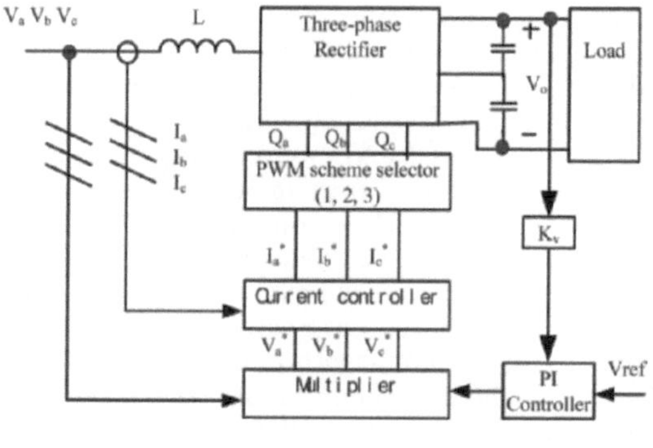

Figure 3(a). The proposed control block diagram of the system.

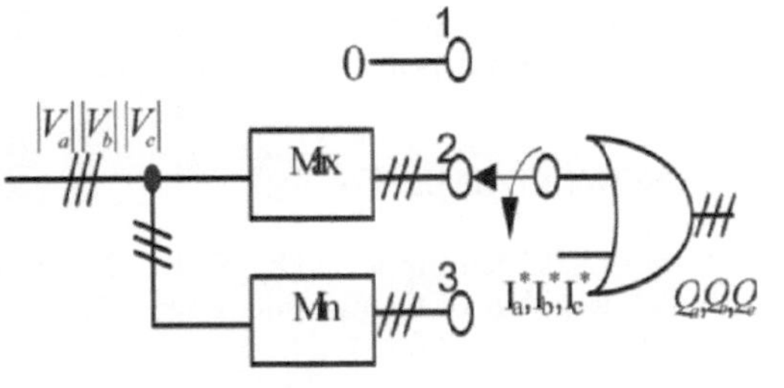

Figure 3(b). PWM scheme selector.

4 SIMULATION AND EXPERIMENTAL RESULTS

4.1 *Simulation*

A comparative simulation has used to check the differences among the three-types of PWM schemes. A 1000 W system shown in Figure 1 was built to prove the feasibility. The system parameters are as follows: the input voltage is 110VLL-rms; the total output voltage is 320 V; each dc output capacitor has the capacitance of 1200 uF; the output power is about 1000 W; the switching frequency is 20 kHz; the source input inductance in each phase is 1 mH. Figure 4 shows the simulation results of the researched circuit with different PWM schemes. Figure 4 shows the simulation results of the researched circuit with different PWM schemes. Figure 4(a) shows the modified PWM switching with the minimum absolute value switch turned on, also the up trace shows the phase switching signal and the low trace shows the three-phase input currents. Figure 4(b) shows the modified PWM switching with the maximum absolute value switch turned on. It shows a better current performance than the case shown in Figure 4(a). The conventional PWM has similar waveforms to Figure 4(b), and the current controller used in above three cases is a P controller of $K_p = 5$.

Even though these three modulation strategies have the similar current waveform performance, however, some different properties exist among these methods. For the conventional modulation strategy, it can regulate a low output voltages near to 0.816 ratio of the root-mean-square value of the source input voltage. But the other two modified strategies can only regulate above 1.414 ratio of the source voltages. On the other hand, the modified strategies can have a better dc balance effect than the conventional strategy, for example with a unbalance capacitance existed in the two dc capacitors will result in an output voltage unbalance of 4% between the split two dc bus voltage for the conventional strategy, and nearly no influence with the modified strategies. Another simulation result shows the modified strategies have a better dc balance effect than the conventional strategy with the case of 20% of dc load unbalance, for example one is 100 Ω, and the other is 120 Ω parallel with the corresponding split two dc capacitors, it shows the modified strategies will result in an output voltage unbalance of 3% between the split two dc bus voltage, while the conventional strategy will result in an output voltage unbalance of 10% between the split two dc bus voltage.

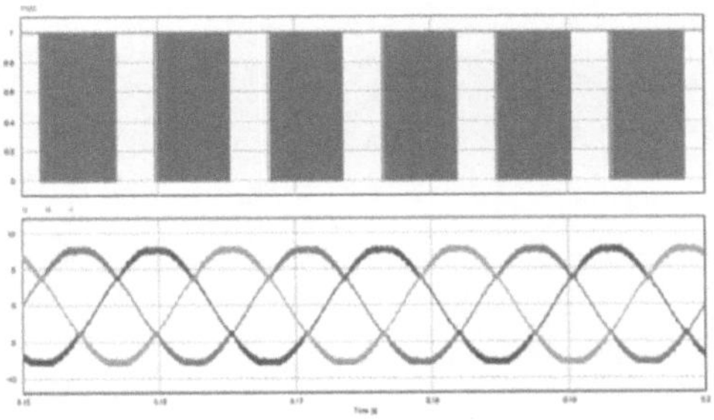

Figure 4(a). Modified PWM with minimum absolute value switching.

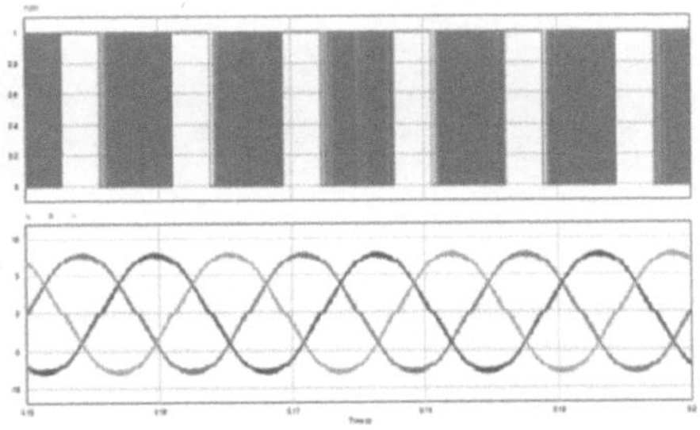

Figure 4(b). Modified PWM with maximum absolute value switching.

4.2 *Experimental results*

A 1000 W prototype shown in Figure 1 was built to prove the feasibility. The system parameters are as follows: the input voltage is 70 V_{LL}rms; the total output voltage is 200 V; each dc output capacitor has the capacitance of 2400 uF; the output power is about 1000 W; the switching frequency is 20 kHz; the source input inductance in each phase is 1 mH. Figure 5 shows the experimental results with respect to the conventional PWM schemes, where the output power is about 1045 W. The up trace shows the total DC output voltages, and the low trace shows the three-phase input currents. It shows a similar result with the simulation shown in Figure 4(a). Figure 6 shows the modified PWM

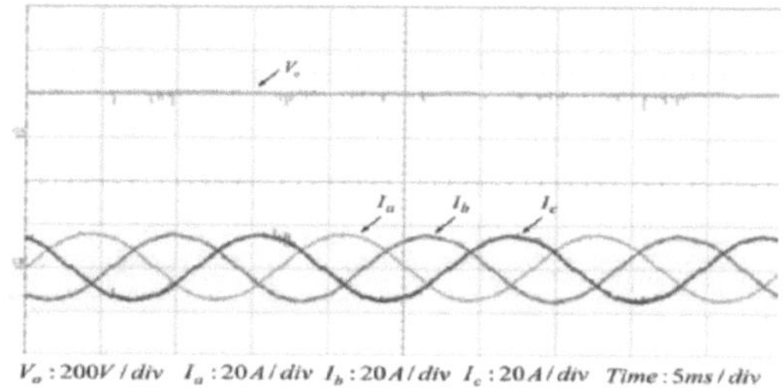

Figure 5. The experimental results with conventional PWM schemes.

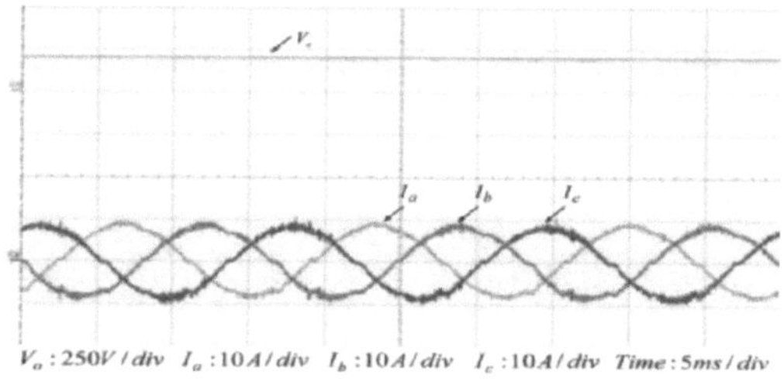

Figure 6. The modified PWM with minimum absolute value switching.

with minimum absolute value switch turned on during the corresponding 60 degree interval, and the load is the same as (a). The up trace shows the total DC output voltage, and the low trace shows the three-phase input currents.

5 CONCLUSION

A comparative research for different pulse-width modulation schemes to achieve the Power-Factor Correction (PFC) function is presented in this paper. It shows the modified PWM scheme with maximum absolute value switch turned on scheme has the best performance than the other two methods and it can perform a lower output voltages ratio than the conventional two level structures of above 1.414 output voltages ratio to the source peak voltage. Prototype results have shown the proposed system to be suitable for the three-phase rectifier with power factor correction function.

ACKNOWLEDGMENTS

This work has been supported by National Science Council, Republic of China under research project NSC100-2622-E-218-005-CC3.

REFERENCES

Drofenik U. and J.W. Kolar, "Comparison of not synchronized saw-tooth carrier and synchronized triangular carrier phase current control for the VIENNA rectifier I," in Proc. IEEE ISIE, Bled, Slovenia, 1999, p. 13.

Liu F. and A.I. Maswood, "A novel variable hysteresis band current control of three-phase three-level unity PF rectifier with constant switching frequency," IEEE Trans. Power Electron., vol. 21, no. 6, pp. 1727–1734, Nov. 2006.

Qiao C. and K.M. Smedley, "Three-phase unity-power-factor star-connected switch (VIENNA) rectifier with unified constant-frequency integration control," IEEE Trans. Power Electron., vol. 18, no. 4, pp. 952–957, Jul. 2003.

Salmon J.C., "Operating a three-phase diode rectifier with a low input current distortion using a series-connected dual boost converter," IEEE Trans. Power Electron., vol. 11, no. 4, pp. 592–603, Jul.

Shaoru Zhang, and Fang Lin Luo, "A new approach for three-phase three-level unity PF rectifier considering alternative performance indices," 2010 the 5th IEEE Conference on Industrial Electronics and Applications (ICIEA), June 2010, pp. 480–485.

Shaoru Zhang, Fang Lin Luo, "A Novel Reference Compensation Current Strategy for Three-phase Three-level Unity PF Rectifier", 4th IEEE Conference on Industrial Electronics and Applications, pp. 178–183, China, 2009.

Siebert A., A. Troedson, and S. Ebner, "AC to DC power conversion now and in the future," IEEE Trans. Ind. Appl., vol. 38, no. 4, pp. 934–940, Jul./Aug. 2002.

Singh B., B.N. Singh, A. Chandra, K. Al-Haddad, A. Pandey, and D.P. Kothari, "A review of three-phase improved power quality ac–dc converters," IEEE Trans. Ind. Electron., vol. 51, no. 3, pp. 641–660, Jun. 2004.

E-Commerce, E-Business and E-Service – Lee (Ed)

Author index